EE 541
USC
Prof. L.W.

ELEMENTS
OF
LINEAR ALGEBRA
AND MATRIX
THEORY

**International Series in
Pure and Applied Mathematics**

William Ted Martin and E. H. Spanier, *Consulting Editors*

AHLFORS · Complex Analysis
BELLMAN · Stability Theory of Differential Equations
BUCK · Advanced Calculus
BUSACKER AND SAATY · Finite Graphs and Networks
CHENEY · Introduction to Approximation Theory
CODDINGTON AND LEVINSON · Theory of Ordinary Differential
  Equations
COHN · Conformal Mapping on Riemann Surfaces
DENNEMEYER · Introduction to Partial Differential Equations
  and Boundary Value Problems
DETTMAN · Mathematical Methods in Physics and Engineering
EPSTEIN · Partial Differential Equations
GOLOMB AND SHANKS · Elements of Ordinary Differential
  Equations
GRAVES · The Theory of Functions of Real Variables
GREENSPAN · Introduction to Partial Differential Equations
GRIFFIN · Elementary Theory of Numbers
HAMMING · Numerical Methods for Scientists and Engineers
HILDEBRAND · Introduction to Numerical Analysis
HOUSEHOLDER · Principles of Numerical Analysis
LASS · Elements of Pure and Applied Mathematics
LASS · Vector and Tensor Analysis
LEPAGE · Complex Variables and the Laplace Transform for
  Engineers
McCARTY · Topology: An Introduction with Applications to
  Topological Groups
MOORE · Elements of Linear Algebra and Matrix Theory
MOURSUND AND DURIS · Elementary Theory and Application
  of Numerical Analysis
NEF · Linear Algebra
NEHARI · Conformal Mapping
NEWELL · Vector Analysis
RALSTON · A First Course in Numerical Analysis
RITGER AND ROSE · Differential Equations with Applications
ROSSER · Logic for Mathematicians
RUDIN · Principles of Mathematical Analysis
SAATY AND BRAM · Nonlinear Mathematics
SIMMONS · Introduction to Topology and Modern Analysis
SNEDDON · Elements of Partial Differential Equations
SNEDDON · Fourier Transforms
STOLL · Linear Algebra and Matrix Theory
STRUBLE · Nonlinear Differential Equations
WEINSTOCK · Calculus of Variations
WEISS · Algebraic Number Theory
ZEMANIAN · Distribution Theory and Transform Analysis

ELEMENTS OF LINEAR ALGEBRA AND MATRIX THEORY

JOHN T. MOORE
The University of Florida
The University of Western Ontario

McGRAW-HILL BOOK COMPANY
New York
St. Louis
San Francisco
Toronto
London
Sydney

**ELEMENTS
OF
LINEAR ALGEBRA
AND MATRIX
THEORY**

Copyright © 1968 by McGraw-Hill, Inc. All Rights Reserved. Printed in the United States of America. No part of this publication may be reproduced, stored in a retrieval system, or transmitted, in any form or by any means, electronic, mechanical, photocopying, recording, or otherwise, without the prior written permission of the publisher.

Library of Congress Catalog Card Number 68-11933

42885

234567890 MAMM 754321069

To
D. E. S.
&
F. A. S.

PREFACE

This book is an outgrowth of a course in linear algebra which I first gave to a class of undergraduates several years ago. In writing it, I have been motivated by the desire to produce a text with the following features: It would be of a size such that a major portion of the contents could be covered in a single academic term; it would have an abundance of problems, with degree of difficulty varying from the routine and near-trivial to the challenging; it would use a notation and terminology as near-standard as possible; and it would present a level of mathematics suitable for students whose major interest is in either pure mathematics or any of the pure or applied sciences. In my desire to limit the size of the book, it is inevitable that I am exposing an "Achilles' heel" by my choice of material for inclusion. However, it is my belief that in a first course one should avoid being encyclopedic but, rather, help to give some sense of direction to the student; and so I take my stand on what I have done—but with apologies to those whose opinions may differ from mine.

Before beginning any writing of my own, I consulted the bulletins of the Committee on the Undergraduate Program in Mathematics (CUPM) and decided to make the course recommendation contained in one of them the nucleus of this book. However, I have not followed these recommendations slavishly but have included additional material and been guided by my own preference with regard to arrangement. The book has been structured, for the most part, as a continuous study, in which each section makes a contribution to the next. At the same time, there are a few sections that are not essential to the main development of the text; in each such case this fact has been made clear by an introductory remark. In order to emphasize the key purpose of a section, I have limited the number of results included in any one of them to a bare minimum.

As was stated above, I have tried to use standard symbolism and terminology, and so I have made free use of the familiar symbols of set theory. To be precise, the following symbols occur without explanation: $\in$ for set membership; $\cup$ for set-theoretic union; $\cap$ for set-theoretic intersection; $\subset$ and $\supset$ for the two kinds of set inclusion; $\times$ for the cartesian product of sets. In addition, it has been convenient to refer to the *cardinality* or *cardinal number* of a set in an intuitive sort of way, and I have used $[a,b]$ to denote the interval of real numbers between a and b, inclusive. Although the book is considered to be self-contained, it is assumed that the reader is familiar with the following basic notions: the *division* and *gcd processes* as they arise in the theory of polynomials; the partitioning of a set effected by an *equivalence relation*, each equivalence class being representable by a single element of the set; and the two methods of proof by *mathematical induction*. If the reader is not familiar with these several concepts, he is urged to acquaint himself with the unfamiliar ones by referring to any book on abstract algebra.

The problem sets have been arranged into three groupings, physically separated from one another by spaces. The problems in the first groupings are basic but quite simple, and some instructors may find them unnecessary for their classes. However, although this may be true in some instances, it is my judgment that a student is often lost at a point where he fails to comprehend something simple—and almost trivial—but also very important. It is the purpose of the early problems to bring to light and, hopefully, to eliminate any such difficulties. The second groupings of problems are complementary to the material in the text, and all of them should be worked out in detail. In this way, students may feel more personally involved with the proofs of the theorems than if these proofs were all presented in final polished form. In the final groupings, I have included other problems of a more general nature, relevant to the particular section under study. Otherwise, the problems have not been arranged according to difficulty. Inasmuch as the first groupings involve problems of such a basic nature, answers or hints have been provided for most of them at the back of the book. The student is thus given an opportunity to work these problems—if he feels the need of doing do—quite independently of whether this is suggested by his instructor and so develop an early confidence in his ability to grasp the concepts being presented.

PREFACE

There are many people to whom I am indebted in the writing of this book. I was first given much encouragement by a number of reviewers of an early portion of the manuscript, all of whom were anonymous except for Prof. George Simmons of Colorado College. To them all I extend my most sincere thanks. At a later stage, much of the manuscript was read by three people who provided many helpful suggestions for its improvement. They are Prof. C. Desoer of the University of California, Berkeley; Prof. D. J. Lewis of the University of Michigan; and Prof. T. L. Wade of Florida State University. I am sincerely grateful to them for this assistance. The manuscript in its near-final form was read by Prof. D. J. Sterling of Bowdoin College, and I wish to express my gratitude to him for his very useful suggestions; I am confident that the book is better because of his pertinent commentary. It is understood, of course, that even though these people have seen part or all of the manuscript, their reading of it does not imply any sort of endorsement; any errors or shortcomings in the book are mine alone. My longtime friend, Prof. W. S. Cannon of Presbyterian College, was of very considerable help in collecting Readings from the *American Mathematical Monthly*. My colleague, Prof. Anne Bode, has been a most gracious consultant on matters which have arisen from time to time, and her generous assistance continued throughout the final stages of the manuscript. Two very good graduate students, Eddy Smet and Jawaid Rizvi, came to my aid at a very busy time and used a portion of a vacation period to supply me with answers for the first groupings of problems. Their further help in reading the galleys was inestimable, and the book is much closer to being free of errors than it would otherwise have been. I offer my most warm and cordial thanks to them all.

JOHN T. MOORE

CONTENTS

Preface vii

1 Finite-dimensional Vector Spaces   1
1.1 Vector Spaces 1
1.2 Directed Line Segments as Vectors 7
1.3 Geometric Vectors and Coordinate Spaces 16
1.4 Subspaces 24
1.5 Solutions of Linear Equations: Gauss Reduction 28
1.6 Solutions of Linear Equations: Determinants 35
1.7 Linear Dependence of Vectors 45
1.8 Basis and Dimension 51
1.9 Two Important Theorems 57

2 Linear Transformations and Matrices   63
2.1 Linear Transformations 63
2.2 Properties of Linear Transformations 69
2.3 Operations on Linear Transformations 73
2.4 Linear Functionals 79
2.5 Annihilators 85
2.6 Nonsingular Linear Transformations 90
2.7 Matrices of Linear Transformations 96
2.8 Matrices as Multiplicative Systems 106
2.9 Change of Basis 119

2.10 Rank 127
2.11 An Important Note of Clarification 134

3 Determinants and Systems of Linear Equations   140
3.1 Matrices and Linear Systems 140
3.2 Elementary Matrices and Inverses 146
3.3 Determinants as Multilinear Functionals 154
3.4 An Alternative Method for Evaluating a Determinant 160
3.5 An Introduction to Alternating Multilinear Forms 168
3.6 Determinants Discovered Anew 173

4 Inner-product Spaces   178
4.1 Inner Products in V_3 178
4.2 General Euclidean Space 186
4.3 The Gram-Schmidt Process of Orthogonalization 193
4.4 Orthogonal Complements 199
4.5 Orthogonal Transformations 206
4.6 Linear Functionals and Adjoints 213
4.7 Inner Products and Positive Operators 221
4.8 Simple Applications of the Distance Function 229

5 Bilinear and Quadratic Forms   236
5.1 Bilinear Functions and Forms 236
5.2 Quadratic Forms 246
5.3 Diagonal Quadratic Forms under Congruence 252
5.4 Invariants of a Symmetric Matrix under Congruence 260
5.5 Eigenvalues and Eigenvectors 265
5.6 Orthogonal Reduction of Quadratic Forms 272

6 Similarity and Normal Operators   282
6.1 The Cayley-Hamilton Theorem 282
6.2 Similarity and Diagonal Matrices 289
6.3 Complex Vector Spaces 297
6.4 The Spectral Theorem 305
6.5 Invariant Subspaces and Primary Decomposition 313
6.6 Nilpotent Operators and T-cyclic Subspaces 321
6.7 The Jordan Canonical Form 327

Appendix 336

Selected Readings 340

Glossary of Symbols 341

Answers 342

Index 363

**ELEMENTS
OF
LINEAR ALGEBRA
AND MATRIX
THEORY**

1
FINITE-DIMENSIONAL VECTOR SPACES

1.1 Vector Spaces

In this study of linear algebra, it is assumed only that the reader has some familiarity with those ubiquitous mathematical objects known as *sets* and their *mappings*. The reader who has escaped these in his mathematical education will find at least some of the related symbols listed at the end of the book. Our basic elements will often be numbers, selected from the sets of natural numbers, integers, rational numbers, real numbers, or complex numbers, designated, respectively, by **N**, **Z**, **Q**, **R**, and **C**. In most instances, except in the final chapter, our numbers will be real and so elements of **R**. We shall assume complete familiarity with all these number systems.

Many physical systems occurring in science can be described conveniently in the language of electrical engineering. The central feature of such a system is a "black box" to which an *input* is applied and from which an *output* leaves, the details of the relationship between the input and output depending on the nature of the "black box" or *system*. Many of the most interesting and common systems are *linear* in the following sense: (1) if an input of quantitative measure A yields an output of measure C, the output that results from an input of rA is rC, for any $r \in$ **R**; (2) if, in addition, an input of B units yields an output of D units, an input of $A + B$ will yield an

2 FINITE-DIMENSIONAL VECTOR SPACES

Fig. 1

output of $C + D$. A study of the mathematical abstractions of these linear systems is the object of a course in *linear algebra*. The pure mathematician needs no such motivation for a study of this sort, but for others it may be encouraging to know that there is a wealth of applications of this undertaking in abstraction. Moreover, in recent times it has become almost a cliché, even in areas of applied science, that the most practical approach to a study of mathematics is an abstract one.

Although the exact meaning of *linear algebra* has varied from time to time, the one just given is probably as comprehensive as is currently possible. In mathematical terms, the input and output referred to above are sets of *vectors*, and the "black box" is a *linear transformation* that transforms input vectors into output vectors. The *vectors* are elements of a mathematical system known as a *vector space*, and so it is a reasonable preview of the content of this book to say that we shall undertake *a study of linear systems through the medium of vector spaces*. We shall now define this all-important concept. In the context of a *real* vector space, about to be defined, it is customary to refer to real numbers, i.e., elements of **R**, as the associated *scalars*.

Definition   *A real vector space V is a set of elements, called vectors, with two operations, called addition (designated by $+$) and multiplication by scalars (designated by juxtaposition), such that the following axioms or conditions are satisfied:*

A1. With every pair $\alpha, \beta \in V$ there is associated a unique vector $\alpha + \beta$, called their *sum*.

A2. Addition is associative: that is, $\alpha + (\beta + \gamma) = (\alpha + \beta) + \gamma$, for arbitrary $\alpha, \beta, \gamma \in V$.

A3. There exists a unique (*zero*) vector 0 with $\alpha + 0 = \alpha$ for any $\alpha \in V$.

A4. With each $\alpha \in V$ there is associated a unique vector $-\alpha$ (called the *additive inverse* of α) such that $\alpha + (-\alpha) = 0$. As in arithmetic, we shall identify the difference $\beta - \alpha$, for vectors α and β, with the sum $\beta + (-\alpha)$.

A5. Addition is commutative: that is, $\alpha + \beta = \beta + \alpha$, for arbitrary $\alpha, \beta \in V$.

M1. With every pair $r \in \mathbf{R}$ and $\alpha \in V$ there is associated a unique vector $r\alpha$, called the *product* of r and α.

M2. Multiplication by scalars is associative: that is, $r(s\alpha) = (rs)\alpha$, for arbitrary $r, s \in \mathbf{R}$ and any vector α.

M3. Multiplication by scalars is distributive with respect to vector addition: that is, $r(\alpha + \beta) = r\alpha + r\beta$, for arbitrary $\alpha, \beta \in V$ and any $r \in \mathbf{R}$.

M4. Multiplication by scalars is distributive with respect to scalar addition: that is, $(r + s)\alpha = r\alpha + s\alpha$, for arbitrary $r, s \in \mathbf{R}$ and any vector α.

M5. $1\alpha = \alpha$, for the real number 1 and any $\alpha \in V$.

We make no claim that the above axioms are logically independent (for example, see Prob. 23) but present them merely as a convenient characterization of the mathematical system to be studied. Before giving any examples of vector spaces, it is appropriate that we make a few further comments on their defining properties.

(i) The vector space that we have defined is *real* because the *multipliers* are real numbers. We refer to these real numbers as the *scalars* associated with the vector space, but in more general circumstances such scalars or multipliers may be elements of any *field F*, and such a vector space is said to be *over F*. Inasmuch as we are not assuming any knowledge of fields, and since little would be gained from such a generalization, *we shall assume that our vector spaces are real (i.e. over* **R**), except in the final chapter.

(ii) Axioms A1 to A4 describe V as an *additive group*, and, if Axiom A5 is included, as an additive *abelian* group. This terminology is convenient and will be familiar to the student of abstract algebra. We shall often refer to this additive structure of V.

(iii) The symbol 0 in this text will have many different meanings, but the context should make clear what is meant at any particular occurrence. For example, 0 designates a vector in Axioms A3 and A4, but the valid (Prob. 16a) equality $0\alpha = 0$ contains the *real number* 0 in its left member and the *vector* 0 in its right member.

(iv) We prefer the terminology *multiplication by scalars* to the more customary *scalar multiplication* in order to avoid any possible confusion with another type of product which is variously called a *scalar product, dot product*, or *inner product*. This type of product will be introduced in Chap. 4.

(v) The student of abstract algebra may be disturbed by our reference to multiplication by scalars as an "operation" in the vector space V, and we have nothing but sympathy for this feeling. Indeed, a (binary) operation defined in an algebraic system regularly associates an element of the system *with each pair of elements of the same system*. In the case of the operation under discussion, however, one member r of the pair (r,α) is from the system of real numbers **R**, and the other member α is from the space V. This type of "operation" is more correctly described as an *endomorphism of the additive group of V*, but it would take us too far astray to make the meaning of this phrase precise. Hence we shall leave the definition of the operation as given, in the belief that the beginning student will be willing to accept it, in spite of our adversely critical comments, and the more experienced student will understand what is meant by this customary terminology.

We now turn to some examples of real vector spaces, and of such there is a great abundance. Several of these examples will recur throughout the book, and we shall continue to use the notation established here. In the discussions of these examples, we shall merely describe the sets of elements and the two basic operations, leaving for the reader the verifications that the resulting systems are vector spaces. Some of these verifications will be requested in the problems, however.

(1) The real numbers **R** and the complex numbers **C** comprise real vector spaces, with ordinary addition and multiplication by reals considered the vector-space operations.

(2) The system of directed line segments in a plane or 3-space, with addition and multiplication by scalars defined as in elementary physics, can be considered to form a real vector space. We shall examine this space of *geometric vectors* in more detail in Sec. 1.2. Although the notion that directed line segments or "arrows" are the progenitors of the modern concept of a vector is now viewed with disfavor in most quarters, it is nonetheless true that a good historical argument can be made to substantiate this claim. However, it is better to base the *algebraic* concept of a vector on algebraic ideas, rather than on ideas which are geometric.

(3) Let P be the set of all real polynomials in an indeterminate or real variable t. Then, if the operations of addition (vector) and multiplication by a real number (scalar) are performed as in college algebra, the set P becomes a real vector space.

(4) If P_n is the set of all real polynomials of degree less than n in an indeterminate or real variable t, and including the "polynomial" 0, this set is a vector space when it is assigned operations like those in **3**. It is convenient to assign the degree $-\infty$ to the number 0, so that P_n may be described as the space of all polynomials in t *of degree less than n*.

(5) A very important vector space has for its elements the set of all n-tuples of real numbers, the operations in this set being defined as follows:

$$(x_1, x_2, \ldots, x_n) + (y_1, y_2, \ldots, y_n) = (x_1 + y_1, x_2 + y_2, \ldots, x_n + y_n)$$

$$r(x_1, x_2, \ldots, x_n) = (rx_1, rx_2, \ldots, rx_n) \quad \text{for any } r \in \mathbf{R}$$

This is the *n-dimensional real coordinate* space and will be denoted by $V_n(\mathbf{R})$ or, more briefly, V_n. The meaning of *n-dimensional* will be clarified in a later section, but tentatively we may associate the *dimension* of V_n with the number n of entries or *components* in the *n*-tuples. We shall see later that if $2 \leq n \leq 3$ this space has a very close relationship with the corresponding geometric space described in **2**. The elements of this space shall be called *coordinate vectors* in the sequel, the significance of the word *coordinate* being clarified later.

(6) All real-valued functions on **R** form a real vector space, with the usual operations for functions f and g:

$$(f+g)(x) = f(x) + g(x)$$
$$(rf)(x) = rf(x) \quad \text{for arbitrary } r \in \mathbf{R}$$

(7) The subset of continuous functions, selected from the system in **6**, forms a real vector space.

(8) A real vector space is formed from the set of all real-valued functions that are defined and integrable on the interval [0,1], the operations being as in **6**.

(9) The set of all solutions of the differential equation $y'' + y' + y = 0$, with the operations on these solutions defined as for functions in **6**, comprises a real vector space.

(10) A rather trivial but ever-present real vector space is one whose single element is the vector **0**. This *zero* space will be denoted by the symbol ☉.

In these examples, we have used the word *real* in each instance in order to emphasize the nature of the multiplying scalars. However, in view of our previously announced plan to restrict our attention, except in the final chapter, to this type of vector space, we shall often omit this adjective in our descriptions. *It should be understood that such vector spaces are considered to be real.* Although there will be some variation in notation for particular spaces, we shall always use Greek letters ($\alpha, \beta, \gamma, \ldots$) to denote abstract vectors and italic capital letters ($X, Y, Z, X_1, Y_1, \ldots$) to denote coordinate vectors.

EXAMPLE 1

Let V be the space of all the real functions (see **6**), with $f, g \in V$, so that $f(x) = 2x + 1$ and $g(x) = 3x^2$, for any $x \in \mathbf{R}$. Then,

$(f+g)(x) = 3x^2 + 2x + 1$ and $(3f)(x) = 3f(x) = 6x + 3$. Moreover, $(f-g)(x) = [f + (-1)g](x) = f(x) + [(-1)g](x) = f(x) - g(x) = 1 + 2x - 3x^2$.

EXAMPLE 2

Let $X = (1,-1,2,4)$ and $Y = (2,-3,4,6)$ be elements of V_4. Then $X + Y = (1,-1,2,4) + (2,-3,4,6) = (3,-4,6,10); 3X = 3(1,-1,2,4) = (3,-3,6,12); X - Y = (1,-1,2,4) - (2,-3,4,6) = (1,-1,2,4) + (-1)(2,-3,4,6) = (1,-1,2,4) + (-2,3,-4,-6) = (-1,2,-2,-2)$.

Problems 1.1

1. State the theorems that must be proved in order to establish Axioms A1 and M1, for **7** and **8** of the vector-space examples.
2. If $X = (1,1,-2,2)$ and $Y = (2,1,-1,0)$ are elements of V_4, determine (a) $X + Y$; (b) $X - Y$; (c) $2X$; (d) $3X - 4Y$.
3. Let $f,g \in V$, where V is the vector space of all real functions on **R**. Then, if $f(x) = x^2 - 1$ and $g(x) = 2x + 1$, describe (a) $2f - g$; (b) $f + 2g$; (c) $3(f - g)$.
4. If $X = (2,1,-1)$, $Y = (1,0,1)$, and $Z = (-1,0,0)$ are elements of V_3, determine (a) $X + Y - Z$; (b) $X + 2Y$; (c) $2X - Y + 2Z$.
5. If x and y designate directed line segments in a plane, recall from elementary physics how one can determine (a) $x + y$; (b) $2x$; (c) $2x + 3y$; (d) $2x - 3y$.
6. Is **R** a real vector space over itself? What about **R** over **Q** and **Q** over **R**? In the cases where a vector space is defined, identify the vectors and scalars.
7. If $\alpha + \beta = \gamma$, for vectors α, β, γ, prove that $\beta = \gamma - \alpha$.
8. Check that it is possible to define a vector space whose single element is 0.
9. Decide whether one should distinguish between V_1 and **R** as real vector spaces.
10. Use whatever formulation of complex numbers is most familiar to you to verify that they comprise a vector space over **R**.
11. Verify that **4** and **5** satisfy all the requirements as examples of a vector space.
12. Verify that the coordinate space V_n satisfies all the conditions of a vector space.
13. Verify that the system of functions given in **6** satisfies all the required conditions of a vector space.

14. Identify the vector 0 for each of the 10 examples of vector spaces described in this section.
15. Prove that the solutions of the differential equation $y'' + y' + y = 0$, with y defined as a function on **R**, form a real vector space.
16. If α is a vector and r is a scalar, regard 0 appropriately and prove that (a) $0\alpha = 0$ [*Hint:* $\alpha = 1\alpha = (1 + 0)\alpha$, and use uniqueness of 0]; (b) $(-1)\alpha = -\alpha$ [*Hint:* Use the uniqueness property in Axiom A4]; (c) $r0 = 0$.
17. If $r\alpha = 0$, for a scalar r and vector α, prove that $r = 0$ or $\alpha = 0$ (or both).

18. If V_2 is defined like the space in **5**, except that $r(x_1,x_2) = (0,rx_2)$, decide whether the new system is a vector space.
19. Prove that the assumption of uniqueness of the vector 0 in Axiom A3 is redundant. [*Hint:* Assume the existence of a vector $0'$ such that $\alpha + 0' = \alpha$, for all $\alpha \in V$.]
20. If $\alpha + 0' = \alpha$, for any particular vector α and some vector $0'$, prove that $0' = 0$. Note that this is a stronger result than that obtained in Prob. 19.
21. Prove that the assumption of uniqueness of the vector $-\alpha$ in Axiom A4 is redundant. [*Hint:* Assume that $\alpha + \alpha' = 0$, for some $\alpha' \in V$, and prove that $\alpha' = -\alpha$.]
22. Establish the uniqueness of the vectors 0 and $-\alpha$ in Axioms A3 and A4 [cf. Probs. 20 and 21] without the use of Axiom A5.
23. Prove that Axiom A5 can be derived from the other axioms, so that it is actually redundant. [*Hint:* Consider $(1 + 1)(\alpha + \beta)$.]
24. Let S be the set of all ordered pairs (x,y) of real numbers, with addition and multiplication by scalars defined as follows:

$$(x_1,y_1) + (x_2,y_2) = (2y_1 + 2y_2, -x_1 - x_2) \qquad r(x_1,y_1) = (2rx_1, -ry_1)$$

Verify that this algebraic system is *not* a vector space.
25. Explain why the solutions of the differential equation $y'' + y' + y = 5$, with y defined as a function on **R**, do not comprise a vector space.

1.2 Directed Line Segments as Vectors

We now proceed to a more detailed discussion of the system of directed line segments, that was briefly described in **2** in Sec. 1.1. In (a) below, we discuss vectors of this type as they appear in elementary science, and in (b) we illustrate their application to problems in euclidean geometry.

8 FINITE-DIMENSIONAL VECTOR SPACES

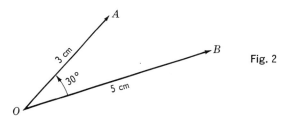

Fig. 2

(a) In elementary science, a directed line segment is often used to represent a measure of some physical quantity, such as velocity, acceleration, or force. It is inherent in the nature of these *vector quantities* that they are not satisfactorily measured by real numbers, as is the case, for example, with the length of a bar or the volume of a can: for somewhere in a measure of a vector quantity a direction must be recorded. If a line segment is to represent a force of 10 lb, it is customary to use a segment 10 units in length whose direction, as indicated by an arrow, will denote the direction of the force. Whether this designated direction is considered actual or relative will depend, of course, on the circumstances of the representation. In Fig. 2, by way of illustration, two forces of magnitudes 3 and 5 lb, which act at the same point but differ in direction by 30°, are satisfactorily represented by the two line segments of lengths 3 and 5 cm inclined to each other at an angle of 30°. In the interest of simplicity, we shall assume that the representing line segments in this discussion are coplanar, but the extension of our remarks to segments in ordinary space should be apparent.

The set of coplanar directed line segments then supplies the basic ingredients of a mathematical system, the line segment directed from a point A to a point B being often designated by AB, with $|AB|$ denoting its length. If this set is to be endowed with the algebraic structure of a vector space, we must clarify our meaning of *equality* ($=$) and describe the two characteristic operations of such a system. In algebraic systems, it is customary to *equate* two elements a and b (and write $a = b$) if and only if a and b are identical elements, but in the context of euclidean geometry, line segments are considered identical only if they contain the same points. For instance, the opposite sides of a parallelogram are certainly equal in length, and they may even be considered to have the same direction, but *as line segments* they would not be identical. However, we *would* regard such distinct line segments as defining or representing *the same vector*, according to the following definition.

Definition   *If AB and CD are directed line segments, they are said to define the same geometric vector—also designated as AB or CD—provided the segments have the same length and direction.*

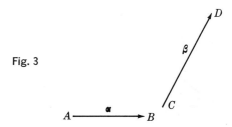

Fig. 3

Expressed otherwise, this means that *a geometric vector may be represented by any one of an infinite number of distinct line segments*, all of which have the same direction and are equal in length (see Prob. 13). It is reasonable to regard two geometric vectors as *parallel* if they are represented by line segments having the same or opposite directions.

It is quite likely that the student of elementary science is familiar with the operations of *addition* and *multiplication by scalars* in the context of geometric vectors. He will know that, when vector quantities are combined, the magnitude of the resultant or *sum* is not necessarily the arithmetic sum of the magnitudes of the components. For example, the sum of a 5-lb force and an 8-lb force is not necessarily 13 lb but may vary from 3 to 13 lb, depending on the directions of the components. The correct way, *as determined by experiment*, to measure the resultant of two vector quantities is to apply the *triangle* or *parallelogram* law to representative geometric vectors.

Addition

If α and β are geometric vectors represented, respectively, by AB and CD in Fig. 3, either of the following two procedures may be used in order to determine the sum $\alpha + \beta$:

(i)   Place the tail of a segment representing β on the head of a segment representing α (Fig. 4), and join the tail of the latter to the head of the former.

(ii)   Represent α and β by segments whose tails coincide (Fig. 5), complete the parallelogram, and draw the diagonal from this common point.

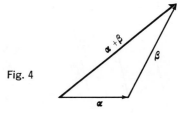

Fig. 4

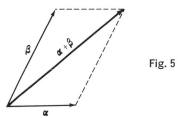

Fig. 5

The method of addition in procedure (i) is called the *triangle law*, and that in procedure (ii) is called the *parallelogram law*.

If two vector quantities have the same direction but their magnitudes differ, it would be only natural to represent them by line segments with the same direction but with lengths proportional to their magnitudes. This leads to the other vector-space operation, as illustrated in Fig. 6.

Multiplication by a Scalar

If α is a geometric vector, represented by the segment AB, and $r \in \mathbf{R}$, then $r\alpha$ is a geometric vector that may be represented by a segment $|r|$ times as long as AB and having the same or opposite direction according as $r > 0$ or $r < 0$. If $r = 0$, then $r\alpha$ is the vector 0, independent of direction but with *length* 0.

If we now agree to regard $-BA$ as an alternative way of denoting the directed line segment AB and *postulate* the existence of the unique vector **0**, we may claim that the system of geometric vectors forms a vector space. The truth of this remark depends on the verification of all the conditions A1 to A5 and M1 to M5 in this system. We leave the details of this to the reader but suggest a few verifications in the problems. This will then justify our terminology of geometric *vectors* as used in this section.

The following point should be considered to be of more than passing interest: Although the conditions of a vector space were *postulated* in an abstract manner, the structural rules for the system of geometric vectors were *imposed*

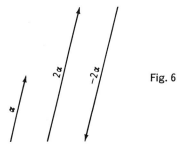

Fig. 6

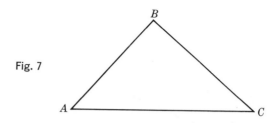

Fig. 7

upon us primarily by our observations of physical quantities. However, the fact that this system is a vector space now makes all the theory of abstract vector spaces applicable in the present context. Since it is likely that the reader is somewhat familiar with applications of geometric vectors to elementary science, we shall omit illustrations of this kind and merely include a few pertinent problems at the close of this section.

(b) In addition to applications in elementary science, it is interesting that geometric vectors can be used to establish many of the propositions of euclidean geometry, in spite of the fact that line segments in this geometry are generally not directed. The basic reason for this is the relationship that exists between the three sides of a triangle, as described by the *triangle law*: for the fact that three line segments comprise the sides of a triangle can be expressed as a vector equation. Although we have already noted that the line segments of geometry are usually not directed, it is clear that they may be given directions that will transform them into acceptable representations of geometric vectors. For example, the relationship between the sides of the triangle in Fig. 7 may be expressed as $AB + BC = AC$, or, alternatively, as $AB - CB + CA = \mathbf{0}$, and one can think of many more.

It should be pointed out that if line segments have any sort of algebraic involvement, as in the preceding equation, they *must* be regarded as directed and, in fact, as representing geometric vectors. When they are not so involved, the context should make clear whether they are directed or undirected segments. To illustrate the application of geometric vectors to geometry, we shall prove a very familiar proposition of Euclid.

EXAMPLE 1

The medians of a triangle intersect at a point located two-thirds of the distance along any median drawn from its associated vertex.

Proof

The situation is illustrated in Fig. 8, where O is the point of intersection of the medians EC and AF. Our procedure will be to obtain two expressions for the *vector* AO and equate them to obtain the desired result. We first let

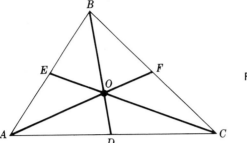

Fig. 8

$AO = r(AF)$ and $EO = s(EC)$, for unknown real numbers r, s, where it should be understood in this proof that we regard all line segments as directed and as representing vectors. From $\triangle ABF$, $AB + BF = AB + \frac{1}{2}BC = AF$, so that $AO = r(AB + \frac{1}{2}BC)$. From $\triangle EBC$, $EC = EB + BC = \frac{1}{2}AB + BC$. Hence, from $\triangle AEO$, $AO = AE + EO = \frac{1}{2}AB + s(EC) = \frac{1}{2}AB + s(\frac{1}{2}AB + BC)$. On equating these two expressions for AO, we obtain $r(AB + \frac{1}{2}BC) = \frac{1}{2}AB + s(\frac{1}{2}AB + BC)$, whence $(\frac{1}{2} + s/2 - r)AB + (s - r/2)BC = 0$. Inasmuch as AB and BC are intersecting segments, it follows that $\frac{1}{2} + s/2 - r = 0$ and $s - r/2 = 0$. Hence $r = \frac{2}{3}$ and $s = \frac{1}{3}$. A similar proof will show that BD also passes through the point O.

Two observations are in order in connection with the method of this example. First, if two geometric vectors are representable by nonparallel line segments, any third coplanar vector can be expressed as a *linear combination* of them. For example, in Fig. 9a, we have represented nonparallel vectors

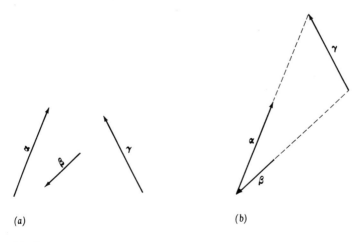

(a)   (b)

Fig. 9

α, β and a third vector γ. In Fig. 9b, we have represented α and β as directed segments with a common endpoint, and it is clear that γ can be represented by the third side of a triangle whose other sides are scalar multiples of α and β. The diagram shows that $\gamma = r\alpha + s\beta$, for certain positive real numbers r, s; if γ had been chosen otherwise, either or both of the numbers r, s might be negative. If the representative segment for γ happens to be parallel to that representing either α or β, no triangle can be formed, as in the illustration; in this case $\gamma = r\alpha = r\alpha + 0\beta$ or $\gamma = 0\alpha + r\beta$, for some real number r. Hence, in every case, γ is a *linear combination* of α and β. The other observation is that if $r\alpha + s\beta = r'\alpha + s'\beta$, for real numbers r, r', s, s' and nonparallel geometric vectors α and β, then $r = r'$ and $s = s'$. This follows because the given equality can be expressed in the form $(r - r')\alpha = (s' - s)\beta$, an assertion that certain scalar multiples of α and β are equal, in contradiction (unless $r = r'$ and $s = s'$) to the assumption that α and β are not parallel.

Any two nonparallel plane vectors, to which the preceding remarks then apply, are called *basis* vectors for the plane. We shall have much to say about basis vectors in a more general environment in the sequel, at which time they will be defined more precisely. In general, as a guide for the solution of problems in plane geometry by vector methods, the following suggestions may be helpful:

(1) Select two intersecting line segments—usually component parts of the figure—as representing plane vectors, and then express some other line segment of the figure in two (or more) ways as linear combinations of the two selected, all segments being assigned directions.

(2) Equate corresponding coefficients of the equivalent expressions.

We close this discussion of plane vectors with another illustration from euclidean geometry.

EXAMPLE 2

> The line segment joining one vertex of a parallelogram with the midpoint of the opposite side trisects the intersecting diagonal and is itself trisected.

Proof
The situation is illustrated in Fig. 10, where E is the midpoint of side AB of the parallelogram $ABCD$. For the proof, we select AD and AB as respective representative directed line segments for basis vectors α and β, and let $AF = s(AC)$ and $EF = r(ED)$, for unknown real numbers r, s. Now $AE + ED = AD = \alpha$, so that $ED = \alpha - AE = \alpha - \frac{1}{2}\beta$, whence $AF = AE + EF = \frac{1}{2}\beta + r(\alpha - \frac{1}{2}\beta)$. Moreover, $AF = s(AC) = s(AD + DC) = s(\alpha + \beta)$.

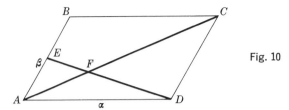

Fig. 10

Hence, $s(\alpha + \beta) = \frac{1}{2}\beta + r(\alpha - \frac{1}{2}\beta)$, $s\alpha + s\beta = r\alpha + \frac{1}{2}(1 - r)\beta$, and so $r = s$ and $s = (1 - r)/2$. From these equalities, it follows that $r = s = \frac{1}{3}$, as asserted.

It should be clear that problems in solid geometry can be solved in a similar manner, but since *three* noncoplanar vectors are necessary to constitute a basis for the space of geometric vectors in ordinary space, the solution of such problems is somewhat more complicated. We shall not illustrate the solution of this type of problem here and shall even add a word of caution concerning the type of problems already considered. Although we have given a "vector" characterization of the geometric concept of parallelism of line segments, we have made no attempt to give similar consideration to segments that are perpendicular or in some other nonparallel relationship to each other. Hence, lest the reader becomes too overenthusiastic about the "vector" method of proving propositions from euclidean geometry, it must be pointed out that other concepts must be introduced before any attempt can be made to solve problems involving measures of angles. This will be postponed until Chap. 4, where we shall study the way in which the angle between two line segments can be described in vector terms.

Problems 1.2

1. Is there a one-to-one correspondence between the directed line segments and the geometric vectors in a plane? Discuss.
2. Explain why it would be appropriate to represent all geometric vectors by line segments that emanate from the same point.
3. Represent two geometric vectors α and β by directed line segments drawn at random in the plane, and then construct a representation of $r\alpha + s\beta$ where (a) $r = \frac{1}{2}$, $s = 2$; (b) $r = -2$, $s = 3$.
4. Represent four geometric vectors by directed line segments drawn at random in the plane, and then determine a representative line segment for their sum.

5. Represent three geometric vectors by directed segments drawn at random in the plane, construct a triangle whose sides are real multiples of the segments, and use the triangle to express one vector *approximately* in terms of the other two.
6. In the space of plane geometric vectors, is the equality $(-1)\alpha = -\alpha$ *provable* for any vector α, as in Prob. 16b of Sec. 1.1?

7. With α and β geometric vectors, explain why $\alpha + \beta = \beta + \alpha$.
8. With α, β, γ geometric vectors, explain why $(\alpha + \beta) + \gamma = \alpha + (\beta + \gamma)$.
9. Explain why $r(s\alpha) = (rs)\alpha$, for any geometric vector α and $r, s \in \mathbf{R}$.
10. Use geometric considerations to show that $(r + s)\alpha = r\alpha + s\alpha$, for real numbers r, s and an arbitrary geometric vector α.

11. An object is acted upon by forces of 20 and 50 lb, the directions of the forces differing by 35°. Use trigonometry to determine the magnitude and direction of the resultant force.
12. A man, who can paddle a canoe at a speed of 4 ft/sec in still water, wishes to reach a point directly across a stream 100 ft wide. If he is able to maintain a constant speed, and if the water is flowing at the rate of 2 ft/sec, in what direction should he head his canoe?
13. In a set S, we say that a *relation* $\mathcal{R}$ has been defined if, for every pair of elements $a, b \in S$, it is possible to decide whether the statement "a is in the relation $\mathcal{R}$ to b" (written $a\mathcal{R}b$) is true or false. The relation $\mathcal{R}$ is an *equivalence* relation if the following properties hold: (i) $a\mathcal{R}a$, for all $a \in S$; (ii) if $a\mathcal{R}b$, then $b\mathcal{R}a$; (iii) if $a\mathcal{R}b$ and $b\mathcal{R}c$, then $a\mathcal{R}c$. Explain why *equality* of geometric vectors is actually an equivalence relation in the set of all directed line segments.
14. Prove that an equivalence relation in a set partitions the set into nonoverlapping subsets of equivalent elements.

Note Use vector methods to establish each of the following propositions:

15. The diagonals of a square are equal in length.
16. The diagonals of a parallelogram bisect each other.
17. The midpoint of the hypotenuse of a right triangle is equidistant from the three vertices.
18. The line segment joining the midpoints of two sides of a triangle is parallel to and half as long as the third side.

19. The line segments joining midpoints of opposite sides of a quadrilateral bisect each other.
20. The diagonals of an isosceles trapezoid are equal in length.
21. If three line segments are drawn through an interior point, parallel to and intersecting the sides of a triangle, the sum of the respective ratios of lengths of the segments to the parallel sides is 2.
22. Let O be any point in the plane of $\triangle ABC$ but exterior to it, with M the centroid of the triangle. Then $|OM| = (|OA| + |OB| + |OC|)/3$.
23. Let P be the point that divides a line segment AB in a given ratio $m:n$. Then, if O is any point in a plane containing AB,

$$|OP| = \frac{n|OA| + m|OB|}{m + n}$$

1.3 Geometric Vectors and Coordinate Spaces

One of the most important concepts in basic abstract algebra is that of *isomorphism*. We shall not give its definition as it applies to general algebraic systems but only for the special case of vector spaces. In intuitive terms, two vector spaces are isomorphic if they are indistinguishable as abstract systems, but the following definition makes the notion more precise.

Definition   *Let V and V' be two real vector spaces. Then V and V' are said to be isomorphic if there exists a one-to-one correspondence $\alpha \leftrightarrow \alpha'$ between the elements α of V and α' of V' such that (i) $\alpha_1 + \alpha_2 \leftrightarrow \alpha_1' + \alpha_2'$ and (ii) $r\alpha_1 \leftrightarrow r\alpha_1'$, for arbitrary $\alpha_1, \alpha_2 \in V$, the corresponding $\alpha_1', \alpha_2' \in V'$, and any $r \in \mathbf{R}$.*

A slight variant of this definition arises if we note that a *one-to-one correspondence* $\alpha \leftrightarrow \alpha'$ between the elements α of V and the elements α' of V' is actually equivalent to the *one-to-one mapping* $\alpha \to \alpha'$ of V onto V' (or the one-to-one mapping $\alpha' \to \alpha$ of V' onto V). We may then say that a one-to-one mapping $\alpha \to \alpha'$ of V onto V' is an *isomorphism* or *isomorphic mapping* if (i) $\alpha_1 + \alpha_2 \to \alpha_1' + \alpha_2'$ and (ii) $r\alpha_1 \to r\alpha_1'$, for any $\alpha_1, \alpha_2 \in V$, the corresponding images $\alpha_1', \alpha_2' \in V'$, and any $r \in \mathbf{R}$. If $\alpha \to \alpha'$ is an isomorphism of V onto V', it is clear that the reverse mapping $\alpha' \to \alpha$ is an isomorphism of V' onto V. When we say that two spaces are *isomorphic* we are indifferent as to which way the mapping takes place. It often happens that an isomorphism maps a space V onto *subset* of a space W, and in this event the mapping is said to be an isomorphism of V *into* W.

We pointed out in Sec. 1.2 that our emphasis on geometric vectors in a *plane* was a matter of convenience and that a similar discussion could be based on line segments in ordinary space or, somewhat trivially, in the line. The geometries of the line, plane, and ordinary space constitute what is ordinarily recognized as the geometry of Euclid. To add variety to our treatments, we first show in this section that the system of geometric vectors in *ordinary space* is isomorphic to the coordinate space V_3, and it will be clear that a similar relationship can be established between the systems of geometric vectors in the plane and line and V_2 and V_1, respectively. In setting up any of these isomorphisms, a key role will be played by a *linear coordinate system*, the basic idea of which will be assumed to be well known to the reader. In ordinary space, four noncoplanar points P_0, P_1, P_2, P_3 are required to determine such a system: one of the points, say P_0, is taken as the *origin*, and the others are *unit* points that determine the unit distances along the *axes* that contain them. If these axes are mutually perpendicular and if all unit segments are the same in absolute length, the coordinate system is said to be *cartesian*. In studies of analytic geometry it is customary to assume that coordinate systems are cartesian but we shall not make that assumption here. Figure 11 illustrates the assignment of coordinates to an arbitrary point P in ordinary space.

In order to establish an *isomorphism* between the system of geometric vectors in ordinary space and the coordinate space V_3, it is necessary to exhibit a one-to-one correspondence between the elements of the two systems and then show that requirements (*i*) and (*ii*) of the definition are satisfied. We first reemphasize a very important feature of the geometric system: *The complete set of geometric vectors can be represented by directed line segments, each of which emanates from the same point, so that any such segment is uniquely*

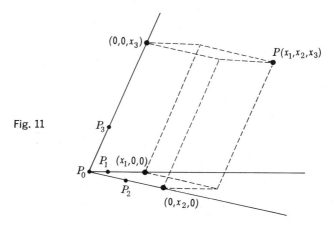

Fig. 11

18 FINITE-DIMENSIONAL VECTOR SPACES

determined by its endpoint. The assumption is now made that the geometric vectors are represented in this manner, and the desired one-to-one correspondence appears quite naturally. We set up a linear coordinate system, using the common point of the representing line segments as origin, and associate each geometric vector with the coordinate triple of the endpoint of the line segment that represents it. Thus, if α is a geometric vector represented by the line segment drawn from the origin to the point whose coordinate triple is (x,y,z), the correspondence is

$$\alpha \leftrightarrow (x,y,z)$$

This correspondence is clearly one-to-one between the set of geometric vectors and the elements of V_3. We again emphasize that α has many directed line segments that *may* represent it, but the one used in this correspondence is that which emanates from the origin of the linear coordinate system. The operations in both vector spaces have already been defined, and it is a simple matter to verify that requirements (i) and (ii) are satisfied, so that the correspondence is an isomorphism. We include only a brief outline of these verifications.

Let α_1 and α_2 be geometric vectors, represented (as shown in Fig. 12) by line segments whose endpoints have coordinate triples (x_1,y_1,z_1) and (x_2,y_2,z_2), respectively. An application of the parallelogram law in the geometric system determines $\alpha_1 + \alpha_2$, and it follows from a simple geometric argument that the coordinate triple of the endpoint of its representing segment is $(x_1 + x_2, y_1 + y_2, z_1 + z_2)$. But this means that, if $\alpha_1 \leftrightarrow (x_1,y_1,z_1)$ and $\alpha_2 \leftrightarrow (x_2,y_2,z_2)$, then $\alpha_1 + \alpha_2 \leftrightarrow (x_1 + x_2, y_1 + y_2, z_1 + z_2)$, so that condition (i) is verified. Figure 13 shows that the coordinate triple of the endpoint of the segment

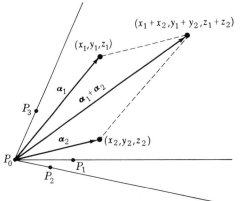

Fig. 12

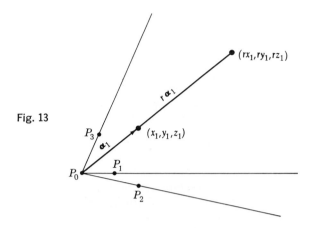

Fig. 13

representing $r\alpha_1$, for $r \in \mathbf{R}$, is (rx_1, ry_1, rz_1); this shows that condition (ii) is also satisfied. The proof of the isomorphism is then complete.

As indicated by an earlier remark, it is clear that the preceding discussion would be altered only superficially if we considered the space of geometric vectors in the *plane* and the associated space V_2 of ordered pairs (x,y). In the analogous correspondence $\alpha \leftrightarrow (x,y)$, the geometric plane vector α, represented by the line segment drawn from a chosen origin to a point P in the plane, is associated with the coordinate pair (x,y) of P. The space of geometric vectors in the *line* is rather uninteresting, but it is not difficult to see that it is isomorphic to $\mathbf{R}$, considered as a real vector space. In this case, the coordinatized geometric space may be identified with the *real line*, and the isomorphism associates each geometric vector drawn from an origin with the real-number coordinate of its endpoint.

It is not possible to generalize the geometry of ordinary space to a space of higher "dimension" if we insist on being able to represent its elements in some pictorial fashion. However, if we abandon this requirement, the fact that we regularly identify a *point* with its coordinate(s) is a simple guide to further generalization. All that is really "geometric" in such a generalization is the language, for we *define* the *points* of such an n-space ($n > 3$) to be the set of all n-tuples of real numbers $(x_1, x_2, \ldots, x_n)$. In this space, the point $(0,0, \ldots, 0)$ may be taken as the *natural* origin, and the points $(1,0,0, \ldots, 0), (0,1,0, \ldots, 0), \ldots, (0,0,0, \ldots, 0,1)$ may be taken as the unit points associated with a *natural* coordinate system. From this foundation of points, a complete geometry of n-space can be developed, including, of course, analogous concepts of lines and planes in this space. A development of this geometry would contain a proof, very similar to that given in this section, that the system of "geometric" vectors in n-space is isomorphic to the coordinate space V_n. We shall not pursue this geometry of n-space further, for we prefer to leave such matters to books with a greater emphasis on geometry. However, we shall see in the next paragraph that the idea of a *point* in n-space is a very convenient one in any discussion of the coordinate space V_n.

We have shown in this section that the system of geometric vectors in ordinary space is isomorphic, and so essentially equivalent, to the coordinate space V_3, and we have indicated that similar results hold for spaces of higher "dimensions." It may have been thought from the above discussion that the principal use of any coordinate space is to provide a means for giving an algebraic representation of a geometric space. However, this is entirely incorrect, for the coordinate spaces V_n are the ones that are of primary importance in this book rather than the spaces of geometry. In Sec. 1.9 we shall derive a result that shows why this is so and also give the important reason why the adjective *coordinate* is attached to a V_n space. It is considered intuitively helpful by many people to have a *geometric* representation of a coordinate space, and, for $n < 4$, the isomorphic space of geometric vectors may be used. However, instead of this representation by directed line segments, *we prefer to use the familiar association of the coordinates of a point with the point itself* and to introduce line segments only if operations in the space are to be performed geometrically. For example, we prefer to regard the geometric representation of $(x,y,z) \in V_3$ as the *point*, with (x,y,z) as its coordinate triple, the line segment joining the origin to the point playing only an "operational" role. In other words, we like to consider the *endpoint* of the associated geometric vector as the geometric correspondent of an element of V_3, as is shown in Fig. 14.

As a further illustration of this, we cite the very familiar case of the *real line*. In this case, the real numbers are considered to be associated in a

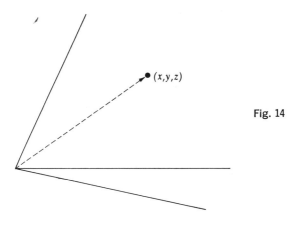

Fig. 14

one-to-one correspondence with the points (*not* line segments) of the line, and the systems of numbers and points are regarded as equivalent. A real number x is associated with the point P whose coordinate is x in the real line, and this number-point association causes no concern *even* when we perform vector-space operations in **R**. For instance, we know (from arithmetic) that $2 + 3 = 5$, and the point representing 5 is just as easy to locate on the line as are the points representing 2 and 3. Only if we wish to perform such an operation *geometrically on the real line itself* is it necessary to introduce line segments joining the origin to points of the line.

A situation, almost as familiar, exists in the case of the complex numbers, for we are quite used to the association of the complex number (x,y) with the point of the complex plane having this as its coordinate pair. Again, the line segments from the origin to the points are useful *only* if we desire to add complex numbers in a geometric fashion, and they fade out of the picture with the completion of the operations. However, what we have said in no way disqualifies a system of geometric vectors, which are represented by directed *line segments*, as a bona fide space isomorphic to the corresponding coordinate space; we are merely stating our preference for the associated space of points as the geometric representation of a coordinate space. If $n > 3$, this is the only natural thing to do because we have *defined* the points of such a general n-space as the n-tuples of V_n.

We are aware of the double use here of an n-tuple $(x_1, x_2, \ldots, x_n)$ as both a *point* in n-space (without algebraic structure) and as an element of V_n (with algebraic structure). However, the context should make clear which meaning is to be understood at any given time, and this double use is in line with common practice. In fact this practice is responsible for the custom of referring to an element of *any* vector space—not just V_n—as a *point* of the space.

Before closing this section, we make a brief comment on some terminology that is common in the literature of science and engineering. The vector quantities of elementary science are often localized to a single point in space, and this point is considered fixed during the course of a given investigation. The associated vectors are then said to be *bound* to this point, whereas vectors not so restricted are *free*. For example, we have been representing the geometric vectors by line segments that emanate from the origin, and these vectors may then be said to be *bound* to the origin. The elements of V_n are all bound in a natural way to the point $(0,0, \ldots, 0)$, but, if so desired, they could be bound to another point $(a_1, a_2, \ldots, a_n)$: for we can write $(x_1, x_2, \ldots, x_n) = (a_1, a_2, \ldots, a_n) + (x_1 - a_1, x_2 - a_2, \ldots, x_n - a_n)$. Note, incidentally, that we have just used $(a_1, a_2, \ldots, a_n)$ both to denote a geometric point and as a vector! In most cases of abstract spaces, the notions of *bound* and *free* do not arise.

Problems 1.3

1. Give separate definitions for *addition* and *multiplication by scalars* in V_2 and V_3, and explain carefully how the elements of these spaces are related, respectively, to the line segments of the plane and ordinary space.
2. Use the points $A(1,-1,2)$, $B(-2,3,5)$, and $C(3,4,-6)$ of ordinary space to verify that the element of V_3, associated with the segment AC by the isomorphism of this section, is equal to the sum of the elements associated with AB and BC.
3. Use the directions given in Prob. 2 for the points $A(2,-5,6)$, $B(4,-2,-4)$, and $C(1,8,7)$.
4. Translate each of the following equalities in V_3 into an equivalent statement about line segments in 3-space: (a) $(1,-2,3) + (0,3,-2) = (1,1,1)$; (b) $2(1,3,-4) = (2,6,-8)$; (c) $2(1,0,0) + 3(0,1,0) - 2(0,0,1) = (2,3,-2)$.
5. Translate each of the following equalities in V_2 into an equivalent statement about line segments in the plane: (a) $(2,1) + (3,2) = (5,3)$; (b) $(1,-1) - 2(3,-2) = (-5,3)$.
6. With operations defined as in V_3, prove that the set of all vector triples $(x_1,x_2,0)$, with $x_1,x_2 \in \mathbf{R}$, comprises a vector space that is isomorphic to V_2.
7. Use the directions given in Prob. 6 but with $(x_1,x_2,0)$ replaced by $(x_1,0,x_3)$.
8. With reference to the examples of vector spaces in Sec. 1.1, describe the "points" of each of the following spaces: (a) P; (b) **C**; (c) the example in **6**; (d) the example in **9**; (e) P_n.
9. Represent each of the following elements of a coordinate space as a point in the appropriate geometric space: (a) $(1,2)$; (b) 3; (c) $(1,-1,2)$; (d) -3; (e) $(0,1,-2)$.

10. Give the geometric proof that line segments in ordinary space, which have the same length and direction, are associated by the isomorphism of this section with the same elements of V_3.
11. Give a geometric characterization of those directed line segments in ordinary space that are associated, by the isomorphism of this section, with the elements (x_1,x_2,x_3) of V_3 such that (a) $x_1 = x_2 = 0$; (b) $x_1 = 0$; (c) $x_1 = x_3 = 0$.
12. Explain why the zero geometric vector in ordinary space is associated by the isomorphism of this section with the element $(0,0,0)$ of V_3.
13. Translate the equality $-\alpha = (-1)\alpha$, with α a geometric vector in ordinary space, into an equivalent statement in V_3.

14. Prove that Axioms A1 to A5 in Sec. 1.1 are satisfied by the elements of V_3.
15. Prove that Axioms M1 to M5 in Sec. 1.1 are satisfied by the elements of V_3.
16. With the numbers 2 and 3 properly associated with *points* of the real line, *use a geometric construction* to verify that $2 + 3$ and 5 are associated with the same point.

17. On the basis of the discussion thus far, give a careful criticism of each of the following "definitions" of a *vector:* (a) an element of a vector space; (b) a directed line segment; (c) an n-tuple of real numbers.
18. Express each of the following elements of V_3 as "bound" to the point $(1,-1,3)$: (a) $(2,1,4)$; (b) $(0,-2,1)$; (c) $(-2,3,5)$.
19. With $A(1,-1,2)$, $B(3,1,-1)$, and $C(2,0,3)$ denoting points of ordinary space, translate the equality $AB + BC + CA = \mathbf{0}$ into an equivalent statement involving elements of V_3.
20. Use properties of V_3 to determine whether the following points in ordinary space are vertices of a parallelogram: $(5,5,1)$, $(3,7,-2)$, $(7,6,0)$, $(5,8,-3)$.
21. If $(-1,3)$ and $(2,-4)$ denote complex numbers, use both an algebraic and a geometric method to determine their sum.
22. If $P(1,-1,1,2)$ and $Q(-2,0,1,4)$ are points in 4-space, and a geometric vector is represented by the line segment PQ, what elements of V_4 are associated with PQ and QP by the isomorphism of this section?
23. If $A(x_1,y_1,z_1)$, $B(x_2,y_2,z_2)$, $C(x_3,y_3,z_3)$, $D(x_4,y_4,z_4)$ are points of ordinary space, give an equality in V_3 that implies that (a) AB is parallel to CD; (b) AB and CD are opposite sides of a parallelogram; (c) the diagonals of the parallelogram in (b) are equal in length.
24. A cartesian coordinate system is established in a plane by an origin P_0 and two unit points P_1, P_2. An additional linear coordinate system is now superimposed on the cartesian system with the points $P'_0 (= P_0)$ and the unit points $P'_1(2,0)$, $P'_2(1,1)$. Determine the coordinates in the new system of the following points identified by their coordinates in the cartesian system: (a) $(0,2)$; (b) $(3,4)$; (c) $(6,7)$; (d) $(0,1)$.
25. Generalize the procedure in Prob. 24 as follows: Let $P'_0 = P_0$, let P'_1 be any point on the line through P_0 and P_1, and select P'_2 so that $P_0P'_2$ makes an acute angle c with P_0P_1. Then determine the coordinates, relative to the linear coordinate system determined by P'_0, P'_1, P'_2, of an arbitrary point whose designation in the original system is (x,y).

26. Give a brief commentary on the distinction that we are making between the "points" of ordinary space and the "points" of V_3. Decide whether it is reasonable to make any distinction if $n > 3$.

1.4 Subspaces

Now that we have completed a brief geometric interlude, we return to a discussion of abstract vector spaces. However, before proceeding to the principal topic of this section, we shall comment briefly on the notation that we have been using to designate the operation of multiplication by scalars. It will doubtless have been noticed—and possibly with some puzzlement—that in a vector space V we have defined a product $r\alpha$, for $r \in \mathbf{R}$ and $\alpha \in V$, but no mention has been made of any meaning of a product αr. The former could be thought of as a multiplication of the vector α *on the left* by r, and it would be natural to regard the latter as a multiplication *on the right* by r. The fact is that *no definition has been given, nor shall we give one, for the symbolic product αr*. In the definition of a vector space, we decide once and for all whether scalars are to be written on the left or the right of the vectors that they multiply, and then *we abide by this decision*. When the scalars are written on the left, as we are doing, the space is sometimes spoken of as a *left vector space*, and in the alternative case a *right vector space*. Inasmuch as we are restricting our attention to left vector spaces, we shall omit the adjective *left* in our discussions.

Next to *isomorphism*, probably the most important concept in the study of any algebraic system is that of a *subsystem*. First, let us consider the subset of V_3 consisting of all vectors of the form $(x_1, 0, x_3)$, with $x_1, x_3 \in \mathbf{R}$. We note that $(x_1, 0, x_3) + (y_1, 0, y_3) = (z_1, 0, z_3)$, where $z_1 = x_1 + y_1$ and $z_3 = x_3 + y_3$, and that $r(x_1, 0, x_3) = (w_1, 0, w_3)$, where $w_1 = rx_1$ and $w_3 = rx_3$, for any $r \in \mathbf{R}$. If we make the identification of $(0,0,0)$ with the zero vector 0, and $(-x_1, 0, -x_3)$ or $-(x_1, 0, x_3)$ with the additive inverse of $(x_1, 0, x_3)$, it is immediate that conditions A1 to A5 and M1 to M5 are satisfied by elements of the subset. Hence this subset of V_3 is a vector space in its own right and provides an example of a *subspace* of a vector space.

Definition *A subspace of a vector space V is a subset of V whose elements satisfy all the requirements of a vector space, with the understanding that the operations and their definitions in the subset are the same as in the space V. The trivial subspaces of any space V are V and $\mathbf{0}$; others are said to be nontrivial.*

A subspace may be said to be *embedded* in the whole space, and the definition just given emphasizes that, insofar as operations are concerned, it makes no difference whether elements are regarded as members of a subspace or of the embedding space: the results are the same. It is quite possible to define operations on a subset of a vector space that are different from those of the space in which it is embedded but which nonetheless make the subset into a vector space. However, in such a case, the system evolved from the subset would not be a subspace of the original. The following theorem shows that the verification that a subset of a vector space is in fact a subspace may be reduced to one simple check.

Theorem 4.1 *A nonempty subset W of a vector space V is a subspace of V if W is closed with respect to the operations of V.*

Proof
The only conditions for a vector space that need a nontrivial check are the existence of the zero vector and inverse vectors in the subset. But W is closed under multiplication by scalars, and so $(-1)\alpha = -\alpha \in W$, for any $\alpha \in W$. Hence the inverse of any vector in W is also in W. In like manner, the closure of W under addition implies that, if $\alpha \in W$, then also $\alpha + (-\alpha) = 0 \in W$. This completes the proof. ∎

EXAMPLE 1

Consider the subset of V_4, consisting of all vectors of the form $a(1,1,-1,1) + b(2,0,2,-3)$, for arbitrary $a,b \in \mathbf{R}$. We see easily that $[a_1(1,1,-1,1) + b_1(2,0,2,-3)] + [a_2(1,1,-1,1) + b_2(2,0,2,-3)] = [a_1 + a_2](1,1,-1,1) + [b_1 + b_2](2,0,2,-3)$ where $a_1 + a_2$ and $b_1 + b_2$ are in $\mathbf{R}$; and that $r[a_1(1,1,-1,1) + b_1(2,0,2,-3)] = ra_1(1,1,-1,1) + rb_1(2,0,2,-3)$, with $ra_1, rb_1 \in \mathbf{R}$. Hence W is closed under the operations of V_4, and it follows by Theorem 4.1 that W is a subspace.

It is easy to use the procedure in Example 1 to form subspaces of any vector space V. If $\alpha_1, \alpha_2, \ldots, \alpha_k$ are k vectors of V, a vector in V of the form $a_1\alpha_1 + a_2\alpha_2 + \cdots + a_k\alpha_k$, with $a_1, a_2, \ldots, a_k$ real numbers, is said to be expressed as a *linear combination* of the given vectors. As in the example, we can then verify that the collection of all such linear combinations makes up a subspace of V. This subspace will be known as the subspace *generated* or *spanned* by $\{\alpha_1, \alpha_2, \ldots, \alpha_k\}$ and will be denoted notationally by $\mathcal{L}\{\alpha_1, \alpha_2, \ldots, \alpha_k\}$. The vectors $\alpha_1, \alpha_2, \ldots, \alpha_k$ are the *generators* or *spanning vectors* of the subspace. It is often useful to describe a subspace, generated by a set of vectors, as the set-theoretic intersection of all subspaces that contain these vectors: for an easy application of Theorem 4.1 shows that this intersection is indeed a subspace and so must be identified with the subspace of all linear combinations of the generators (Prob. 10).

If S and T are subspaces of a space V, the intersection $S \cap T$ is also a subspace, but it is easy to see that the set-theoretic union $S \cup T$ is generally *not* a subspace. However, if we define $S + T$ to be the set of all $s + t$, with $s \in S$ and $t \in T$, this *sum* may be shown to be a subspace of V (Prob. 13). This subspace is actually the intersection of all subspaces of V that contain both S and T. If $S \cap T = \varnothing$, and $S + T = V$, each of S and T may be said to be the *complement* of the other in V.

Problems 1.4

1. Decide whether the space of continuous functions on **R** is a subspace of all real functions on **R**, as these are defined in **6** and **7** of Sec. 1.1. What about the differentiable functions?
2. Prove that any two subspaces of a vector space have a nonempty intersection that is also a subspace. Generalize this result to any number of subspaces.
3. If P_4 is the vector space of real polynomials, as described in **4** of Sec. 1.1, describe the subspace generated by $\{1, t, t^2\}$.
4. Decide whether the subset of elements (x_1, x_2, x_3) of V_3, subject to the indicated requirement, is a subspace:
   (a) $x_1 = x_3$
   (b) $x_1 + x_2 = 0$
   (c) $x_1 - x_2 + x_3 = 0$
   (d) $x_2 x_3 = 0$
   (e) $x_1 = x_3 = 0$
   (f) $2x_1 + 3x_2 = 0$
   (g) $x_1 \in \mathbf{Z}$
   (h) $x_1, x_2, x_3 \in \mathbf{Z}$
5. If $X_1 = (1, -2, 0, 1)$, $X_2 = (1, 1, 1, 0)$, $X_3 = (3, 0, 2, 1)$, $X_4 = (0, 3, 1, -1)$ are elements of V_4, prove that $\mathcal{L}\{X_1, X_2\} = \mathcal{L}\{X_3, X_4\}$.
6. Verify that $\mathcal{L}\{1, \sqrt{3}\}$ and $\mathcal{L}\{1 - \sqrt{3}, 1 + \sqrt{3}\}$ are the same subspace of **R**, with **R** considered a vector space over **Q**.
7. Identify (a) the two subspaces of **R** over **R** and (b) three non-isomorphic subspaces of V_2. Give a geometric interpretation to each of these subspaces.
8. Give a geometric description of all subspaces of V_3.
9. If $\alpha_1, \alpha_2, \ldots, \alpha_k \in V$, use Theorem 4.1 to verify that $\mathcal{L}\{\alpha_1, \alpha_2, \ldots, \alpha_k\}$ is a subspace of the space V.
10. With $\alpha_1, \alpha_2, \ldots, \alpha_k$ as in Prob. 9, explain why $\mathcal{L}\{\alpha_1, \alpha_2, \ldots, \alpha_k\}$ must be the intersection of all subspaces of V that contain the given vectors.

11. Prove that the subset of vectors $(x_1, x_2, \ldots, x_n)$ such that $a_1 x_1 + a_2 x_2 + \cdots + a_n x_n = 0$, for fixed real numbers $a_1, a_2, \ldots, a_n$, comprises a subspace of V_n.
12. Define vector-space operations on a subset of V_4 in such a way that the subset becomes a vector space but not a subspace of V_4.
13. If S and T are subspaces of a vector space V, prove that $S + T$ is also a subspace of V.
14. Let V be the space of all real functions f on the interval $[0,1]$, with operations defined as in **6** of Sec. 1.1. Then determine whether the subset of V, subject to the indicated condition, is a subspace of V:
    (a) $f(0) = 0$
    (b) $2f(0) = 1$
    (c) $f(1) = 0$
    (d) $f(x) = f(1 - x)$
    (e) $f(x) > 0$
    (f) $f(1) + f(-1) = 0$
    (g) $f(x) \geq 0$
    (h) $f(x) = ax^2 + bx + c$ with $a (\neq 0), b, c \in \mathbf{R}$

15. Determine the smallest (in the sense of set inclusion) subspace of V_4 containing the following vectors: (a) $(0,0,0,0)$, $(0,1,0,0)$; (b) $(1,0,1,0)$, $(0,0,1,0)$; (c) $(1,0,0,0)$, $(0,1,0,0)$, $(0,0,1,0)$.
16. If $\alpha_1, \alpha_2, \alpha_3$ are vectors such that $\alpha_1 + \alpha_2 + \alpha_3 = 0$, prove that $\mathcal{L}\{\alpha_1, \alpha_2\} = \mathcal{L}\{\alpha_2, \alpha_3\}$.
17. If W_1 and W_2 are subspaces of a vector space V, prove that $W_1 \cup W_2$ is also a subspace if and only if $W_1 \subset W_2$ or $W_2 \subset W_1$.
18. If W_1, W_2, W_3 are subspaces of a vector space V, show that it is not necessarily true that $W_1 \cap (W_2 + W_3) = (W_1 \cap W_2) + (W_1 \cap W_3)$.
19. Change the left member of the "equality" in Prob. 18 so that the new statement is true.
20. If S is the subspace of V_3 consisting of all vectors $(x_1, x_2, 0)$, and T is the subspace generated by $(1,3,0)$ and $(2,1,3)$, determine the subspaces $S \cap T$ and $S + T$.
21. Try to decide whether a nontrivial subspace can have a unique complement.
22. A polynomial $p(t)$, in an indeterminate t, is said to be *even* if $p(-t) = p(t)$ and *odd* if $p(-t) = -p(t)$. Prove that the set P_o of all odd polynomials and the set P_e of all even polynomials form subspaces of the space P of all polynomials, as described in **3** of Sec. 1.1.
23. Prove that the subspaces in Prob. 22 are complements of each other and that $P = P_o + P_e$.

1.5 Solutions of Linear Equations: Gauss Reduction

At this stage it is necessary to interrupt our discussion of vector spaces for a brief study of systems of linear equations. Our objective in this and the following section is very limited, for we shall include only a bare minimum of essentials needed before the important results in Sec. 1.7. To be specific, we need to know *how to solve a given system of linear homogeneous equations* or, even more frequently, *whether a given system of linear homogeneous equations has a solution that is nontrivial.* A more thorough and theoretical treatment of linear equations is the subject of Chap. 3.

For our study of linear equations, both here and in Chap. 3, we shall be concerned with a system of equations of the following form, in which the number n of unknowns is not necessarily the same as the number m of equations:

(1)
$$\begin{aligned} a_{11}x_1 + a_{12}x_2 + \cdots + a_{1n}x_n &= b_1 \\ a_{21}x_1 + a_{22}x_2 + \cdots + a_{2n}x_n &= b_2 \\ &\cdots \\ a_{m1}x_1 + a_{m2}x_2 + \cdots + a_{mn}x_n &= b_m \end{aligned}$$

The coefficients a_{ij} and b_i ($i = 1, 2, \ldots, m$; $j = 1, 2, \ldots, n$) are, of course, real numbers, and the following rectangular arrays of numbers are known, respectively, as the *matrix of coefficients* and the *augmented coefficient matrix* of the system of equations:

$$\begin{bmatrix} a_{11} & a_{12} & \cdots & a_{1n} \\ a_{21} & a_{22} & \cdots & a_{2n} \\ \cdots & \cdots & \cdots & \cdots \\ a_{m1} & a_{m2} & \cdots & a_{mn} \end{bmatrix} \qquad \begin{bmatrix} a_{11} & a_{12} & \cdots & a_{1n} & b_1 \\ a_{21} & a_{22} & \cdots & a_{22} & b_2 \\ \cdots & \cdots & \cdots & \cdots & \cdots \\ a_{m1} & a_{m2} & \cdots & a_{mn} & b_m \end{bmatrix}$$

In the sequel, we shall have much to say about *matrices*, but for the present they are merely arrays of numbers which conveniently record the coefficients of systems of linear equations. Our real concern at this point in the text is with systems in which $b_1 = b_2 = \cdots = b_m = 0$, otherwise known as *homogeneous* systems. However, since the treatment of the *nonhomogeneous* systems, where this condition is not satisfied, is just as easy, we shall investigate this more general situation.

By a *solution* of system (1), we mean an ordered set $\{c_1, c_2, \ldots, c_n\}$ of real numbers such that the equations of the system reduce to arithmetic identities if we let $x_1 = c_1$, $x_2 = c_2$, $\ldots$, $x_n = c_n$. For a given system of equations, there are three solution possibilities: there is a unique solution; there are many sets of solutions; there are no solutions, and in this case the system is *inconsistent*. A homogeneous system with n unknowns *always* has the *trivial* solution $x_1 = x_2 = \cdots = x_n = 0$, but other solutions may also exist. The *Gauss reduction method* for solving a system of equations is a method that is guaranteed to produce *all* existing solutions of a given system. We now examine this method.

Equations or systems of equations are said to be *equivalent* if they have the same solutions. The method of Gauss reduction consists in making successive replacements of the given system of equations by equivalent systems until ultimately there is obtained one that is essentially in solved form. The permissible operations on the equations (i.e., those which do not change the solution set) of a system are called *elementary (row) operations* and are three in number:

Type 1. The interchange of any two equations of the system

Type 2. The multiplication of both members of an equation of the system by a nonzero real number r: that is, a replacement of both members of the equation by r-multiples of themselves

Type 3. The addition of an r-multiple of both members of an equation to the corresponding members of another equation of the system

It is a simple matter to verify that each of these operations, when applied to a system of equations, yields an equivalent system. As a consequence of this, it is then possible to use any finite sequence of these operations on a given system of equations with the assurance that the final system will have the same solutions as the original.

It should be clear that, in performing any of these elementary operations, we are really concerned only with the coefficients and so with the augmented matrix of the system. It follows that we can perform elementary operations on a system of equations by performing the equivalent operations on the horizontal arrays or *rows* of its augmented matrix. For example, if we wish to solve the system

$$\begin{aligned} 2x + y - 2z &= 1 \\ x - 2y + z &= 5 \\ 3x + y - z &= 4 \end{aligned}$$

it is more convenient to work with the associated augmented matrix

$$\begin{bmatrix} 2 & 1 & -2 & 1 \\ 1 & -2 & 1 & 5 \\ 3 & 1 & -1 & 4 \end{bmatrix}$$

In this matrix, the first three vertical arrays, or *columns*, are the respective coefficients of x, y, z, and the final column consists of the *constants* or right members of the equations. Instead of performing the elementary operations on the *equations* of the system, we may then perform the equivalent operations on the *rows* of the matrix to produce a *row-equivalent* matrix. In the above example, it is possible to use a sequence of these operations (to be discussed in more detail later) to obtain the matrix

$$\begin{bmatrix} 1 & 0 & 0 & 2 \\ 0 & 1 & 0 & -1 \\ 0 & 0 & 1 & 1 \end{bmatrix}$$

If we now interpret the columns of this final matrix as respective coefficients of x, y, z, we see that the corresponding "reduced" system of equations is

$$\begin{aligned} x & = 2 \\ y & = -1 \\ z & = 1 \end{aligned}$$

It is clear that this system actually *exhibits* the solution—unique in this case— of the given system as $x = 2$, $y = -1$, $z = 1$.

The final matrix above, from which we were able to obtain the solution of the system of equations so easily, is in *row-echelon* or *Hermite normal* form according to the following definition. A *nonzero* row has at least one nonzero entry, and the *leading entry* of any nonzero row, as referred to in the definition, is the first nonzero number in the row as one reads its entries from left to right.

Definition   *A matrix is in row-echelon or Hermite normal form if the following conditions are satisfied:*
    (1) *The zero rows lie below the nonzero rows.*
    (2) *The leading entry of any nonzero row is* 1.
    (3) *The column that contains the leading entry of any row has* 0 *for all other entries.*
    (4) *If the leading entry of the ith row is in the t_ith column, then $t_1 < t_2 < \cdots < t_r$, where r is the number of nonzero rows.*

From a purely visual perspective, a matrix in this normal form exhibits a sequence of leading 1's which tend to follow a path downward to the right. For example, in addition to the final matrix in the example above, the matrix on the left below is in this form; the one on the right is not.

$$\begin{bmatrix} 1 & 0 & 0 & 0 & 2 \\ 0 & 0 & 1 & 0 & 1 \\ 0 & 0 & 0 & 1 & 2 \\ 0 & 0 & 0 & 0 & 0 \end{bmatrix} \quad \begin{bmatrix} 1 & 0 & 0 & 0 & 1 \\ 0 & 0 & 1 & 0 & 2 \\ 0 & 1 & 0 & 1 & 1 \\ 0 & 0 & 0 & 0 & 0 \end{bmatrix}$$

We have just seen how easy it is to solve a system of equations if its associated augmented matrix is in row-echelon form. As a further illustration of this simple process, let us suppose that the above matrix on the left is the augmented matrix of a system of equations in x_1, x_2, x_3, x_4. A proper interpretation of the columns of the matrix then yields the solution at once: $x_1 = 2$, $x_3 = 1, x_4 = 2$, and $x_2 = c$ is an arbitrary real number. Our earlier example illustrated the case of a unique solution, but this example shows how multiple solutions—in fact, an infinite number—may arise.

We now show that it is always possible, with a finite number of steps, to reduce any given matrix to Hermite normal form by elementary (row) operations and thence to obtain an immediate solution to the related system of equations. In our discussion, we refer to the original general system (1) of equations. Either $a_{11} \neq 0$, or a Type 1 operation can be used to replace it by a nonzero element. (If all elements of the first column were 0, the associated unknown would not appear in the equations.) The nonzero element can now be reduced to 1 by using its inverse as a multiplier of all elements of the first row (Type 2 operation). All other elements of the first column can now be reduced to 0 by subtracting the correct multiple of the first row from the other rows of the matrix in turn. This "sweep-out" process reduces the first column to its desired form. If the remaining $m - 1$ rows contain only 0's, the reduction is complete. If not, we consider the next column, after the first, that contains a nonzero entry in other than the first row. A Type 1 operation will put this number in the second row, if it is not there already, and multiplication of the second row (Type 2 operation) by the inverse of this element will reduce the element to 1. The "sweep-out" process can now be applied by using an appropriate sequence of Type 3 operations to reduce all entries of this column to 0, except, of course, for the pivotal 1 in the second row. The second stage of the reduction is now complete, and the process can be continued until the matrix is in the desired normal form. We note that this normal form must be unique, for, if there were two such equivalent forms, it would be possible to transform one into the other by means of a sequence of elementary (row) operations. In view of the nature of the Hermite normal form (Prob. 10), this is clearly impossible.

EXAMPLE 1

Solve the system of equations

$$2x + 2y + 2z = 5$$
$$x + 2y - z = 4$$

Solution

The augmented matrix of the system is

$$\begin{bmatrix} 2 & 2 & 2 & 5 \\ 1 & 2 & -1 & 4 \end{bmatrix}$$

We now subject this matrix to the indicated sequence of transformations, the number in parentheses above the arrow in each case designating the type of operation involved at that stage:

$$\begin{bmatrix} 2 & 2 & 2 & 5 \\ 1 & 2 & -1 & 4 \end{bmatrix} \xrightarrow{(1)} \begin{bmatrix} 1 & 2 & -1 & 4 \\ 2 & 2 & 2 & 5 \end{bmatrix} \xrightarrow{(3)} \begin{bmatrix} 1 & 2 & -1 & 4 \\ 0 & -2 & 4 & -3 \end{bmatrix}$$

$$\xrightarrow{(2)} \begin{bmatrix} 1 & 2 & -1 & 4 \\ 0 & 1 & -2 & \frac{3}{2} \end{bmatrix} \xrightarrow{(3)} \begin{bmatrix} 1 & 0 & 3 & 1 \\ 0 & 1 & -2 & \frac{3}{2} \end{bmatrix}$$

The associated system of equations is

$$x + 3z = 1$$
$$y - 2z = \frac{3}{2}$$

and so $x = -3z + 1$, and $y = 2z + \frac{3}{2}$, with z arbitrary. If we let $z = c$, for any $c \in \mathbf{R}$, the final solution can be expressed in the form

$$x = -3c + 1$$
$$y = 2c + \frac{3}{2}$$
$$z = c$$

There are then infinitely many solutions to the given system of equations.

EXAMPLE 2

Solve the "system" consisting of the single equation

$$2x - 3y + z - w = 4$$

Solution
In this case, the augmented matrix is [2 -3 1 -1 4], and this simple matrix may be put into Hermite normal form by merely dividing each element by 2 (Type 2 operation). The result is [1 $-\frac{3}{2}$ $\frac{1}{2}$ $-\frac{1}{2}$ 2], and its associated equation is $x - \frac{3}{2}y + \frac{1}{2}z - \frac{1}{2}w = 2$. We could, of course, have obtained this equation very easily without use of a matrix! Inasmuch as

SEC. 1.5 SOLUTIONS OF LINEAR EQUATIONS: GAUSS REDUCTION   33

we may assume that y, z, w are arbitrary, the desired solution may be expressed as

$$x = \tfrac{3}{2}c_1 - \tfrac{1}{2}c_2 + \tfrac{1}{2}c_3 + 2$$
$$y = c_1$$
$$z = c_2$$
$$w = c_3$$

Although the system of equations in Example 1 has what may be described as a *one-parameter family* of solutions, the system in Example 2 has a three-parameter family of solutions, the *parameters* being the arbitrary constants. As a further comment, it is no accident that the simple matrix in Example 2 looks very much like a vector from V_5. More will be said about this later.

We close this section with an example of a homogeneous system, the type of system of primary interest at this point in the text.

EXAMPLE 3

Solve the following system of equations:

$$2x + 3y - z = 0$$
$$x - y + 2z = 0$$
$$x + 2y - z = 0$$

Solution
The solution consists in exhibiting how the augmented matrix of the system can be put step-by-step into Hermite normal form.

$$\begin{bmatrix} 2 & 3 & -1 & 0 \\ 1 & -1 & 2 & 0 \\ 1 & 2 & -1 & 0 \end{bmatrix} \xrightarrow{(1)} \begin{bmatrix} 1 & -1 & 2 & 0 \\ 2 & 3 & -1 & 0 \\ 1 & 2 & -1 & 0 \end{bmatrix} \xrightarrow{(3)} \begin{bmatrix} 1 & -1 & 2 & 0 \\ 0 & 5 & -5 & 0 \\ 1 & 2 & -1 & 0 \end{bmatrix} \xrightarrow{(3)}$$

$$\begin{bmatrix} 1 & -1 & 2 & 0 \\ 0 & 5 & -5 & 0 \\ 0 & 3 & -3 & 0 \end{bmatrix} \xrightarrow{(2)} \begin{bmatrix} 1 & -1 & 2 & 0 \\ 0 & 5 & -5 & 0 \\ 0 & 6 & -6 & 0 \end{bmatrix} \xrightarrow{(3)} \begin{bmatrix} 1 & -1 & 2 & 0 \\ 0 & 5 & -5 & 0 \\ 0 & 1 & -1 & 0 \end{bmatrix} \xrightarrow{(1)}$$

$$\begin{bmatrix} 1 & -1 & 2 & 0 \\ 0 & 1 & -1 & 0 \\ 0 & 5 & -5 & 0 \end{bmatrix} \xrightarrow{(3)} \begin{bmatrix} 1 & 0 & 1 & 0 \\ 0 & 1 & -1 & 0 \\ 0 & 5 & -5 & 0 \end{bmatrix} \xrightarrow{(3)} \begin{bmatrix} 1 & 0 & 1 & 0 \\ 0 & 1 & -1 & 0 \\ 0 & 0 & 0 & 0 \end{bmatrix}$$

It follows that $x = -z$ and $y = z$, with z arbitrary. Hence, we may express the complete solution in the form: $x = -c$, $y = z = c$.

Problems 1.5

1. Write the rows and columns of the augmented matrix in Example 3.
2. Explain why a homogeneous system of linear equations always has the trivial solution.
3. Give an example of a system of two linear equations in two unknowns for which there is no solution.
4. Verify that $x = 1$, $y = -1$, $z = 1$ is a solution of the following system of equations:

$$2x + y - z = 0$$
$$3x - z = 2$$
$$x + y + z = 1$$

5. Verify that $x = 2$, $y = 0$, $z = 1$ is a solution of the following system of equations:

$$x + 2y + z = 3$$
$$2x - 2z = 2$$
$$3y + 2z = 2$$

6. Verify that $x = 2$, $y = 1$, $z = 2$ is a solution of the following system of equations:

$$2x + 2y - 3z = 0$$
$$-x + 4y - z = 0$$
$$3x + 2y - 4z = 0$$

Also check that $x = 2c$, $y = c$, $z = 2c$ is a solution for any $c \in \mathbf{R}$.

7. Write the augmented matrix of the "system" $2x - 3y + z = 4$, and use this simple matrix to determine all solutions.

8. Explain why elementary operations of Types 1 and 2 preserve the solutions of a given system of linear equations.
9. Explain why an elementary operation of Type 3 preserves the solutions of a system of linear equations.
10. Explain in detail why a Hermite normal form for a matrix is unique.
11. Give a matrix with two rows and two columns that is in Hermite normal form.
12. Describe a characteristic feature of the Hermite normal form of the augmented matrix of a system of linear equations for which there is no solution.

13. Prove that the system

$$ax + by = 0$$
$$cx + dy = 0$$

has only the trivial solution $x = y = 0$, if $ad - bc \neq 0$.

Note In Probs. 14 to 16 the given matrices are the Hermite normal form of the augmented coefficient matrices for systems of linear equations in x, y, z. Solve the system in each case.

14. $\begin{bmatrix} 1 & 0 & 0 & 4 \\ 0 & 1 & 2 & 3 \end{bmatrix}$

15. $\begin{bmatrix} 1 & 0 & 0 & 2 \\ 0 & 1 & 0 & 4 \\ 0 & 0 & 1 & 3 \end{bmatrix}$

16. $[1 \quad 2 \quad 3 \quad 4]$

Note In Probs. 17 to 19 reduce the given matrices to Hermite normal form.

17. $\begin{bmatrix} 2 & 1 & 4 & 2 \\ 4 & 3 & 4 & 1 \end{bmatrix}$

18. $\begin{bmatrix} 3 & 1 & 6 & 4 \\ 2 & 4 & 1 & 2 \\ 3 & 1 & 0 & 2 \end{bmatrix}$

19. $\begin{bmatrix} 2 & 4 & 1 \\ 1 & 3 & 4 \end{bmatrix}$

Note In each of Probs. 20 to 23 solve the given system of equations by reducing its augmented coefficient matrix to Hermite normal form.

20. $2x - y + z = 0$
    $x + 2y - 3z = 0$

21. $3x + y - 2z = 0$
    $2x - y - 2z = 0$
    $x + y - 2z = 0$

22. $4x - 3y + z = 8$

23. $2x - y + 2z + w = 0$
    $x + y + z - w = 0$
    $2x + 4z - w = 0$

1.6 Solutions of Linear Equations: Determinants

In Sec. 1.5 we presented an infallible method for finding all existing solutions of any system of linear equations. However, this method of Gauss is rather involved, and fortunately we can sometimes achieve our purpose with less work: for, in the application of linear equations that we need at this time, often a mere knowledge of the *existence* or *nonexistence* of solutions is sufficient. Our objective in this section is then a very practical one: *to find a means for deciding on the existence of nontrivial solutions of a system of homogeneous linear equations.*

We shall examine only the most common case, in which the number of equations is equal to the number of unknowns, leaving the general decision problem for the methods of Chap. 3. Our present method is based on *determinants* and is essentially a development of what is known as *Cramer's rule*. The reader who is already familiar with this well-known technique may omit this section (except for a glance at the statement of the final result) with no loss of continuity. The theory of determinants is a very beautiful one, and there are several different approaches to it, some of which will be discussed in Chap. 3. However, in view of our strictly practical interest in determinants here, our viewpoint at times will be more heuristic than theoretical, with special attention to situations where determinants are of actual use. As was the case with the Gauss method in Sec. 1.5, there are no added difficulties here if we consider a system of nonhomogeneous equations, even though our present interest is in the homogeneous case.

Before we define determinants, it is convenient to know something about *permutations*, a word for which there are two common usages. In college algebra, a *permutation* is usually considered an *arrangement* of a given set of objects; in modern abstract algebra it is a one-to-one *mapping* of a finite set into itself. These two viewpoints are actually equivalent, for suppose the "objects" are the natural numbers $\{1,2,\ldots,n\}$. Then any *arrangement* $i_1, i_2, \ldots, i_n$ of these numbers determines a *mapping* π such that $\pi(k) = i_k$, $1 \leq k \leq n$; and a *mapping* π of $\{1,2,\ldots,n\}$ onto itself determines an *arrangement* of these numbers in which their natural order is disturbed by replacing each number k by $\pi(k)$. Although permutations may involve arbitrary finite sets of objects, our interest will be restricted to permutations of the natural numbers $\{1,2,\ldots,n\}$.

It is customary to specify a permutation π by writing the elements of the permuted set in two rows: the top row contains the elements in any order, although the natural order is preferred if the elements are real numbers; the bottom row contains the elements written so that $\pi(k)$ is directly below the permuted element k. For example, if π is the permutation of the set $\{1,2,3,4,5\}$ such that $\pi(1) = 4$, $\pi(2) = 5$, $\pi(3) = 2$, $\pi(4) = 1$, $\pi(5) = 3$, we may represent π in many ways such as

$$\pi = \begin{pmatrix} 1 & 2 & 3 & 4 & 5 \\ 4 & 5 & 2 & 1 & 3 \end{pmatrix} = \begin{pmatrix} 2 & 3 & 5 & 4 & 1 \\ 5 & 2 & 3 & 1 & 4 \end{pmatrix} = \begin{pmatrix} 1 & 5 & 3 & 2 & 4 \\ 4 & 3 & 2 & 5 & 1 \end{pmatrix}$$

We have defined a permutation as a mapping or function and so the product $\pi_2\pi_1$ of two permutations π_1, π_2 is defined (or undefined) as in the case of functions. That is, $\pi_2\pi_1$ will mean the mapping π_1 *followed by* the mapping π_2. (Note that the reverse order may be understood in other books.) Thus if $\pi_1 = \begin{pmatrix} 1 & 2 & 3 \\ 3 & 1 & 2 \end{pmatrix}$ and $\pi_2 = \begin{pmatrix} 1 & 2 & 3 \\ 3 & 2 & 1 \end{pmatrix}$, then $\pi_2\pi_1 = \begin{pmatrix} 1 & 2 & 3 \\ 1 & 3 & 2 \end{pmatrix}$, whereas $\pi_1\pi_2 = \begin{pmatrix} 1 & 2 & 3 \\ 2 & 1 & 3 \end{pmatrix} \neq \pi_2\pi_1$. The *identity* permutation (1) leaves all elements of the given set fixed; the permutation π^{-1} is the *inverse* of π and reverses or "undoes" the effect of π, so that $\pi\pi^{-1} = \pi^{-1}\pi = (1)$. It may be of passing interest to note that the set of all permutations of a set of n elements forms a nonabelian group, called the *symmetric* group on n symbols and is usually denoted by S_n.

Our present interest in permutations centers around the notion of an *inversion*. If $i < j$, where i and j are natural numbers from the set $\{1,2, \ldots, n\}$, we say that π performs an *inversion* on the set if $\pi(i) > \pi(j)$. For example, if $\pi = \begin{pmatrix} 1 & 2 & 3 \\ 3 & 1 & 2 \end{pmatrix}$, π performs two inversions on $\{1,2,3\}$: for $1 < 2$ and $1 < 3$, whereas $\pi(1) > \pi(2)$ and $\pi(1) > \pi(3)$. If we write the top row of numbers of an expressed permutation in their natural order, it is easy to determine the number of inversions in the permutation by merely counting the number of natural-order inversions in the bottom row. For example, in the case of π above, we note that 3 is out of natural order with respect to both 1 and 2, so that two inversions are present. As a slightly more complicated example, consider the permutation $\begin{pmatrix} 1 & 2 & 3 & 4 & 5 \\ 2 & 4 & 1 & 5 & 3 \end{pmatrix}$. As we examine the bottom row from left to right, we notice that 2 is out of order only with respect to 1 (one inversion); 4 is out of order with respect to 1 and 3 (two inversions); 1 is ordered naturally with respect to 5 and 3 (zero inversions); 5 is out of order with respect to 3 (one inversion). The total number of inversions in the permutation is then $1 + 2 + 0 + 1 = 4$.

Definition   *A permutation π is said to be even or odd according as the number $k(\pi)$ of its inversions is an even or odd integer.*

It is seen that the permutation discussed just prior to the definition is then even. We shall use the abbreviation sgn π (*signum* of π) for $(-1)^{k(\pi)}$, so that sgn $\pi = 1$ if π is even and sgn $\pi = -1$ if π is odd. It should be clear that sgn π (or the *parity* of π) is uniquely determined by the permutation.

We now turn to the central matter of this section and again consider the system of equations (1) in Sec. 1.5, but with $m = n$. In this case, the matrix of coefficients has n rows and n columns, is called an $n \times n$ (n by n) *square matrix*, and is abbreviated for convenience to $A = [a_{ij}]$. The latter symbolism for the matrix A indicates that a_{ij} is the element at the intersection of the ith row and jth column of A. Each square matrix has associated with it a real number according to the following definition.

Definition *The determinant of the $n \times n$ matrix $A = [a_{ij}]$ is the real number designated det A or $|A|$, where*

$$\det A = |A| = \sum_{\pi} (sgn\ \pi) a_{1\pi(1)} a_{2\pi(2)} \cdots a_{n\pi(n)}$$

and the sum is taken over all permutations π of $\{1, 2, \ldots, n\}$. For computational purposes, it is often convenient to use the following more expressive symbolism and write

$$|A| = \begin{vmatrix} a_{11} & a_{12} & \cdots & a_{1n} \\ a_{21} & a_{22} & \cdots & a_{2n} \\ \cdot & \cdot & \cdot & \cdot \\ a_{n1} & a_{n2} & \cdots & a_{nn} \end{vmatrix}$$

An examination of the expression for det A shows it to be a sum of terms, each a product of n numbers taken *one from each row and column* of A and with the appropriate sign attached. If $A = [a_{ij}]$ is the 1×1 matrix whose single element is a, it is immediate that det $A = a$. An examination of the inversions in the set of second indices for each term also shows quite easily (Prob. 13) that

$$\begin{vmatrix} a_{11} & a_{12} \\ a_{21} & a_{22} \end{vmatrix} = a_{11} a_{22} - a_{12} a_{21}$$

and

$$\begin{vmatrix} a_{11} & a_{12} & a_{13} \\ a_{21} & a_{22} & a_{23} \\ a_{31} & a_{32} & a_{33} \end{vmatrix} = a_{11} a_{22} a_{33} + a_{12} a_{23} a_{31} + a_{13} a_{21} a_{32} \\ - a_{12} a_{21} a_{33} - a_{13} a_{22} a_{31} - a_{11} a_{23} a_{32}$$

For a general $n \times n$ matrix, there are $n!$ terms in the *expansion* of its determinant; and so, as n increases, the computational work involved in such an expansion becomes excessive.

In practice, one seldom uses the actual definition to compute the determinant of a matrix, because there are other more efficient methods. The one described here (but not always the best one) makes use of what are called *cofactors*. In the expansion for det A, where $A = [a_{ij}]$, if we collect all terms involving the element a_{ij} and factor it out, the *other factor* is the *cofactor* of a_{ij}. That is, if we express det A in the form $a_{ij}A_{ij} +$ (terms not involving a_{ij}), the number A_{ij} is the *cofactor* of a_{ij}. A glance at the general expansions given above shows, for example, that the cofactor of a_{11} in the 2×2 case is a_{22}, and the cofactors of a_{11} and a_{21} in the 3×3 case are, respectively, $a_{22}a_{33} - a_{23}a_{32}$ and $a_{13}a_{32} - a_{12}a_{33}$. Even more concretely, if

$$A = \begin{vmatrix} 2 & 3 & -4 \\ 1 & 0 & 2 \\ 3 & -2 & 1 \end{vmatrix}$$

the cofactors of the numbers in the a_{11} and a_{23} positions are, respectively, 4 and 13.

We have noted earlier that each term of the expansion of det A contains exactly one factor from each row and each column of A; or, stated otherwise, each row and each column of A has exactly one of its elements as a factor in each term of this expansion. Hence, for any i, $1 \leq i \leq n$, we have the following expansion of det A *in terms of the ith row*: det $A = \sum_{j=1}^{n} a_{ij}A_{ij}$. In like manner, we may express det A *in terms of the kth column* as follows: det $A = \sum_{j=1}^{n} a_{jk}A_{jk}$. Either of these formulas gives an expansion of det A in terms of the elements of a row or column and its associated cofactors, but there is one difficulty: How do we determine the cofactors, except by "brute force"? A theoretical development leading to a satisfactory answer to this query, although not difficult, is quite tedious for the general $n \times n$ matrix. Hence, since our present interest in determinants is strictly utilitarian, we shall be heuristic and state a rule for the determination of cofactors. The rule is easy to apply to $n \times n$ matrices if $n \leq 4$, and *the reader should check its validity for such cases*, making use of the actual definition of det A. As a matter of fact, the method of cofactors is a very inefficient way of determining $|A|$ if $n > 4$, and other methods—not to be discussed here—are much to be preferred. Here is the rule:

> *In order to determine the cofactor A_{ij} of a_{ij} in the matrix $A = [a_{ij}]$, delete the ith row and jth column of A, compute the determinant (called a "minor") of the remaining "submatrix," and multiply the result by* $(-1)^{i+j}$.

We notice that this rule describes the cofactor of an element of an $n \times n$ matrix as the determinant of an $(n - 1) \times (n - 1)$ matrix. The validity of the rule can be readily checked for the general 2 × 2 and 3 × 3 matrices, whose determinant expansions were given earlier. For example, in the 3 × 3 case, if we desire A_{12}, we delete the first row and second column of A and obtain the minor

$$\begin{vmatrix} a_{21} & a_{23} \\ a_{31} & a_{33} \end{vmatrix} = a_{21}a_{33} - a_{31}a_{23}$$

Since $(-1)^{1+2} = -1$, $A_{12} = a_{31}a_{23} - a_{21}a_{33}$.

EXAMPLE 1

Determine the cofactor of the element in the position of a_{23} in the matrix

$$A = \begin{bmatrix} 2 & -1 & 1 \\ 1 & 4 & 0 \\ 2 & 0 & -2 \end{bmatrix}$$

Solution
An application of the rule gives

$$A_{23} = (-1)^{2+3} \begin{vmatrix} 2 & -1 \\ 2 & 0 \end{vmatrix} = -(0+2) = -2$$

EXAMPLE 2

Determine $|A|$, where A is the matrix in Example 1.

Solution
We apply the rule and expand $|A|$ in terms of the elements of the first row — the usual practice. Then $|A| = 2A_{11} + (-1)A_{12} + 1(A_{13})$, where

$$A_{11} = (-1)^2 \begin{vmatrix} 4 & 0 \\ 0 & -2 \end{vmatrix} = -8 \qquad A_{12} = (-1)^3 \begin{vmatrix} 1 & 0 \\ 2 & -2 \end{vmatrix} = 2$$

$$A_{13} = (-1)^4 \begin{vmatrix} 1 & 4 \\ 2 & 0 \end{vmatrix} = -8$$

Hence, $|A| = 2(-8) + (-1)2 + 1(-8) = -16 - 2 - 8 = -26$.

Before reaching the *raison d'être* of this section, the solution of equations, we call attention to some simple properties of the determinant function.

(1) If all elements of any row or column of a matrix A are 0, then $|A| = 0$. This follows directly from our earlier comment that each term of an expansion of $|A|$ contains one element from each row and column of A as a factor.

(2) *If two rows (columns) of a matrix A are interchanged, the algebraic sign of its determinant is changed.* An interchange of two rows of A is equivalent to interchanging two row indices in the set of elements of A. If we designate this interchange as the permutation σ, the interchange of rows has the effect of replacing each permutation π in the expansion of $|A|$ by $\sigma\pi$. Since σ is an odd permutation, it follows that the sign of each term of $|A|$ is changed, as asserted. A similar argument gives the parallel result for columns.

(3) *If two rows (columns) of a matrix A are identical, then $|A| = 0$.* An interchange of these two rows (columns) would result in the same matrix, whereas its determinant has been changed in sign by (2). Hence $|A| = 0$.

(4) *If the elements of any row (column) of a matrix A are multiplied by the corresponding cofactors of a different row (column), the sum of these products is 0.* In symbols, this means that $\sum_{j=1}^{n} a_{ij}A_{kj} = 0$, if $k \neq i$, and $\sum_{i=1}^{n} a_{ij}A_{ik} = 0$, if $k \neq j$. We outline the argument for *rows*, the central point in either argument being that the cofactors of the elements of a row or column are independent of these actual elements. If we replace the elements of the kth row by the corresponding elements of the ith row, (2) implies that the determinant of this altered matrix is 0. But now the expansion of this determinant in terms of the kth row gives the desired result: $\sum_{j=1}^{n} a_{ij}A_{kj} = 0$, $k \neq i$ (Prob. 17).

Finally, with our preliminary discussions now completed, we are in a position to obtain the important result of this section very easily. We refer again to the system of equations listed as (1) in Sec. 1.5, but with $m = n$, writing them in abbreviated form as

$$\sum_{j=1}^{n} a_{ij}x_j = b_i \qquad i = 1, 2, \ldots, n$$

The matrix of coefficients of the system is $A = [a_{ij}]$ and, as above, we let A_{ij} designate the cofactor of the element a_{ij} in A. Then, for any fixed k,

$$\sum_{i=1}^{n} A_{ik} \left(\sum_{j=1}^{n} a_{ij}x_j \right) = \sum_{j=1}^{n} \left(\sum_{i=1}^{n} A_{ik}a_{ij} \right) x_j = |A|x_k = \sum_{i=1}^{n} A_{ik}b_i$$

where we have made a critical use of Property (4) and the expansion of $|A|$ in terms of its kth column. But then, if $|A| \neq 0$, we have the following result known as *Cramer's rule*:

$$x_k = \frac{\sum_{i=1}^{n} A_{ik}b_i}{|A|} \qquad k = 1, 2, \ldots, n$$

As a mnemonic device, it may be noted that, although the denominator of the expression on the right side is simply the determinant of the coefficient matrix A, its numerator is the cofactor expansion of the determinant of the matrix obtained from A by replacing its kth column by the column of constants b_i.

We have already remarked that Cramer's rule is quite cumbersome to apply if $n > 4$, but it is very convenient if $n \leq 3$. We shall not illustrate its use here, for our immediate concern is not so much the actual solution of equations as the question of the existence of solutions. We note from the form of Cramer's rule that $|A| \neq 0$ is *sufficient* for the existence of a unique solution of the system of equations. In the case of homogeneous equations, $b_i = 0$ (for $i = 1, 2, \ldots, n$) and here, if $|A| \neq 0$, the *only* solution is the trivial one: $x_1 = x_2 = \cdots = x_n = 0$. Inasmuch as this is the desired culmination point of the section, we restate this result in the form of a theorem.

Theorem 6.1 *A system of n linear homogeneous equations in n unknowns has only the trivial solution if $|A| \neq 0$, where A is the matrix of coefficients of the system.*

EXAMPLE 3

Determine, without solving, whether the following system of equations has a nontrivial solution:

$$\begin{aligned} 2x - y + z &= 0 \\ x + 4y &= 0 \\ 2x \quad\quad - 2z &= 0 \end{aligned}$$

Solution
The matrix A of coefficients of this system is

$$\begin{bmatrix} 2 & -1 & 1 \\ 1 & 4 & 0 \\ 2 & 0 & -2 \end{bmatrix}$$

and we discovered in Example 2 that $|A| = -26 \neq 0$. It then follows from Theorem 6.1 that the only solution is the trivial one: $x = y = z = 0$.

Problems 1.6

1. Explain why the arrangement of the numbers in the top row of a symbolized permutation makes no difference to the permutation.

SEC. 1.6 SOLUTIONS OF LINEAR EQUATIONS: DETERMINANTS   43

2. Write out each of the following indicated sums:

   (a) $\sum_{i=1}^{3} \left(\sum_{j=1}^{3} a_{ij}x_j \right)$; (b) $\sum_{i=1}^{3} \left(\sum_{j=1}^{2} a_{ij}x_j \right) x_i$.

3. Use the two-row scheme to express the permutation π, where
   (a) $\pi(1) = 4$, $\pi(2) = 3$, $\pi(3) = 1$, $\pi(4) = 2$; (b) $\pi(4) = 5$, $\pi(5) = 1$, $\pi(3) = 2$, $\pi(2) = 3$, $\pi(1) = 4$.

4. If the permutation π is expressed as $\pi = \begin{pmatrix} 1 & 2 & 3 & 4 & 5 & 6 \\ 4 & 3 & 1 & 2 & 6 & 5 \end{pmatrix}$, find $\pi(1)$, $\pi(3)$, $\pi(5)$, and $\pi(6)$.

5. If $\pi_1 = \begin{pmatrix} 1 & 2 & 3 & 4 & 5 & 6 \\ 2 & 1 & 5 & 3 & 4 & 6 \end{pmatrix}$ and $\pi_2 = \begin{pmatrix} 2 & 3 & 4 & 1 & 5 & 6 \\ 5 & 4 & 3 & 1 & 6 & 2 \end{pmatrix}$, find $\pi_1\pi_2$ and $\pi_2\pi_1$.

6. Find π_1^{-1} and π_2^{-1} for the permutations described in Prob. 5.

7. Count the number of inversions in each of the following permutations, regarded as arrangements: (a) 245163; (b) 451263; (c) 4312.

8. Verify by counting that the permutations 142536 and 42536, regarded as arrangements, have the same number of inversions.

9. If
$$A = \begin{bmatrix} 1 & 2 & 1 \\ 3 & 1 & 2 \\ -1 & 4 & 5 \end{bmatrix}$$
find (a) A_{11}; (b) A_{22}; (c) A_{13}; (d) A_{31}.

10. With A as in Prob. 9, use the first row and the cofactors of its elements to determine det A.

11. Check your result in Prob. 10 by making a new evaluation of det A, using the second column and its set of cofactors.

12. If
$$A = \begin{bmatrix} 2 & 3 & -5 \\ -1 & 2 & 0 \\ 4 & -2 & 1 \end{bmatrix}$$
use the cofactor expansion with the elements of the first row to determine $|A|$, and check your result with another expansion.

13. Use the *definition* of a determinant to verify the expansions given for general 2×2 and 3×3 matrices.

14. Verify the cofactor rule for the expansions of the determinants of a general 2×2 and a general 3×3 matrix.

15. Verify the cofactor rule for the expansion of the determinant of a general 4×4 matrix.
16. Fill in the details of the argument needed to establish Property (3) of the determinant function, as outlined in the text.
17. Explain the last equality in the outline of the argument given in the text to establish Property (4) of the determinant function.
18. In the text derivation of Cramer's rule, explain the sequence of equalities leading to the final result.

Note In Probs. 19 to 23, use Theorem 6.1 to decide, if possible, whether the given systems of equations have only the trivial solution. [Note that no decision can be made, on the basis of the theorem, if $|A| = 0$.]

19. $2x + 5y = 0$
    $x - 3y = 0$

20. $3x + 2y + z = 0$
    $x + 4y + z = 0$
    $2x + y + 4z = 0$

21. $x - 2y + 3z = 0$
    $2x + y - z = 0$
    $x - 7y + 10z = 0$

22. $2x_1 + 3x_2 - x_3 = 0$
    $x_1 - 2x_2 = 0$
    $3x_1 + x_2 - x_3 = 0$

23. $3x_1 - x_2 + 2x_3 = 0$
    $x_1 + 2x_2 - 5x_3 = 0$
    $2x_1 + 5x_2 + x_3 = 0$

24. Find $|A|$, where

$$A = \begin{bmatrix} 2 & 0 & 1 & 5 \\ 4 & 0 & -1 & 6 \\ 2 & 5 & 0 & 1 \\ -2 & 0 & 1 & 4 \end{bmatrix}$$

If A is the matrix of coefficients of a system of linear homogeneous equations, can this result be used to make a decision on the existence of any nontrivial solutions?

25. Express

$$\begin{vmatrix} 1 & x_1 & x_1^2 \\ 1 & x_2 & x_2^2 \\ 1 & x_3 & x_3^2 \end{vmatrix}$$

as a polynomial in x_1, x_2, x_3.

26. Use Cramer's rule to solve the following system of equations:

$$\begin{aligned} 2x + y - z &= 4 \\ x + 2y + 3z &= -8 \\ -x + 2y + z &= 16 \end{aligned}$$

27. If the product $\pi_1\pi_2$ of two permutations π_1 and π_2 is defined, prove that sgn $\pi_1\pi_2 =$ (sgn π_1)(sgn π_2). [*Hint:* Express π_1 in the form $\begin{pmatrix} \cdots & \pi_2(i) & \cdots & \pi_2(k) & \cdots \\ \pi_1\pi_2(i) & \cdots & \pi_1\pi_2(k) & \cdots \end{pmatrix}$.]

28. Prove that a permutation π has the same parity as π^{-1}.

29. Under what circumstances would it be simpler *not* to use the *first* row of a matrix A in a cofactor expansion of det A?

1.7 Linear Dependence of Vectors

We now return to the mainstream of our study of vectors with the introduction of an essential idea. In Sec. 1.2, we saw in a very intuitive way that, if given any three nonzero coplanar line segments, no two of which are parallel, it is possible to "fit together" certain scalar multiples of them to form the sides of a triangle. In Fig. 9, this construction was applied to segments representing *geometric vectors* α, β, γ, and from this we were able to conclude that $\gamma = a\alpha + b\beta$, with $a, b \in \mathbf{R}$. We also noted that, even if γ happens to be parallel to α or β, it is true that $\gamma = a\alpha + 0\beta$ or $\gamma = 0\alpha + b\beta$, and so in every case the relationship $\gamma = a\alpha + b\beta$ is valid. That is, if α and β are given as any nonparallel plane vectors, any third vector γ can be expressed as a linear combination of them, and so α and β *generate* the whole space of plane vectors containing them. A similar result occurs for the case of any three nonzero noncoplanar space vectors no two of which are parallel: If α, β, γ are three such vectors, any fourth space vector δ can be expressed in the form $\delta = a\alpha + b\beta + c\gamma$, for real numbers a, b, c, and so α, β, γ are *generators* of the space of geometric space vectors.

The equation $\gamma = a\alpha + b\beta$ can be written in the equivalent form $a\alpha + b\beta + (-1)\gamma = 0$, and the equation $\delta = a\alpha + b\beta + c\gamma$ can be expressed as $a\alpha + b\beta + c\gamma + (-1)\delta = \mathbf{0}$. The second forms of these equations provide an intuitive lead to the definition of *dependence* of vectors, but in this definition it is important to understand that we are thinking not only of geometric vectors but of elements of an arbitrary vector space.

Definition   *The vectors $\alpha_1, \alpha_2, \ldots, \alpha_k$ are said to be linearly dependent if there exist real numbers $a_1, a_2, \ldots, a_k$, not all 0, such that $a_1\alpha_1 + a_2\alpha_2 + \cdots + a_k\alpha_k = 0$. If the latter equal-*

ity is satisfied only when $a_1 = a_2 = \cdots = a_k = 0$, the vectors are linearly independent. Any set of vectors is linearly independent (dependent) if every (some) finite subset is linearly independent (dependent). As an aid in the general statement of certain theorems, it is convenient to regard the empty set as linearly independent.

EXAMPLE 1

The vectors $(1,2,-1,1)$, $(3,-2,-2,0)$, $(3,-10,-1,-3)$ of V_4 are *linearly dependent* because $3(1,2,-1,1) - 2(3,-2,-2,0) + (3,-10,-1,-3) = (0,0,0,0) = 0$. That is, $a_1(1,2,-1,1) + a_2(3,-2,-2,0) + a_3(3,-10,-1,-3) = O$, where $a_1 = 3, a_2 = -2, a_3 = 1$. On the other hand, the vectors $(1,0,0,0)$, $(0,1,0,0)$, $(0,0,1,0)$ are *linearly independent*, because if $a_1(1,0,0,0) + a_2(0,1,0,0) + a_3(0,0,1,0) = 0 = (0,0,0,0)$, it follows that $a_1 = a_2 = a_3 = 0$.

We note in passing that any collection of vectors that includes the vector 0 is linearly dependent (Why?), whereas a single nonzero vector makes up a linearly independent set. The following theorem is useful in many considerations of dependency for sets of vectors.

Theorem 7.1 *A set of nonzero vectors $\{\alpha_1, \alpha_2, \ldots, \alpha_n\}$ in any indexed order, is linearly dependent if and only if some vector $\alpha_k (k \geq 2)$ is a linear combination of the vectors $\alpha_1, \alpha_2, \ldots, \alpha_{k-1}$ with smaller indices.*

Proof
If $\alpha_k = a_1\alpha_1 + a_2\alpha_2 + \cdots + a_{k-1}\alpha_{k-1}$, for $k \leq n$, then $a_1\alpha_1 + a_2\alpha_2 + \cdots + a_{k-1}\alpha_{k-1} + (-1)\alpha_k + 0\alpha_{k+1} + \cdots + 0\alpha_n = 0$, where we note that the coefficient (-1) of α_k is not 0. Hence, $\{\alpha_1, \alpha_2, \ldots, \alpha_n\}$ is a linearly dependent set. Conversely, if $\{\alpha_1, \alpha_2, \ldots, \alpha_n\}$ is a linearly dependent set of vectors, there exist coefficients $a_1, a_2, \ldots, a_n$, not all 0, such that $a_1\alpha_1 + a_2\alpha_2 + \cdots + a_n\alpha_n = 0$. If a_k is the nonzero coefficient of largest index in this equation, then $\alpha_k = (-a_1/a_k)\alpha_1 + (-a_2/a_k)\alpha_2 + \cdots + (-a_{k-1}/a_k)\alpha_{k-1}$, so that the condition of the theorem is fulfilled. (Why is $k \neq 1$?) ∎

The result of the preceding theorem can now be used to replace a set of generators of a vector space by a linearly independent subset, which generates the same space. Since the subset is usually *proper*, we then shall have obtained a simpler description of the space.

Theorem 7.2 *If $\alpha_1, \alpha_2, \ldots, \alpha_n$ are nonzero generators of a vector space V, there exists a linearly independent subset of these generators that also generates V.*

Proof

If the given vectors are linearly independent, there is nothing to prove, and so let us assume them to be linearly dependent. We now make a critical examination of each vector from α_1 to α_n in their indexed order. Since $\alpha_1 \neq 0$, we retain it in the subset. If α_2 is a linear combination (multiple) of α_1, we reject it; otherwise α_2 is retained. If α_3 is a linear combination of α_1 and α_2 (which implies that it is a linear combination of the members of the subset), it is rejected; otherwise α_3 is retained. We continue this examination of the vectors in turn, rejecting a vector if it is a linear combination of the preceding vectors, and otherwise including it in the subset. If the vectors comprising the subset are $\alpha_1, \alpha_{i_2}, \alpha_{i_3}, \ldots, \alpha_{i_k}$, for certain indices $i_2, i_3, \ldots, i_k$ between 2 and n, it is apparent that $V = \mathcal{L}\{\alpha_1, \alpha_{i_2}, \alpha_{i_3}, \ldots, \alpha_{i_k}\}$. This follows because any vector $\alpha \in V$ can be expressed as $\alpha = a_1\alpha_1 + a_2\alpha_2 + \cdots + a_n\alpha_n$, and any of the original generators, not present in the subset, can be replaced by some finite linear combination of the vectors of this subset. Moreover, by Theorem 7.1 and the construction of the subset, these latter vectors are linearly independent. ∎

EXAMPLE 2

If $V = \mathcal{L}\{\alpha_1, \alpha_2, \alpha_3, \alpha_4\}$, where $\alpha_1 = (1,0,0,0)$, $\alpha_2 = (1,1,1,0)$, $\alpha_3 = (0,1,1,0)$, $\alpha_4 = (3,1,1,0)$, let us find a linearly independent subset of generators of V. We examine the given generators, in the manner outlined in Theorem 7.2, and retain α_1 and α_2 since $\alpha_1 \neq 0$ and α_2 is certainly not a multiple of α_1. However, a close examination of α_3 shows that $\alpha_3 = \alpha_2 + (-1)\alpha_1$, and so we reject α_3 from the subset. Likewise, we see that $\alpha_4 = 2\alpha_1 + \alpha_2$, so that α_4 is also rejected, thereby leaving α_1 and α_2 to comprise the linearly independent subset of generators. Hence, $V = \mathcal{L}\{\alpha_1, \alpha_2\}$.

It is an elementary matter to establish directly the linear dependence or independence of two vectors, but with three or more it is usually more convenient to reduce the problem to one on the solvability of a system of linear equations. It was with this application in mind that we included the preceding two sections, and we illustrate with some examples.

EXAMPLE 3

Decide whether the vectors $\alpha_1 = (2,-1,1)$, $\alpha_2 = (0,2,-1)$, $\alpha_3 = (1,-1,0)$ are linearly independent in V_3.

Solution

The equation $a_1\alpha_1 + a_2\alpha_2 + a_3\alpha_3 = a_1(2,-1,1) + a_2(0,2,-1) + a_3(1,-1,0) = 0$ is equivalent to $(2a_1 + a_3, -a_1 + 2a_2 - a_3, a_1 - a_2) = (0,0,0)$. But then

$$\begin{aligned} 2a_1 + a_3 &= 0 \\ -a_1 + 2a_2 - a_3 &= 0 \\ a_1 - a_2 &= 0 \end{aligned}$$

and the question is now whether this system of equations has a nontrivial solution. If A is the matrix of coefficients of the system,

$$A = \begin{bmatrix} 2 & 0 & 1 \\ -1 & 2 & -1 \\ 1 & -1 & 0 \end{bmatrix}$$

and we find easily that $|A| = -3 \neq 0$. Hence, the system of equations has only the trivial solution $a_1 = a_2 = a_3 = 0$, and the vectors $\alpha_1, \alpha_2, \alpha_3$ are linearly independent.

EXAMPLE 4

Decide whether the following vectors in V_3 are linearly independent or dependent: $\alpha_1 = (2,1,1)$, $\alpha_2 = (3,-1,2)$, $\alpha_3 = (-1,2,-1)$.

Solution

The equation $a_1\alpha_1 + a_2\alpha_2 + a_3\alpha_3 = a_1(2,1,1) + a_2(3,-1,2) + a_3(-1,2,-1) = 0 = (0,0,0)$ reduces to the following linear system:

$$\begin{aligned} 2a_1 + 3a_2 - a_3 &= 0 \\ a_1 - a_2 + 2a_3 &= 0 \\ a_1 + 2a_2 - a_3 &= 0 \end{aligned}$$

There are many nontrivial solutions to this system, since it may be seen to be the same except for symbolism, as that in Example 3 of Sec. 1.5; and so the vectors $\alpha_1, \alpha_2, \alpha_3$ are linearly dependent. In fact, if we refer to the solution in the earlier example and let $a_1 = -1$, $a_2 = a_3 = 1$, we see that $-\alpha_1 + \alpha_2 + \alpha_3 = 0$, so that $\alpha_1 = \alpha_2 + \alpha_3$.

SEC. 1.7 LINEAR DEPENDENCE OF VECTORS

The method of Example 4 leads directly to the following important theorem.

Theorem 7.3 The n vectors $Y_1 = (a_{11}, a_{21}, \ldots, a_{n1})$, $Y_2 = (a_{12}, a_{22}, \ldots, a_{n2})$, $\ldots$, $Y_n = (a_{1n}, a_{2n}, \ldots, a_{nn})$ of V_n are linearly independent if $|A| \neq 0$, where A is the $n \times n$ matrix whose ith column may be identified with the n-tuple Y_i, $i = 1, 2, \ldots, n$.

Proof
The equation $c_1 Y_1 + c_2 Y_2 + \cdots + c_n Y_n = 0$ is equivalent to $(a_{11}c_1 + a_{12}c_2 + \cdots + a_{1n}c_n, a_{21}c_1 + a_{22}c_2 + \cdots + a_{2n}c_n, \ldots, a_{n1}c_1 + a_{n2}c_2 + \cdots + a_{nn}c_n) = (0, 0, \ldots, 0)$, and this in turn is equivalent to the following system of linear equations:

$$\begin{aligned} a_{11}c_1 + a_{12}c_2 + \cdots + a_{1n}c_n &= 0 \\ a_{21}c_1 + a_{22}c_2 + \cdots + a_{2n}c_n &= 0 \\ &\cdots \\ a_{n1}c_1 + a_{n2}c_2 + \cdots + a_{nn}c_n &= 0 \end{aligned}$$

It is now a consequence of Theorem 6.1 that this system has only the trivial solution $c_1 = c_2 = \cdots = c_n = 0$ and the vectors are linearly independent if $|A| \neq 0$, where A is the coefficient matrix of the system; and the columns of this matrix are the ordered n-tuples comprising the given vectors $Y_1, Y_2, \ldots, Y_n$. ∎

Although the more complete ("if and only if") theorem will be established in Chap. 3, it is generally the above result that is useful.

EXAMPLE 5

A vector space V is described as being generated by the five polynomial vectors $\alpha_1 = 2$, $\alpha_2 = 3 + t$, $\alpha_3 = 2t$, $\alpha_4 = t + 2t^2$, $\alpha_5 = t^2$, where t is an indeterminate over **R**. Find a linearly independent subset of generators of V.

Solution
Since $\alpha_1 = 2 \neq 0$, we retain α_1. It is clear that $\alpha_2 = 3 + t$ is not a multiple of 2, and so we retain α_2. We note, however, that $\alpha_3 = 2t = 2(3 + t) - 3(2) = 2\alpha_2 - 3\alpha_1$, and so we reject α_3. Since it is easy to see that $\alpha_4 = t + 2t^2$ is not a linear combination of α_1 and α_2 (in view of the term in t^2 in α_4) we retain α_4. In the case of α_5, a little computation shows that $\alpha_5 = t^2 = \frac{1}{2}(t + 2t^2) - \frac{1}{2}(3 + t) + (\frac{3}{4})2 = \frac{1}{2}\alpha_4 - \frac{1}{2}\alpha_2 + \frac{3}{4}\alpha_1$, and so we reject α_5. (The computation involved here is left to the reader as Prob. 9.) Hence $V = \mathcal{L}\{\alpha_1, \alpha_2, \alpha_4\}$, where these three vectors are linearly independent.

Problems 1.7

1. Explain why a subset of linearly independent vectors must be linearly independent.
2. Replace the following "vector" equation by the equivalent system of three equations in three unknowns: $x(2,1,1) + y(-2,3,0) + z(-1,1,2) = 0$.
3. Show that the vectors $(1,0,0)$, $(0,1,1)$ do not generate V_3.
4. Decide whether the following vectors are linearly dependent, and, if so, express one vector as a linear combination of the others: (a) $(2,1,0,-1)$, $(1,1,0,1)$; (b) $(1,1,1,3)$, $(2,2,2,6)$; (c) $(2,1,0)$, $(1,1,-2)$, $(0,0,0)$; (d) $(0,1,0,0)$, $(1,0,1,0)$, $(0,2,0,0)$.
5. Use the directions in Prob. 4 for the following vectors: (a) $(1,1,-2,1)$, $(2,-1,0,1)$, $(-4,5,-4,-1)$; (b) $(2,0,-1,0)$, $(1,1,1,1)$, $(1,-1,-2,-1)$; (c) $(1,0,0)$, $(0,1,0)$, $(0,0,1)$; (d) $(2,-1)$, $(3,1)$, $(-1,1)$.

6. Explain why any set of vectors that includes 0 is dependent.
7. With reference to the proof of Theorem 7.2, if α_k is in the space $\mathcal{L}\{\alpha_1, \alpha_2, \ldots, \alpha_{k-1}\}$, explain why α_k must also be a linear combination of the vectors of this set *that have been retained*.
8. If $\{\alpha_1, \alpha_2, \ldots, \alpha_r\}$ is a set of linearly dependent vectors, explain why it is not necessarily the case that $\alpha_1 \in \mathcal{L}\{\alpha_2, \alpha_3, \ldots, \alpha_r\}$.
9. Complete the computation with respect to α_5 in Example 5.
10. If two polynomials in an indeterminate t are equal, what is the basis for the "rule" that corresponding coefficients of the two polynomials may be equated?

11. Use the method of Theorem 7.2 to find a linearly independent subset of generators of $\mathcal{L}\{\alpha_1, \alpha_2, \alpha_3\}$, where
    (a) $\alpha_1 = (1,1,-1,-2)$, $\alpha_2 = (0,1,1,3)$, $\alpha_3 = (-1,1,3,4)$
    (b) $\alpha_1 = (1,1,0)$, $\alpha_2 = (2,2,0)$, $\alpha_3 = (-1,1,0)$
    (c) $\alpha_1 = (2,1,0,0,3)$, $\alpha_2 = (-1,1,0,0,1)$, $\alpha_3 = (6,0,0,0,4)$
    (d) $\alpha_1 = (-1,2)$, $\alpha_2 = (-2,-3)$, $\alpha_3 = (3,2)$
12. Decide whether the following vectors in V_4 are linearly independent: $(1,2,-1,1)$, $(2,0,1,0)$, $(1,1,0,-1)$.
13. Prove that a set of vectors is linearly dependent if and only if it contains a proper subset that generates the same space.
14. If every vector of a vector space has a *unique* representation as a linear combination of certain generating vectors, prove that these generating vectors are linearly independent.

15. If α_1, α_2, α_3 are linearly independent vectors, determine whether the following are linearly dependent or independent: (a) $\alpha_1 + \alpha_2$, α_2, $\alpha_2 - \alpha_3$; (b) α_1, $\alpha_2 + \alpha_3$, $\alpha_1 - \alpha_3$; (c) $\alpha_1 - \alpha_2$, α_2, $\alpha_1 + \alpha_2$; (d) $2\alpha_1 + \alpha_2$, α_2, $\alpha_1 - 2\alpha_2$.

16. Prove that (a,b) and (c,d) are linearly dependent in V_2 if and only if $ad - bc = 0$.

17. If P_4 is the space of all real polynomials in an indeterminate t, of degree 3 or less, find a set of generators of P_4 that contains: (a) 1, t; (b) 1, $2 + t^2$; (c) $1 + t$, $t^3 - 1$; (d) $2t + 1$, $t^2 + 1$.

18. If α_1, α_2, α_3 are vectors such that $a_1\alpha_1 + a_2\alpha_2 + a_3\alpha_3 = 0$, $a_1 a_3 \neq 0$, prove that $\mathcal{L}\{\alpha_1, \alpha_2\} = \mathcal{L}\{\alpha_2, \alpha_3\}$.

19. Prove the converse to Prob. 14: If $\alpha_1, \alpha_2, \ldots, \alpha_k$ are linearly independent vectors, prove that $a_1\alpha_1 + a_2\alpha_2 + \cdots + a_k\alpha_k = b_1\alpha_1 + b_2\alpha_2 + \cdots + b_k\alpha_k$ implies that $a_1 = b_1$, $a_2 = b_2, \ldots, a_k = b_k$.

20. If $\{\alpha_1, \alpha_2, \ldots, \alpha_r\}$ is a linearly independent set of vectors, and the set $\{\alpha_1, \alpha_2, \ldots, \alpha_r, \alpha\}$ is linearly dependent, prove that $\alpha \in \mathcal{L}\{\alpha_1, \alpha_2, \ldots, \alpha_r\}$.

21. If f_i is the function defined on $[0,1]$ so that $f_i(x) = 0$ (for $i = 1, 2, \ldots, n$), except that $f_i(x) = 1$ for $(i-1)/n \leq x < i/n$ (for $i = 1, 2, \ldots, n-1$) and $f_n(x) = 1$ for $(n-1)/n \leq x \leq 1$, describe the space $V = \mathcal{L}\{f_1, f_2, \ldots, f_n\}$.

22. Decide whether f is in the vector space V of Prob. 21, where (a) $f(x) = 1$, $0 \leq x \leq 1$; (b) $f(x) = 1$, $0 \leq x < \frac{1}{2}$; $f(x) = -1$, $\frac{1}{2} \leq x \leq 1$; (c) $f(x) = x$, $0 \leq x \leq 1$.

23. If V' and V'' are subspaces of a vector space V, such that $V = V' + V''$ and $V' \cap V'' = \mathcal{O}$, prove that, for each vector $\alpha \in V$, there exist *unique* vectors $\alpha' \in V'$ and $\alpha'' \in V''$ such that $\alpha = \alpha' + \alpha''$.

24. What can you say—if anything—about the cardinality of a maximal set of linearly independent functions in the set of all real functions on the interval $[0,1]$?

1.8 Basis and Dimension

It is easy to see that there is nothing unique about the number of generators of a vector space. In fact, if we so wish, we may consider *all* elements of a vector space to constitute a generating set—which is clearly maximal for the space. For a less extreme example, we note that $(1,1,0,0) + (1,0,1,0) = (2,1,1,0)$, from which it follows that $\mathcal{L}\{(1,1,0,0), (1,0,1,0), (2,1,1,0)\} = \mathcal{L}\{(1,1,0,0), (1,0,1,0)\}$. The main difference in the indicated generating sets is that one is linearly independent, whereas the other is not. We shall show in this section that, if a space can be generated by a finite number of generators (i.e., it is *finitely generated*), the *number* of *linearly independent* generators is always the same. We shall sometimes use a terminology introduced earlier and refer to generators of a vector space as *spanning* vectors or vectors that *span* the space. We now state and prove the fundamental theorem on linear independence.

Theorem 8.1 If $\beta_1, \beta_2, \ldots, \beta_r \in V = \mathcal{L}\{\alpha_1, \alpha_2, \ldots, \alpha_n\}$, where the β_i are linearly independent, then $r \leq n$.

Proof
Since β_1 is a linear combination of $\alpha_1, \alpha_2, \ldots, \alpha_n$, the set $\{\beta_1, \alpha_1, \alpha_2, \ldots, \alpha_n\}$ of vectors is linearly dependent but spans V. It then follows from Theorem 7.1 that some vector of this ordered set is a linear combination of the vectors that precede it in the ordering. We know that $\beta_1 \neq 0$, since β_1 is a member of a linearly independent subset, and so we conclude that some vector $\alpha_i (1 \leq i \leq n)$ must be dependent on its predecessors. On discarding this vector α_i, we obtain the subset $\{\beta_1, \alpha_1, \ldots, \alpha_{i-1}, \alpha_{i+1}, \ldots, \alpha_n\}$, which must also span V. We now add β_2 to this ordered generating set, *as its first member*, and repeat the argument. The set $\{\beta_2, \beta_1, \alpha_1, \ldots, \alpha_{i-1}, \alpha_{i+1}, \ldots, \alpha_n\}$ clearly spans V and is linearly dependent, since $\beta_2 \in V$ and V is spanned by $\{\beta_1, \alpha_1, \ldots, \alpha_{i-1}, \alpha_{i+1}, \ldots, \alpha_n\}$. Again, by Theorem 7.1, and since $\{\beta_2, \beta_1\}$ is a linearly independent subset, one of the vectors α_j $(j \neq i)$ must be dependent on its predecessors in the ordered set of generators of V to which β_2 has just been added. We discard α_j and obtain (say, with $j > i$) the new subset $\{\beta_2, \beta_1, \alpha_1, \ldots, \alpha_{i-1}, \alpha_{i+1}, \ldots, \alpha_{j-1}, \alpha_{j+1}, \ldots, \alpha_n\}$ which also spans V. This type of argument can be repeated r times, when all the vectors $\beta_1, \beta_2, \ldots, \beta_r$ will have been adjoined. Inasmuch as one vector of the original set of generators has been discarded at each step, the number of original generators must have been at least r. That is, $n \geq r$, as asserted. ∎

If $\{\alpha_1, \alpha_2, \ldots, \alpha_r\}$ and $\{\beta_1, \beta_2, \ldots, \beta_s\}$ are two linearly independent sets of generators for a vector space V, it follows from Theorem 8.1 that $r \geq s$ and $s \geq r$. Hence, the following corollary results.

Corollary 1 *Any two sets of linearly independent generators of a finitely generated vector space contain the same number of vectors.*

Definition *A basis for a vector space is a linearly independent subset of vectors that spans the space. If the space is finitely generated, the number of elements in a basis is called the dimension of the space, and the space is said to be finite-dimensional. Since the zero space is the intersection of all spaces containing (and so is spanned by) the empty set—which we have* defined to be linearly independent—this trivial space may be said to have the empty set as basis with 0 as its dimension.

The concept of *dimension* may be applied to spaces that are not finitely generated. However, although spaces of this kind will occur in examples in the sequel, our principal concern will be with finite-dimensional spaces. We shall often use dim V to denote the (finite) dimension of the vector space V.

Corollary 2 If $V = \mathcal{L}\{\alpha_1, \alpha_2, \ldots, \alpha_n\}$, and dim $V = r$, then $r \leq n$.

Corollary 3 If dim $V = n$, for a vector space V, any maximal subset of linearly independent vectors of V contains n elements.

Although the notion of *ordering* is quite irrelevant in the development of the concept of a basis for a vector space, it is nonetheless true that much of this book will be concerned with bases whose members have been ordered by some indexing device. Such a basis will be called an *ordered basis*. It may be recalled that this type of index ordering was used for an arbitrary set of vectors in the important Theorem 7.1.

Let us now take another look at one of our key vector spaces, the space V_n, in the light of our present discussion. Since $(x_1, x_2, \ldots, x_n) = x_1(1,0,0, \ldots, 0) + x_2(0,1,0, \ldots, 0) + \cdots + x_n(0,0,0, \ldots, 1)$, it is clear that the vectors $E_1 = (1,0,0, \ldots, 0), E_2 = (0,1,0, \ldots, 0), \ldots, E_n = (0,0,0, \ldots, 1)$ span V. Moreover, these vectors are linearly independent, because $a_1(1,0,0, \ldots, 0) + a_2(0,1,0, \ldots, 0) + \cdots + a_n(0,0,0, \ldots, 1) = (a_1, a_2, \ldots, a_n) = 0 = (0,0,0, \ldots, 0)$ would imply that $a_1 = a_2 = \cdots = a_n = 0$. Hence V_n has finite dimension n, and the vectors $E_1, E_2, \ldots, E_n$ form a basis for the space. We shall use this basis very often in the sequel, and it will be convenient to refer to it as the $\{E_i\}$ basis for the space. It is a consequence of the definition of a basis that it is possible to express an arbitrary vector $\alpha \in V_n$ in the form $\alpha = a_1 E_1 + a_2 E_2 + \cdots + a_n E_n$, where $a_1, a_2, \ldots, a_n$ are *unique* real numbers (Prob. 8). In our earlier discussion of V_n, with $n \leq 3$, we made a rather loose identification of the notion of *dimension* with the number of components in the vector n-tuples. Since we have now shown that the dimension of the space V_n is indeed n, our previous intuitive use of the word *dimension* has been justified. It is easy to attach geometric significance to an $\{E_i\}$ basis, for these basis vectors can be regarded as *unit* vectors associated geometrically with a linear coordinate system in n-space. For example, in 3-space, we may associate E_1, E_2, E_3 with the *unit* points $(1,0,0), (0,1,0), (0,0,1)$, respectively, or, if desired, with the line segments joining the origin with these points.

Although the number of vectors in a basis of a given vector space is unique, there is nothing unique about the actual vectors that make up a basis. The following theorem is of frequent use in this connection.

Theorem 8.2   *Any linearly independent subset of vectors can be expanded to form a basis for a finite-dimensional space.*

Proof
Let $\{\alpha_1, \alpha_2, \ldots, \alpha_n\}$ be a basis for the space V, and $\{\beta_1, \beta_2, \ldots, \beta_r\}$ a linearly independent subset of V. The collection of vectors $\{\beta_1, \beta_2, \ldots, \beta_r, \alpha_1, \alpha_2, \ldots, \alpha_n\}$ certainly spans V, and, by Theorem 7.2, we can extract from it a linearly independent subset that also spans V. If we follow the procedure of the theorem cited, a vector is discarded if and only if it is a linear combination of the preceding vectors in the ordered set. Hence no β_i will be discarded, and the linearly independent subset that remains will contain all the original vectors. This subset will then be a basis of the desired kind. ∎

Corollary   *A set of n vectors from an n-dimensional vector space V forms a basis if either they span V or they are linearly independent.*

Proof
If they span V, it follows from Theorem 7.2 that a subset will form a basis; and, since the dimension of the space is n, the original set must be a basis. On the other hand, if the vectors of the given set are linearly independent, it follows from Theorem 8.2 that the set can be expanded to a basis. Again, the dimension of the space assures that the original set must be a basis. ∎

EXAMPLE 1

Find a basis for V_3 that contains the vectors $(1,0,1)$ and $(0,1,1)$.

Solution
Any vector that is a linear combination of $(1,0,1)$ and $(0,1,1)$ has the form $a_1(1,0,1) + a_2(0,1,1)$ or $(a_1, a_2, a_1 + a_2)$, for $a_1, a_2 \in \mathbf{R}$. The vector $(1,1,-1)$, for instance, is not of this form, and so we may include it to obtain the linearly independent set $\{(1,0,1), (0,1,1), (1,1,-1)\}$. By the corollary, this set is then a basis for V_3.

EXAMPLE 2

Find a basis for the subspace V of V_4, consisting of all vectors (x_1, x_2, x_3, x_4) such that $x_1 + x_2 - x_3 = 0$.

Solution

It is clear that the vectors $(1,1,2,0)$, $(1,0,1,0)$, $(0,0,0,1)$ are elements of V. Moreover, it is clear that the second is not a multiple of the first, nor is the third a linear combination of the first two, so that the vectors are linearly independent. Any basis of V_4 must contain four vectors, and it is apparent, in view of the condition imposed on V, that V is a *proper* subspace of V_4. Hence the three vectors that we have exhibited make up a basis for V.

In Example 2, since there were *three* vectors of V_4 under examination, it was not possible to use the determinant method of Sec. 1.6 to establish the linear independence of the vectors. We could have used the solution method of Sec. 1.5, but for this simple example it was easier to use the direct intuitive approach. However, our final example illustrates the usefulness of determinants in this context.

EXAMPLE 3

Find a basis for V_4 that contains the vectors $(1,-1,2,3)$ and $(3,-2,1,5)$.

Solution

All that is required is the discovery of two other 4-tuples with the property that the determinant associated with the four vectors is nonzero. The most direct approach is simply to pick two vectors, *somewhat* at random but preferably with components of 0 or 1 for ease of calculation, and check the determinant. We do this by picking the vectors $(1,0,0,0)$ and $(0,1,0,0)$, and an easy calculation shows that

$$\begin{vmatrix} 1 & 3 & 1 & 0 \\ -1 & -2 & 0 & 1 \\ 2 & 1 & 0 & 0 \\ 3 & 5 & 0 & 0 \end{vmatrix} = 7 \neq 0$$

Hence the four vectors form a basis for V_4.

Problems 1.8

1. Find a basis for V_2 that includes the vector (*a*) $(1,1)$; (*b*) $(-1,0)$; (*c*) $(-2,-3)$.
2. With reference to Prob. 1, express the basis elements in (*a*) in terms of those in (*b*), and conversely.

3. Find a basis for V_3 that includes (a) (1,1,0), (0,1,0); (b) (1,1,1), (0,−1,1).
4. With reference to Prob. 3, express each vector of both bases in terms of the $\{E_i\}$ basis.
5. Expand the set $\{(1,1,0), (−1,2,1)\}$ to a basis for V_3.

6. Prove the first statement in the proof of Theorem 8.1.
7. With reference to the proof of Theorem 8.1, explain why the set $\{\beta_1, \alpha_1, \ldots, \alpha_{i-1}, \alpha_{i+1}, \ldots, \alpha_n\}$ spans V.
8. Explain why the representation of a vector as a linear combination of the elements of *any* basis is unique. [Cf. Prob. 19, Sec. 1.7.]

9. Decide whether the vectors (1,0,0,1), (1,−1,0,0), (0,0,1,0), (1,−1,−1,0) form a basis for V_4.
10. Show that the vectors (1,1,0), (1,1,−2), (1,0,−1) form a basis for V_3, and express each member of the $\{E_i\}$ basis in terms of them.
11. Find two bases for V_4 that have no elements in common.
12. In the space of geometric vectors in ordinary space, what are the geometric requirements for a basis? Can two of the representing line segments be parallel, even if oppositely directed? Perpendicular?
13. Find two bases, with no common elements, for the space P_4 of real polynomials of degree less than 4 in an indeterminate t.
14. Use your knowledge of differential equations to find a basis for the vector space of real solutions of $y'' + 2y' + y = 0$.
15. The complex numbers $a + bi$, with $a,b \in \mathbf{R}$, form a two-dimensional vector space over $\mathbf{R}$ with $\{1,i\}$ as a basis. Find another basis for this space, and express the elements of each basis in terms of the others.
16. Let V be the subspace of V_4 consisting of all vectors (x_1, x_2, x_3, x_4) such that $x_1 = x_3$. Prove that $\{(1,0,1,0), (0,1,0,0), (0,0,0,1)\}$ is a basis for V.
17. Let P_4 be the vector space of real polynomials of degree less than 4 in an indeterminate t.
    (a) Expand $\{1, 2 − t, t^2 + 1\}$ to a basis for P_4.
    (b) Express $2 − 3t + t^2 + 4t^3$ in terms of the basis in (a).
18. Find a basis for the space in Prob. 17 that includes the elements 1 and $t^3 + t$.
19. What can be said about the dimension of a subspace of a finite-dimensional vector space?

20. Let V be the subspace of V_4, consisting of all vectors (x_1, x_2, x_3, x_4) such that $x_1 = x_3 = -x_4$. Find a basis for V, and include a proof that it is in fact a basis.
21. Prove that the vector space P of all real polynomials in an indeterminate t is not finite-dimensional.
22. Prove that the vector space of all real functions defined on the interval [0,1] is not finite-dimensional.
23. If V_1 and V_2 are two subspaces of a vector space V, such that dim V_1 = dim V_2 and $V_1 \subset V_2$, prove that $V_1 = V_2$.
24. It is true that *every* vector space has a basis, but have we established this result in this section? Comment on the matter of the existence of a basis for a vector space.

1.9 Two Important Theorems

The basic concept of *isomorphic* systems, although probably familiar to most readers of this book, was reviewed in Sec. 1.3 as it applies to vector spaces. We now establish a very important result concerning isomorphism which will justify the central role of the space V_n in any study of finite-dimensional vector spaces.

Theorem 9.1   *Every vector space V of dimension $n > 0$ is isomorphic to V_n.*

Proof

We must find a one-to-one mapping of V onto V_n such that conditions (i) and (ii) of the definition in Sec. 1.3 hold. Since the dimension of V is n, any basis of V has n elements, and we select $\{\alpha_1, \alpha_2, \ldots, \alpha_n\}$ as such a basis. For any $\alpha \in V$, $\alpha = a_1\alpha_1 + a_2\alpha_2 + \cdots + a_n\alpha_n$ for *unique* real numbers $a_1, a_2, \ldots, a_n$, and we may define the following *one-to-one* mapping of V onto V_n:

$$\alpha = a_1\alpha_1 + a_2\alpha_2 + \cdots + a_n\alpha_n \to (a_1, a_2, \ldots, a_n)$$

This mapping associates each vector of V with the element of V_n that is its n-tuple of coefficients *relative to the selected basis* $\{\alpha_i\}$. Since $r\alpha = r(a_1\alpha_1 + a_2\alpha_2 + \cdots + a_n\alpha_n) = ra_1\alpha_1 + ra_2\alpha_2 + \cdots + ra_n\alpha_n$, for any $r \in \mathbf{R}$, $r\alpha \to (ra_1, ra_2, \ldots, ra_n) = r(a_1, a_2, \ldots, a_n)$. Also, if $\beta = b_1\alpha_1 + b_2\alpha_2 + \cdots + b_n\alpha_n$, $\alpha + \beta = (a_1\alpha_1 + a_2\alpha_2 + \cdots + a_n\alpha_n) + (b_1\alpha_1 + b_2\alpha_2 + \cdots + b_n\alpha_n) = (a_1 + b_1)\alpha_1 + (a_2 + b_2)\alpha_2 + \cdots + (a_n + b_n)\alpha_n$, so that

$$\alpha + \beta \to (a_1 + b_1, a_2 + b_2, \ldots, a_n + b_n)$$
$$= (a_1, a_2, \ldots, a_n) + (b_1, b_2, \ldots, b_n)$$

Thus conditions (i) and (ii) in the definition of an isomorphism are seen to hold, and this shows that $\alpha \to (a_1, a_2, \ldots, a_n)$ is an isomorphic map of V onto V_n and that V and V_n are isomorphic spaces. ∎

The isomorphism between V_n and a space V of dimension n implies that these spaces are algebraically indistinguishable: whether we deal with a vector $\alpha \in V$ or its image $(a_1, a_2, \ldots, a_n) \in V_n$ is immaterial insofar as algebraic properties are concerned. Since the image in V_n of α is the n-tuple of coefficients or *coordinates* of α, relative to the basis selected for V, it is only natural to refer to V_n as a *coordinate space*. This is the explanation promised earlier in Sec. 1.1 for this name for V_n. However, in spite of the obvious importance of coordinate spaces, it is still often desirable to work with a given n-dimensional space in preference to its isomorphic image V_n.

EXAMPLE 1

The space P_n of all polynomials of degree less than n in an indeterminate t is isomorphic to V_n. This follows immediately from Theorem 9.1, because it is clear that $\{1, t, t^2, \ldots, t^{n-1}\}$ is a basis for P_n. If this basis is used, the isomorphic mapping of P_n onto V_n is seen to be described as follows:

$$p(t) = a_0 + a_1 t + \cdots + a_{n-1} t^{n-1} \to (a_0, a_1, \ldots, a_{n-1})$$

The other major result in this section is a theorem on *dimension*, a word we have already found convenient to abbreviate to dim. If S and T are subspaces of a vector space V, we have defined $S + T$ to be the subspace generated by S and T and consisting of all vectors of the form $\alpha + \beta$, with $\alpha \in S$ and $\beta \in T$. It was noted that the set-theoretic intersection $S \cap T$ is also a subspace of V, and the following theorem connects the dimensions of these two subspaces with those of S and T.

Theorem 9.2 *If S and T are subspaces of a finite-dimensional vector space, then $\dim(S + T) + \dim(S \cap T) = \dim S + \dim T$.*

Proof
Let $\{\alpha_1, \alpha_2, \ldots, \alpha_k\}$ be a basis for $S \cap T$. By Theorem 8.2, these vectors can be supplemented by other vectors $\beta_1, \beta_2, \ldots, \beta_{r-k}$ to make up a basis for S. In like manner, the basis for $S \cap T$ can be supplemented by vectors $\gamma_1, \gamma_2, \ldots, \gamma_{m-k}$ to make up a basis for T. We shall now show that $\{\beta_1, \beta_2, \ldots, \beta_{r-k}, \alpha_1, \alpha_2, \ldots, \alpha_k, \gamma_1, \gamma_2, \ldots, \gamma_{m-k}\}$ constitutes a basis for $S + T$, and to this end we verify that these vectors are linearly independent and span the space.

The vectors in question are certainly in $S + T$. Moreover, since any vector of $S + T$ has the form $\beta + \gamma$, where β is in $\mathcal{L}\{\beta_1, \beta_2, \ldots, \beta_{r-k}, \alpha_1, \alpha_2,$

..., α_k} and γ is in $\mathcal{L}\{\gamma_1,\gamma_2, \ldots, \gamma_{m-k},\alpha_1,\alpha_2, \ldots, \alpha_k\}$, it is clear that $\beta + \gamma$ is in $\mathcal{L}\{\beta_1,\beta_2, \ldots, \beta_{r-k},\alpha_1,\alpha_2, \ldots, \alpha_k,\gamma_1,\gamma_2, \ldots, \gamma_{m-k}\}$. Hence it is true that the vectors span the space $S + T$.

To show linear independence, let us suppose that $b_1\beta_1 + b_2\beta_2 + \cdots + b_{r-k}\beta_{r-k} + a_1\alpha_1 + a_2\alpha_2 + \cdots + a_k\alpha_k + c_1\gamma_1 + c_2\gamma_2 + \cdots + c_{m-k}\gamma_{m-k} = 0$, for real numbers $a_1, a_2, \ldots, a_k, b_1, b_2, \ldots, b_{r-k}, c_1, c_2, \ldots, c_{m-k}$. This equation can be rewritten in the form $b_1\beta_1 + b_2\beta_2 + \cdots + b_{r-k}\beta_{r-k} = -(a_1\alpha_1 + a_2\alpha_2 + \cdots + a_k\alpha_k + c_1\gamma_1 + c_2\gamma_2 + \cdots + c_{m-k}\gamma_{m-k})$, and we note that the right member is in T. Hence the left member is in both T and S and so is in $S \cap T$. It follows that $b_1\beta_1 + b_2\beta_2 + \cdots + b_{r-k}\beta_{r-k} = d_1\alpha_1 + d_2\alpha_2 + \cdots + d_k\alpha_k$, for real numbers $d_1, d_2, \ldots, d_k$. But the vectors appearing in this equation make up a basis for S, and so $b_1 = b_2 = \cdots = b_{r-k} = 0 = d_1 = d_2 = \cdots = d_k$. The original equation now becomes $a_1\alpha_1 + a_2\alpha_2 + \cdots + a_k\alpha_k + c_1\gamma_1 + c_2\gamma_2 + \cdots + c_{m-k}\gamma_{m-k}$; and, since $\{\alpha_1, \ldots, \alpha_k, \gamma_1, \ldots, \gamma_{m-k}\}$ is a basis set for T, these vectors are linearly independent and so $a_1 = a_2 = \cdots = a_k = c_1 = c_2 = \cdots = c_{m-k} = 0$. We have now shown that all coefficients of the originally assumed linear combination of vectors are 0, and so these vectors are linearly independent and form a basis for $S + T$.

The theorem now follows by simple arithmetic, because $\dim (S + T) = r - k + k + m - k = r + m - k$, $\dim (S \cap T) = k$, $\dim S = r$ and $\dim T = m$. Since $(r + m - k) + k = r + m$, it follows that $\dim (S + T) + \dim (S \cap T) = \dim S + \dim T$, as asserted. ∎

EXAMPLE 2

As an illustration of the force of Theorem 9.2, we can prove that any two-dimensional subspaces of V_3 must have a nonzero intersection. If S and T are two such subspaces, $\dim (S + T) = 2$ or $\dim (S + T) = 3$. But, by the theorem, $\dim (S + T) + \dim (S \cap T) = \dim S + \dim T = 2 + 2 = 4$, so that $\dim (S \cap T) = 4 - \dim (S + T)$. Hence, $\dim (S \cap T) = 1$ or $\dim (S \cap T) = 2$.

We have seen that, if S and T are subspaces of a vector space V, so is the sum $S + T$. It is also true that any space V (with $\dim V > 1$) can be decomposed nontrivially into a sum of this kind. Let $\{\alpha_1, \alpha_2, \ldots, \alpha_n\}$ be a basis of V. If we now define $S = \mathcal{L}\{\alpha_1, \alpha_2, \ldots, \alpha_k\}$ and $T = \mathcal{L}\{\alpha_{k+1}, \alpha_{k+2}, \ldots, \alpha_n\}$, $1 \leq k < n$, we can express V in a nontrivial manner as $V = S + T$. If the vectors $\alpha_1, \alpha_2, \ldots, \alpha_n$ are merely *generating* elements, and so *possibly* dependent, we may still write $V = S + T$, but there is a difference: in this case it *may* be that $S \cap T = \varnothing$, whereas in the former case this *must* be so. The distinction between these two cases is often of importance, and so we need the following definition.

Definition If $V = S + T$, where S and T are subspaces of V such that $S \cap T = \mathbf{0}$, we say that V is the *direct sum* of S and T and write $V = S \oplus T$. (See Prob. 23 for a generalization of this concept.)

Two comments are in order with respect to direct sums. First, if $V = S \oplus T$, the representation of $\alpha \in V$ as $\alpha = \alpha_S + \alpha_T$ ($\alpha_S \in S$, $\alpha_T \in T$), is unique (Prob. 10). Each component in such a representation is the *complement* (see Sec. 1.4) of the other in V, but it is also convenient and more descriptive to speak of α_S as the *projection of α on S along T* and α_T as the *projection of α on T along S*. The reader should be cautioned, however, that the *mappings* $\alpha \to \alpha_S$ and $\alpha \to \alpha_T$ will also be referred to as *projections* of V on S and T, respectively.

The other comment is that, whereas the representation of any vector is unique with respect to any *given* direct decomposition, the direct decomposition of a vector space into subspaces is not unique, even if one of the subspaces is designated. For example, if $V = S \oplus T$, where $S = \mathcal{L}\{(1,0,1)\}$ and $T = \mathcal{L}\{(-1,-1,0)\}$, it is evident that we may replace T by $\mathcal{L}\{(0,-1,1)\}$ so that $V = \mathcal{L}\{(1,0,1)\} \oplus \mathcal{L}\{(0,-1,1)\}$, and we note that $\mathcal{L}\{(0,-1,1)\} \neq T$.

EXAMPLE 3

If $V = S \oplus T$, where $S = \mathcal{L}\{(1,-2,1),(-1,1,0)\}$ and $T = \mathcal{L}\{(0,1,1)\}$, find the projections of $(1,-1,4)$ on S and T.

Solution
The problem is to determine real numbers a, b, c such that $a(1,-2,1) + b(-1,1,0) + c(0,1,1) = (1,-1,4)$. This is equivalent to the solution of the following system of equations:

$$\begin{aligned} a - b &= 1 \\ -2a + b + c &= -1 \\ a + c &= 4 \end{aligned}$$

On solving, we find that $a = 2$, $b = 1$, $c = 2$ and so $(1,-1,4) = 2(1,-2,1) + (-1,1,0) + 2(0,1,1) = (1,-3,2) + (0,2,2)$. Since $(1,-3,2) \in S$ and $(0,2,2) \in T$, these are the desired projections on S and T, respectively.

Problems 1.9

1. Use Theorem 9.2 to investigate the dimension of the intersection of any two three-dimensional subspaces of V_5.

2. If $V = S \oplus T$, where $S = \mathcal{L}\{(2,1,0)\}$ and $T = \mathcal{L}\{(1,-1,1)\}$, find the projection of $(1,2,-1)$ on S and T.
3. Explain why the space V in Example 3 must be identical with V_3.
4. Explain why two vector spaces that are isomorphic to a third space must be isomorphic to each other.
5. If V is the space of all real quadratic polynomials, exhibit an isomorphic mapping of V onto V_3.
6. If $V = S \oplus T$, where $S = \mathcal{L}\{(1,1,0)\}$ and $T = \mathcal{L}\{(0,1,1)\}$, find the projection on S and T of (a) $(2,1,-1)$; (b) $(-2,-1,1)$; (c) $(3,1,-2)$.
7. If $V = S \oplus T$, what can be said about a vector $\alpha \in V$ if (a) the projection of α on S is 0; (b) the projection of α on T is 0?

8. Explain why the mapping $\alpha \to (a_1, a_2, \ldots, a_n)$ in Theorem 9.1 is one-to-one of V onto V_n.
9. In view of Theorem 9.1, there is an isomorphic mapping of any n-dimensional vector space V onto V_n. But can there be more than one such mapping? Explain.
10. If $V = S \oplus T$, explain why the representation of a vector $\alpha \in V$ in the form $\alpha = \alpha_S + \alpha_T$, with $\alpha_S \in S$ and $\alpha_T \in T$, is unique.
11. Prove that the uniqueness property of Prob. 10, as applied to a decomposition $V = S + T$, would *require* that $V = S \oplus T$.
12. If S and T are subspaces of V, is there any difference between $S \oplus T$ and $T \oplus S$, as we have defined these sums? Explain.

13. If $V = S \oplus T$, where $S = \mathcal{L}\{(1,0,0,1),(0,1,-1,1)\}$ and $T = \mathcal{L}\{(0,1,-1,2)\}$, find the projection of $(1,1,1,1)$ on (a) S; (b) T.
14. With reference to V of Prob. 6, express V in the form $V = S_1 \oplus T_1$, where (a) $S_1 = S$, $T_1 \neq T$; (b) $S_1 \neq S$, $T_1 = T$; (c) $S_1 \neq S$, $T_1 \neq T$.
15. If **C** is the two-dimensional vector space of complex numbers (see Prob. 15 of Sec. 1.8), exhibit two isomorphic mappings of **C** onto V_2.
16. If $S = \mathcal{L}\{(1,1,0),(0,1,1)\}$ and $T = \mathcal{L}\{(-1,1,2),(0,1,3)\}$, find $S \cap T$ and $S + T$.
17. If S and T are two-dimensional subspaces of V_4, determine $\dim (S + T)$, where (a) $S \cap T = \mathcal{O}$; (b) $S \neq T$, but the given bases of S and T have one vector in common; (c) $S + T \neq V$.

18. If $S = \mathcal{L}\{(1,2,0),(-1,1,2)\}$, find distinct subspaces T_1 and T_2 of V_3 such that $V_3 = S \oplus T_1 = S \oplus T_2$.

19. If $S = \mathcal{L}\{(1,1,2,0),(-2,1,2,0)\}$ and $T = \mathcal{L}\{(2,0,1,-1),(-3,2,0,4)\}$, show that the sum $S + T$ is not direct.

20. Prove that the images of linearly independent vectors under any isomorphic mapping are also linearly independent.

21. Use the result in Prob. 20 to show that an isomorphism between two vector spaces maps any basis of one space onto a basis of the other.

22. Let P be the space of all real polynomials $p(t)$ in an indeterminate t, and $P' = \{p(t) \in P | p(1) = 0\}$ and $P'' = \{p(t) \in P | p(2) = 0\}$. Describe $P' + P''$ and $P' \cap P''$, without any reference to bases. Are these subspaces finite-dimensional?

23. The notion of a direct sum can be extended in a natural way to include a finite number of subspaces as summands: The sum $V_1 + V_2 + \cdots + V_n$ is *direct*, and we write $V_1 \oplus V_2 \oplus \cdots \oplus V_n$, provided $V_i \cap \left(\sum_{j \neq i} V_j \right) = \mathcal{O}$, for each $i = 1, 2, \ldots, n$. Prove that the sum $V_1 + V_2 + \cdots + V_n$ is direct if and only if $\dim (V_1 + V_2 + \cdots + V_n) = \dim V_1 + \dim V_2 + \cdots + \dim V_n$.

24. Explain why a sum of two one-dimensional subspaces is always either direct or redundant (i.e., one summand may be omitted). Discuss the cases where one summand has dimension 2 and the other dimension 1. Give illustrations in V_3, including geometric representations.

25. If $V_3 = S \oplus T$, for nontrivial subspaces S and T, verify that the geometric representation of one of the subspaces is a plane and the other a line.

26. Describe the geometric representations of the following subspaces of V_3: (a) $\mathcal{L}\{(1,0,2)\} \oplus \mathcal{L}\{(-1,1,0)\}$; (b) $\mathcal{L}\{(1,1,1)\} \oplus \mathcal{L}\{(2,1,-1)\} \oplus \mathcal{L}\{(0,1,2)\}$.

Selected Readings

Cheema, M. S.: Integral Solutions of a System of Linear Equations, *Amer. Math. Monthly,* **73**(5): 487–490 (May, 1966).

Kurosh, A. G.: "General Algebra," secs. 33.4–33.6, Chelsea Publishing Company, New York, 1963.

Stoll, R. R.: Elementary Operations and Systems of Linear Equations, *Amer. Math. Monthly,* **69**(1): 42–44 (January, 1962).

2

LINEAR TRANSFORMATIONS AND MATRICES

2.1 Linear Transformations

A great deal of modern mathematics is concerned with studies of algebraic systems and their mappings, with particular attention to invariants under the mappings. These will be matters for much discussion in this volume. A *mapping* or *function* is a one-way correspondence which associates each element of one system with a unique element of another—or possibly the same—system. The isomorphisms in Sec. 1.3 are special cases of the general concept. The mappings of real numbers, as defined by algebraic equations, are commonplace in studies of analytic geometry and calculus, and the association of a point in a plane with its coordinate pair (x,y) and the association of a geometric vector in 3-space with an element (x,y,z) of V_3 are other familiar illustrations of the concept. If T is a mapping of a set A into a set B, the essential idea is that T associates with each $a \in A$ (the *domain* of T) some unique element $b \in B$, the set TA of image points being called the *range* of T. We may describe this by writing T$a = b$ or $a \xrightarrow{T} b$; in the symbolism of the sets A and B, we may write T$A \subset B$, $A \to B$, or T: $A \to B$. It would be well for the reader to become familiar with all these methods of indicating a mapping. As with the isomorphic mappings of Theorem 9.1 (Chap. 1), the mapping T is said to be *onto* B if T$A = B$. (Such a mapping is sometimes called *surjective*, but we prefer the simpler language.) Thus, in the case of a mapping of A onto B, *every* element of the set B is the image of at least one element of the domain A.

One of the most familiar examples of mappings, apart from those already mentioned, occurs in geometry under the heading *rotations of the plane*. In this geometric setting, a point $P(x,y)$ is mapped onto a point $P^*(x^*,y^*)$ by a rotation of the plane through an angle of θ about the origin of some cartesian coordinate system. Analytic geometry shows that $x^* = x \cos \theta - y \sin \theta$ and $y^* = x \sin \theta + y \cos \theta$. In view of the usual association between the points of the plane and the vectors of V_2, a rotation can be regarded as a mapping of V_2 onto itself in which $(x,y) \to (x^*,y^*)$. It is easy to check that this mapping is actually an isomorphism of V_2 with itself, a type of mapping also known as an *automorphism* of V_2:

(1) It is one-to-one of V_2 onto V_2.
(2) If $(x_1,y_1) \to (x_1^*,y_1^*)$ and $(x_2,y_2) \to (x_2^*,y_2^*)$, then $(x_1 + x_2, y_1 + y_2) \to (x_1^* + x_2^*, y_1^* + y_2^*)$.
(3) $r(x,y) \to r(x^*,y^*)$, for any $r \in \mathbf{R}$.

The verification of these properties is left to the reader, but their validity should be intuitively evident from Fig. 15.

The kind of mapping that we shall now define is somewhat less restrictive than that of the preceding illustration but includes it as a special case.

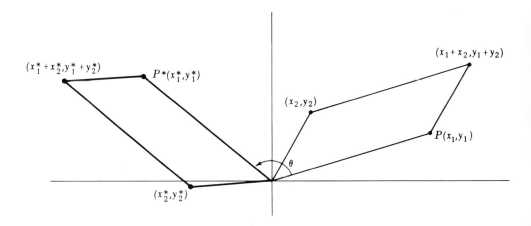

Fig. 15

Definition   *A mapping T of a vector space V into a vector space W is called a linear transformation of V into W if, for any vectors $\alpha, \beta \in V$ and an arbitrary real number r, the following hold:*

(1) $T(\alpha + \beta) = T\alpha + T\beta$
(2) $T(r\alpha) = r(T\alpha)$

If $W = V$, the transformation T is often called an *operator on V*. Associated with any linear transformation T: $V \to W$ are two very important spaces: the *rank space* or *range*, as previously defined for a mapping T, and denoted by R_T; the *kernel* or the *null space* of T, denoted by N_T, and defined as $\{\alpha \in V | T\alpha = 0\}$. These spaces will play an essential role in the sequel, and there will be repeated references to them.

It is easy to see (Prob. 12) that the two requirements of a linear transformation T may be replaced by the single condition that

$$T(r_1\alpha + r_2\beta) = r_1(T\alpha) + r_2(T\beta)$$

for any vectors $\alpha, \beta \in V$ and arbitrary real numbers r_1, r_2. It is implicit in the definition that a linear transformation must map the zero of the object space onto the zero of the image space (Prob. 7). Note, however, that the general definition does not require that the mapping be one-to-one and that the two spaces V and W may possibly be distinct.

EXAMPLE 1

If we consider the mapping T: $V_3 \to V_2$, defined so that $T(x,y,z) = (x,z)$, it is easy to check that T is linear. If $(x_1,y_1,z_1) \to (x_1,z_1)$ and $(x_2,y_2,z_2) \to (x_2,z_2)$, we see that $(x_1,y_1,z_1) + (x_2,y_2,z_2) = (x_1 + x_2, y_1 + y_2, z_1 + z_2) \to (x_1 + x_2, z_1 + z_2) = (x_1,z_1) + (x_2,z_2)$. Moreover, $r(x,y,z) = (rx,ry,rz) \to (rx,rz) = r(x,z)$, for any $r \in \mathbf{R}$, and the check is complete.

EXAMPLE 2

As an example of a mapping that is not linear, we may examine T: $V_2 \to V_2$, where $(x,y) \to (x + 1, y)$. If $(x_1,y_1) \to (x_1 + 1, y_1)$ and $(x_2,y_2) \to (x_2 + 1, y_2)$, then $(x_1,y_1) + (x_2,y_2) = (x_1 + x_2, y_1 + y_2) \to (x_1 + x_2 + 1, y_1 + y_2)$. But we note that $(x_1 + x_2 + 1, y_1 + y_2) \neq (x_1 + 1, y_1) + (x_2 + 1, y_2)$, and so T is not linear.

66 LINEAR TRANSFORMATIONS AND MATRICES

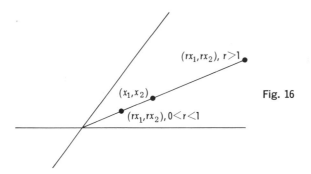

Fig. 16

We conclude this section with a few more examples of linear transformations, drawn from diverse areas of mathematics.

EXAMPLE 3

If T is the operator defined on V_n by $T(x_1, x_2, \ldots, x_n) = (rx_1, rx_2, \ldots, rx_n)$ for a real number r, it is easy to verify that T is linear. If we look at the geometric representation of V_n, for the case $n = 2$, we see that T maps each point of the plane onto a point whose distance from the origin is $|r|$ times as great as initially. A transformation of this kind is often called a *dilation* if $r > 1$ and a *contraction* if $0 < r < 1$, for geometric reasons, as illustrated in Fig. 16.

EXAMPLE 4

If T is the operator defined on V_2 by $T(x_1, x_2) = (x_1 + rx_2, x_2)$, for some real number $r \neq 0$, it is again easy to verify that T is linear. In this case, the geometric representation is a mapping or "shifting" of the points of the plane in a direction parallel to the x_1 axis and a distance proportional to $|x_2|$. This type of transformation is often called a *shear*, for obvious geometric reasons, and is illustrated in Fig. 17.

EXAMPLE 5

If V is the vector space of all continuous functions on **R**, we can define the operator $T: V \to V$ so that $(Tf)(x) = \int_0^x f(x)\, dx$. It is known from elementary calculus that Tf is also continuous on **R** and

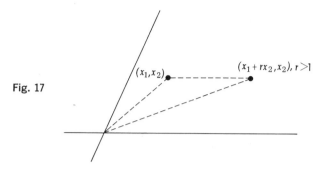

Fig. 17

that $\int_0^x [r_1 f_1(x) + r_2 f_2(x)]\, dx = r_1 \int_0^x f_1(x)\, dx + r_2 \int_0^x f_2(x)\, dx$.
Hence T is linear.

EXAMPLE 6

Another familiar example of a linear operator occurs in the space P of all real polynomials in an indeterminate t. Since the derivative polynomial $p'(t)$ is in P, for any $p(t) \in P$, we can define T: $P \to P$ so that $Tp(t) = p'(t)$. It is well known from calculus that $[r_1 p_1(t) + r_2 p_2(t)]' = r_1 p_1'(t) + r_2 p_2'(t)$, for any polynomials $p_1(t)$, $p_2(t) \in P$ and real numbers r_1, r_2, and so we have verified that T is linear.

Problems 2.1

1. Decide what is probably meant by an "invariant" of an operator on a vector space?
2. Explain why the equation $y = 2x^2 + x - 1$ defines a mapping of **R**, and decide whether the mapping is onto **R**.
3. Verify that the operator T in Example 3 is linear.
4. Verify that the operator T in Example 4 is linear.
5. Decide which of the following mappings are linear transformations: (a) $(x_1, x_2, x_3) \to (x_1, x_2 - 2, x_3)$; (b) $(x_1, x_2) \to (0,0)$; (c) $(x_1, x_2) \to (\sin x_1, x_2)$.
6. Decide which of the following mappings are linear transformations: (a) $(x_1, x_2, x_3) \to (x_2, x_1 - x_2, x_3 + x_1)$; (b) $(x_1, x_2) \to (x_1, x_1 x_2)$; (c) $(x_1, x_2, x_3) \to (0, x_2^2, x_3)$.

68   LINEAR TRANSFORMATIONS AND MATRICES

7. Explain why the definition of a linear transformation T requires that T0 = 0, with 0 designating indifferently the zero of any vector space.

8. Give a geometric proof of the formulas connecting (x,y) with (x^*,y^*), where (x^*,y^*) is the image of (x,y) after a rotation of the plane about the origin.

9. Explain why the mapping $(x,y) \to (x^*,y^*)$, induced by the rotation in Prob. 8, is one-to-one.

10. Give a geometric argument to show that the mapping in Prob. 9 is an isomorphism of V_2 with itself.

11. Prove that the mapping in Prob. 9 is an isomorphism, using a nongeometric argument.

12. Show why the two defining requirements for a linear transformation T are equivalent to the single condition that $T(r_1\alpha + r_2\beta) = r_1(T\alpha) + r_2(T\beta)$, for any vectors α, β and arbitrary real numbers r_1, r_2.

13. Prove that the following mappings are linear transformations of V_2, and describe the effect of each on the geometric representation of the vectors of V_2 as (1) points of 2-space; (2) geometric vectors: (a) $T(x_1,x_2) = (x_2,x_1)$; (b) $T(x_1,x_2) = (x_1,3x_2)$; (c) $T(x_1,x_2) = (x_2 - x_1, x_1 - x_2)$.

14. Let V be a vector space with the direct decomposition $V = S \oplus T$, so that $\alpha = \alpha_S + \alpha_T$, for any vector $\alpha \in V$ and associated vectors $\alpha_S \in S$ and $\alpha_T \in T$. Prove that the projection T on S, defined by $T\alpha = \alpha_S$, is a linear transformation of V.

15. Prove that the range and kernel of a linear transformation T are, in fact, vector spaces.

16. If the decomposition of V in Prob. 14 were not direct, explain why the corresponding mapping T would not even be defined.

17. We have noted that the differentiation operator induces a linear transformation on the space of all real polynomials in an indeterminate t. Can a similar remark be made for the vector space of real functions generated by (a) $\{1, e^x, e^{2x}, \ldots\}$; (b) $\{1, \sin x, \sin 2x, \ldots\}$; (c) $\{1, \sin^m x, \cos^n x,$ with m, n nonnegative integers$\}$?

18. By the *equation* of the line segment joining two points p_1 and p_2 of a plane, we mean an expression that describes any point p of the segment in terms of p_1 and p_2. If we regard $p_1 = (x_1,y_1)$ and $p_2 = (x_2,y_2)$ indifferently as elements of V_2 or points of a plane, show that the equation of the line seg-

ment joining these two points is $p = (1 - t)p_1 + tp_2$, $0 \leq t \leq 1$.

19. Use the result in Prob. 18 to prove that any linear operator on V_2, with the latter interpreted geometrically, maps each line segment of the plane onto either another line segment or a single point.
20. The equation (see Prob. 18) of a plane line segment is $p = (1 - t)p_1 + tp_2$, $0 \leq t \leq 1$, where $p_1 = (1, -1)$ and $p_2 = (2, 1)$. Find the equation of the segment into which this segment is carried by each of the linear transformations in Prob. 13, and in each case give a sketch of the original segment and its transform.

2.2 Properties of Linear Transformations

A linear transformation is basically a mapping with an additional characteristic property that makes it linear. This property is the one which we have described as follows for a transformation T of a vector space V:

$$T(r_1\alpha_1 + r_2\alpha_2) = r_1(T\alpha_1) + r_2(T\alpha_2)$$

for any vectors $\alpha_1, \alpha_2 \in V$ and arbitrary real numbers r_1, r_2. By an immediate generalization of this result (Prob. 1) we see that if $\alpha_1, \alpha_2, \ldots, \alpha_n$ are any n vectors of V, and $r_1, r_2, \ldots, r_n$ are arbitrary real numbers, then $T(r_1\alpha_1 + r_2\alpha_2 + \cdots + r_n\alpha_n) = r_1(T\alpha_1) + r_2(T\alpha_2) + \cdots + r_n(T\alpha_n)$. This property makes possible the very important result that a linear transformation is completely described by its effect on a set of basis vectors.

Theorem 2.1 *If $\{\alpha_1, \alpha_2, \ldots, \alpha_n\}$ is a basis for a vector space V, there is one and only one linear transformation T of V with $T\alpha_1$, $T\alpha_2, \ldots, T\alpha_n$ preassigned. Moreover, $R_T = \mathcal{L}\{T\alpha_1, T\alpha_2, \ldots, T\alpha_n\}$.*

Proof
If $\alpha \in V$, we can write $\alpha = r_1\alpha_1 + r_2\alpha_2 + \cdots + r_n\alpha_n$, for real numbers $r_1, r_2, \ldots, r_n$. We now extend the mapping T from the basis $\{\alpha_i\}$ to all V by *defining* $T\alpha = r_1(T\alpha_1) + r_2(T\alpha_2) + \cdots + r_n(T\alpha_n)$, where we note that this mapping is consistent with the given mappings of the basis vectors (Prob. 6). To show that this extended mapping T is linear on V, we first take another vector $\beta = s_1\alpha_1 + s_2\alpha_2 + \cdots + s_n\alpha_n \in V$, with $s_1, s_2, \ldots, s_n \in \mathbf{R}$. Then $T(\alpha + \beta) = T[(r_1\alpha_1 + r_2\alpha_2 + \cdots + r_n\alpha_n) + (s_1\alpha_1 + s_2\alpha_2 + \cdots + s_n\alpha_n)] = T[(r_1 + s_1)\alpha_1 + (r_2 + s_2)\alpha_2 + \cdots + (r_n + s_n)\alpha_n] = (r_1 + s_1)(T\alpha_1) + (r_2 + s_2)(T\alpha_2) + \cdots + (r_n + s_n)(T\alpha_n) = [r_1(T\alpha_1) +$

$r_2(T\alpha_2) + \cdots + r_n(T\alpha_n)] + [s_1(T\alpha_1) + s_2(T\alpha_2) + \cdots + s_n(T\alpha_n)] = T\alpha + T\beta$. Moreover, for any $r \in \mathbf{R}$, $T(r\alpha) = T(rr_1\alpha_1 + rr_2\alpha_2 + \cdots + rr_n\alpha_n) = rr_1(T\alpha_1) + rr_2(T\alpha_2) + \cdots + rr_n(T\alpha_n) = r[r_1(T\alpha_1) + r_2(T\alpha_2) + \cdots + r_n(T\alpha_n)] = r(T\alpha)$. Hence T is linear, and there remains to show that there exists no other linear mapping of V with the prescribed mapping of the basis elements. If we suppose the existence of an additional linear transformation T', where $T'\alpha_i = T\alpha_i$, $i = 1, 2, \ldots, n$, the linearity of T' requires that $T'\alpha = T'(r_1\alpha_1 + r_2\alpha_2 + \cdots + r_n\alpha_n) = r_1(T'\alpha_1) + r_2(T'\alpha_2) + \cdots + r_n(T'\alpha_n) = r_1(T\alpha_1) + r_2(T\alpha_2) + \cdots + r_n(T\alpha_n) = T\alpha$. Since this means that T and T' map an arbitrary vector of V onto the same element of the image space, we conclude that T' = T and so T is uniquely determined by $T\alpha_1, T\alpha_2, \ldots, T\alpha_n$. Since $T\alpha = r_1(T\alpha_1) + r_2(T\alpha_2) + \cdots + r_n(T\alpha_n)$, it follows that $R_T = \mathcal{L}\{T\alpha_1, T\alpha_2, \ldots, T\alpha_n\}$, and the proof is complete. ∎

In the applications of Theorem 2.1, we note that no restrictions have been placed on $T\alpha_1, T\alpha_2, \ldots, T\alpha_n$; these image vectors may be all alike, all different, or some alike and some different. They may even be all 0 or, at the other extreme, they may comprise a basis of the image space. If $T\alpha_1 = T\alpha_2 = \cdots = T\alpha_n = 0$, $T\alpha = 0$ for any $\alpha \in V$, and we write T = 0. It is noteworthy that $\{\alpha_1, \alpha_2, \ldots, \alpha_n\}$ may be *any* basis of the space V, so that a linear transformation can be defined in terms of an arbitrary basis.

EXAMPLE 1

Describe the linear transformation T of V_3 that maps the natural basis vectors E_1, E_2, E_3 onto (2,0,0), (0,1,1), (1,1,1), respectively.

Solution
If $X = (x_1, x_2, x_3) = x_1 E_1 + x_2 E_2 + x_3 E_3$ is an arbitrary vector in V_3, we define T so that $TX = x_1(TE_1) + x_2(TE_2) + x_3(TE_3) = x_1(2,0,0) + x_2(0,1,1) + x_3(1,1,1) = (2x_1 + x_3, x_2 + x_3, x_2 + x_3)$. By way of illustration, we see that $T(1,0,1) = (3,1,1)$ and $T(-1,1,-1) = (-3,0,0)$. It is easy to check (Prob. 7) by the method of Sec. 1.6 that the image vectors TE_1, TE_2, TE_3 do not constitute a basis for V_3.

In Sec. 2.1, we defined the *range* and *null space* of a linear transformation T but left to the reader in Prob. 15 of that section the actual verification that they are vector spaces. The following theorem connects the dimensions of these two important spaces with the dimension of the original space.

Theorem 2.2   *Let* T *be a linear transformation of a finite-dimensional vector space* V *into a vector space* W. *Then, if* N_T *and* R_T *are the null space and range of* T, *respectively*, dim V = dim N_T + dim R_T.

Proof

If $T = 0$, dim $V =$ dim N_T and dim $R_T = 0$, so the theorem holds. Thus we may assume $T \neq 0$ and dim $R_T > 0$. The proof is much like that of Theorem 9.2 (Chap. 1): we find bases for N_T and R_T, and the result follows by arithmetic. We suppose that $\{\alpha_1, \alpha_2, \ldots, \alpha_n\}$ is a basis for V and, for convenience, let $T\alpha_i = \beta_i$, $i = 1, 2, \ldots, n$. Then, as a result of Theorem 2.1, we know that $R_T = \mathcal{L}\{\beta_1, \beta_2, \ldots, \beta_n\}$, where dim $R_T = r \leq n$. Moreover, we may assume that the α_i and β_i have been labeled so that $\{\beta_1, \beta_2, \ldots, \beta_r\}$ is a basis of R_T. Then $\beta_{r+i} = a_{i1}\beta_1 + a_{i2}\beta_2 + \cdots + a_{ir}\beta_r$ or, expressed otherwise, $T\alpha_{r+i} = a_{i1}(T\alpha_1) + a_{i2}(T\alpha_2) + \cdots + a_{ir}(T\alpha_r)$, for real numbers $a_{i1}, a_{i2}, \ldots, a_{ir}$, $i = 1, 2, \ldots, n - r$. If we now define $\gamma_i = \alpha_{r+i} - (a_{i1}\alpha_1 + a_{i2}\alpha_2 + \cdots + a_{ir}\alpha_r)$, $i = 1, 2, \ldots, n - r$, it is easy to verify (Prob. 8) that $T\gamma_i = 0$. Hence $\gamma_i \in N_T$, and we shall show that $\{\gamma_i\}$, $i = 1, 2, \ldots, n - r$, is a basis for this null space.

In order to show linear independence, let us suppose that $c_1\gamma_1 + c_2\gamma_2 + \cdots + c_{n-r}\gamma_{n-r} = 0$, for real numbers $c_1, c_2, \ldots, c_{n-r}$. If we replace each γ_i by the expression defining it, we obtain $c_1\alpha_{r+1} + c_2\alpha_{r+2} + \cdots + c_{n-r}\alpha_n + (d_1\alpha_1 + d_2\alpha_2 + \cdots + d_r\alpha_r) = 0$, where $d_1, d_2, \ldots, d_r$ are real numbers which replace expressions involving $c_1, c_2, \ldots, c_{n-r}$ and $a_{i1}, a_{i2}, \ldots, a_{ir}$. Since $\{\alpha_i\}$ is a basis of V, all coefficients in the preceding equation are 0 and, in particular, $c_1 = c_2 = \cdots = c_{n-r} = 0$. Hence $\gamma_1, \gamma_2, \ldots, \gamma_{n-r}$ are linearly independent.

We must now show that $\gamma_1, \gamma_2, \ldots, \gamma_{n-r}$ span N_T. To this end, we examine an arbitrary vector $\gamma \in N_T$, so that $T\gamma = 0$; and we may assume that $\gamma = b_1\alpha_1 + b_2\alpha_2 + \cdots + b_n\alpha_n$, with $b_1, b_2, \ldots, b_n \in \mathbf{R}$. Since $\gamma, \gamma_1, \gamma_2, \ldots, \gamma_{n-r}$ are in N_T, it is immediate that $T(\gamma - b_{r+1}\gamma_1 - b_{r+2}\gamma_2 - \cdots - b_n\gamma_{n-r}) = 0$. Since (Prob. 8) the coefficient of α_i, for $i > r$, is 0 in the expression of $\gamma - b_{r+1}\gamma_1 - b_{r+2}\gamma_2 - \cdots - b_n\gamma_{n-r}$ in terms of the $\{\alpha_i\}$ basis, we may write this formally as $a_1\alpha_1 + a_2\alpha_2 + \cdots + a_r\alpha_r$, with $a_1, a_2, \ldots, a_r \in \mathbf{R}$. But then $T(a_1\alpha_1 + a_2\alpha_2 + \cdots + a_r\alpha_r) = 0 = a_1\beta_1 + a_2\beta_2 + \cdots + a_r\beta_r$ and, since $\beta_1, \beta_2, \ldots, \beta_r$ are linearly independent, it follows that $a_1 = a_2 = \cdots = a_r = 0$. Hence, $\gamma = b_{r+1}\gamma_1 + b_{r+2}\gamma_2 + \cdots + b_n\gamma_{n-r}$, and $N_T = \mathcal{L}\{\gamma_1, \gamma_2, \ldots, \gamma_{n-r}\}$. Inasmuch as simple arithmetic shows that $(n - r) + r = n$, it follows that dim N_T + dim R_T = dim V, as asserted by the theorem. (See Prob. 18 for a suggested alternative proof.) ∎

EXAMPLE 2

A linear transformation T maps V_4 so that $TE_1 = TE_3 = (1,1,1)$, $TE_2 = (1,1,0)$, $TE_4 = (0,0,0)$. Find the dimension of N_T, and use this information to describe the subspace completely.

72 LINEAR TRANSFORMATIONS AND MATRICES

Solution
The vectors $(1,1,1)$, $(1,1,0)$ are linearly independent, and $(0,0,0)$ is, of course, a dependent vector. Hence dim $R_T = 2$ by Theorem 2.1, and it follows from Theorem 2.2 that dim $N_T = 2$. Since $E_4 \in N_T$, and (Prob. 9) $E_1 - E_3 = (1,0,0,0) - (0,0,1,0) = (1,0,-1,0) \in N_T$ (and these vectors are clearly linearly independent) it follows immediately that $N_T = \mathcal{L}\{E_4,(1,0,-1,0)\} = \mathcal{L}\{(0,0,0,1),(1,0,-1,0)\}$.

If T is a linear transformation of a vector space V into a vector space W, dim R_T and dim N_T are often called the *rank* and *nullity*, respectively, of T. In this language, the statement of Theorem 2.2 may be phrased as follows:
Rank T + nullity T = dim V.

Problems 2.2

1. If T is a linear transformation of a vector space V, prove that $T(r_1\alpha_1 + r_2\alpha_2 + \cdots + r_n\alpha_n) = r_1(T\alpha_1) + r_2(T\alpha_2) + \cdots + r_n(T\alpha_n)$, for any vectors $\alpha_1, \alpha_2, \ldots, \alpha_n$ and real numbers $r_1, r_2, \ldots, r_n$.
2. If a linear transformation T is defined on V so that $T\alpha_i = 0$, for each vector α_i of a basis of V, describe T and the spaces N_T and R_T.
3. If $T: V_3 \to V_2$ is defined by $TE_1 = (1,2)$, $TE_2 = (1,-1)$, and $TE_3 = (1,1)$, determine $T\alpha$ for an arbitrary vector $\alpha = (x_1,x_2,x_3) \in V_3$.
4. Describe N_T and R_T, for the transformation T in Prob. 3, and check their dimensions by Theorem 2.2.
5. If T is a linear transformation of a vector space V, where dim $V = 6$, determine the rank of T, given that the nullity of T is (a) 0; (b) 1; (c) 2; (d) 6.

6. Explain what is meant, in the proof of Theorem 2.1, by the remark that the extended mapping T is "consistent with the given mappings of the basis vectors."
7. With reference to Example 1, verify that the vectors TE_1, TE_2, TE_3 do not form a basis of V_3.
8. Supply the arguments, in the proof of Theorem 2.2, that $T\gamma_i = 0$ and that the coefficient of α_i, $i > r$, in the new expression for $\gamma - b_{r+1}\gamma_1 - b_{r+2}\gamma_2 - \cdots - b_n\gamma_{n-r}$ is 0.
9. With reference to Example 2, verify that $E_1 - E_3 \in N_T$.
10. If $\alpha_1, \alpha_2, \ldots, \alpha_n$ are generators, but not basis vectors, of a vector space V, explain why it is not necessarily the case

that T is defined by an assignment of values for $T\alpha_1$, $T\alpha_2$, ..., $T\alpha_n$.

11. If $V = W_1 \oplus W_2$, and T is the projection of V on W_1, determine N_T and R_T. Use Theorem 2.2 to check their dimensions.
12. If T is a linear operator on a finite-dimensional vector space V such that N_T is identical with R_T, prove that dim V is an even integer. Illustrate with $V = V_2$.
13. A linear transformation T on V_2 is defined by $TE_1 = (1,2)$ and $TE_2 = (-1,1)$. Express T in terms of the basis $\{(1,1),(2,1)\}$.
14. Let T be a mapping of a vector space V into a vector space W. Then prove that T is linear if $T(\alpha + r\beta) = T\alpha + r(T\beta)$, for any $\alpha, \beta \in V$ and arbitrary $r \in \mathbf{R}$.
15. Let V and W be n-dimensional vector spaces. Then, if there exist linear transformations $T_1: V \to W$ and $T_2: W \to V$ such that $(T_2T_1)\alpha = T_2(T_1\alpha) = \alpha$, for any $\alpha \in V$, prove that T_1 and T_2 are isomorphisms onto W and V, respectively.
16. If T is a linear operator on a vector space V, such that $R_T \cap N_T = \mathfrak{O}$, prove that $T(T\alpha) = 0$ implies that $T\alpha = 0$.
17. Prove the converse to Prob. 16.
18. Use the following suggestions for an alternative proof of Theorem 2.2: Let U be the complement (see Sec. 1.9) of N_T in V, and obtain a basis for V by forming the union of bases for U and N_T; if $\{\beta_1, \beta_2, ..., \beta_s\}$ is a basis for U, show by direct computation and the definition of N_T that $\{T\beta_1, T\beta_2, ..., T\beta_s\}$ is a basis for R_T.
19. If T is an operator on V_2, defined by $TE_1 = (a,b)$ and $TE_2 = (c,d)$, determine TX, where $X = (x_1, x_2)$.
20. If the rank of a linear transformation T of a vector space V is 1, prove that $T(T\alpha) = c(T\alpha)$, for any $\alpha \in V$ and some associated real number c.
21. Decide whether there exists a linear transformation T of V_2 such that $T(1,2) = (1,1)$, $T(-2,-2) = (1,0)$, $T(3,1) = (2,-3)$.
22. If T is a linear transformation of a vector space V, prove that vectors $\alpha_1, \alpha_2, ..., \alpha_n$ are linearly independent if $T\alpha_1, T\alpha_2, ..., T\alpha_n$ are linearly independent. Is the converse true?

2.3 Operations on Linear Transformations

In this section we shall see how it is possible to make algebraic systems from linear transformations of vector spaces. If T_1 and T_2 are linear transformations of a vector space V, we can define $T_1 + T_2$ as follows:

$$(T_1 + T_2)\alpha = T_1\alpha + T_2\alpha \quad \text{for any } \alpha \in V$$

One can see that $T_1 + T_2$ is also linear on V. For, if $\alpha, \beta \in V$ and $r_1, r_2 \in \mathbf{R}$, $(T_1 + T_2)(r_1\alpha + r_2\beta) = T_1(r_1\alpha + r_2\beta) + T_2(r_1\alpha + r_2\beta) = r_1(T_1\alpha) + r_2(T_1\beta) + r_1(T_2\alpha) + r_2(T_2\beta) = r_1(T_1\alpha + T_2\alpha) + r_2(T_1\beta + T_2\beta) = r_1[(T_1 + T_2)\alpha] + r_2[(T_1 + T_2)\beta]$, as required.

EXAMPLE 1

Let us suppose that T_1 and T_2 are defined on V_3 so that $T_1(x_1, x_2, x_3) = (x_1 - x_2, 2x_2)$ and $T_2(x_1, x_2, x_3) = (2x_1 + x_2, x_2 - x_3)$. Then, according to the definition just given for the sum of two transformations, $(T_1 + T_2)(x_1, x_2, x_3) = (x_1 - x_2, 2x_2) + (2x_1 + x_2, x_2 - x_3) = (3x_1, 3x_2 - x_3)$.

If the range of T_1 is a subspace of the domain of T_2, it is possible to define the product T_2T_1 as follows:

$$(T_2T_1)\alpha = T_2(T_1\alpha) \quad \text{for any } \alpha \text{ in the domain of } T_1$$

An important special case arises when $T_1 = T_2 = T$ so that TT is defined. We denote this product T^2 and, by induction, T^n is defined for any positive integer n; we define T^0 to be the operator I that maps each vector in the domain of T onto itself. The product T_2T_1 is illustrated in Fig. 18 and it may be regarded as the resultant of T_1 followed by T_2. We note that, whereas the domain of $T_1 + T_2$ is the common domain of T_1 and T_2, the domain of T_2T_1 is the domain of T_1, and the domain of T_1T_2 is the domain of T_2. The proof that T_2T_1 is linear, *provided it is defined*, is similar to the corresponding proof for the sum. If α, β are in the domain of T_2T_1 and $r_1, r_2 \in \mathbf{R}$, then $(T_2T_1)(r_1\alpha + r_2\beta) = T_2[T_1(r_1\alpha + r_2\beta)] = T_2[r_1(T_1\alpha) + r_2(T_1\beta)] = r_1[T_2(T_1\alpha)] + r_2[T_2(T_1\beta)] = r_1[(T_2T_1)\alpha] + r_2[(T_2T_1)\beta]$. A very important case when the product T_2T_1 *is*

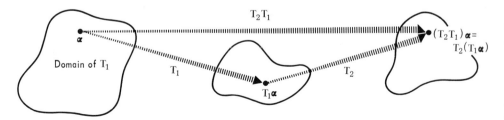

Fig. 18

defined occurs whenever *both* T_1 and T_2 are *operators* or linear transformations of a vector space V into itself. It may be useful to point out that the definitions given for the sum and product of linear transformations may be applied to *arbitrary mappings* of vectors spaces, subject only to the same conditions for these definitions to be meaningful.

In many important circumstances, the operation of multiplication of linear transformations is not commutative, even when both products are defined. For example, let D be the operator that transforms any polynomial $p(t)$ in the vector space P of all such polynomials into its derivative $p'(t)$, and let T be defined on P so that $Tp(t) = tp(t)$. Then, $(TD)p(t) = Tp'(t) = tp'(t)$, and $(DT)p(t) = D[tp(t)] = tp'(t) + p(t)$, so that $DT \neq TD$.

EXAMPLE 2

Let T_1 and T_2 be operators on V_3, defined so that $T_1(x_1,x_2,x_3) = (x_1 + x_2, 0, x_1 - x_3)$ and $T_2(x_1,x_2,x_3) = (0, x_1 + x_3, x_2 - x_1)$. Then, by the definition of a product of transformations, $(T_2T_1)(x_1,x_2,x_3) = T_2(x_1 + x_2, 0, x_1 - x_3) = (0, 2x_1 + x_2 - x_3, -x_1 - x_2)$ and $(T_1T_2)(x_1,x_2,x_3) = T_1(0, x_1 + x_3, x_2 - x_1) = (x_1 + x_3, 0, x_1 - x_2)$. Again, $T_2T_1 \neq T_1T_2$.

It is easy to see that the transformation that maps each element of a vector space V onto the zero of any space is linear, and we have already (page 70) denoted this *zero* transformation by 0, regardless of the nature of the image space (another use of 0). If we define $-T$ to be the transformation associated with a transformation T in such a way that $(-T)\alpha = -(T\alpha)$, for any α in the domain of T, it is evident (Prob. 9) that $-T$ is linear. Moreover, $T + (-T) = 0$, so that $-T$ acts like an "additive inverse" of T. In fact, since it is easy to see (Prob. 7) that $(T_1 + T_2) + T_3 = T_1 + (T_2 + T_3)$, for any three transformations T_1, T_2, T_3 of a vector space, it follows from a remark in Sec. 1.1 that the collection of all linear transformations of a vector space into any given space comprises an *additive abelian group*. The *additive* operation of this group is the operation of addition as it has been defined in this section for linear transformations.

If we consider the set of operators or linear transformations of a vector space V *into itself*, the preceding remarks apply and so this subset of linear transformations of V makes up an additive abelian group. But, in this case, it is possible to go farther. It is a consequence of our definitions of the operations that $(T_1T_2)T_3 = T_1(T_2T_3)$, $(T_1 + T_2)T_3 = T_1T_3 + T_2T_3$, and $T_1(T_2 + T_3) = T_1T_2 + T_1T_3$, for any three operators T_1, T_2, T_3 on V (Probs. 8 and 10), so that multiplication of operators is an operation that is associative and also distributive with respect to addition. The *identity* operator I has already been defined (see page 74) on V so that $I\alpha = \alpha$, for any $\alpha \in V$, and this transformation is clearly linear. It is then possible to assert that the collection of all linear transformations of a vector space into itself (or operators on the space) forms a *ring with identity*, under the operations of addition and multiplication as defined in this section. For those unfamiliar with the algebraic system known as a *ring*, we list its properties in the following definition.

Definition *A ring is a nonempty set of elements, endowed with two binary operations (addition, designated by $+$; multiplication, designated by $\cdot$ or juxtaposition), subject to the following conditions:*

(1) *The system is an abelian group under addition, i.e., an additive abelian group.*

(2) *Multiplication is an associative operation; that is, $a(bc) = (ab)c$, for arbitrary elements a, b, c of the system.*

(3) *Multiplication is distributive with respect to addition; that is, $a(b + c) = ab + ac$ and $(a + b)c = ac + bc$, for arbitrary elements a, b, c of the system.*

If the following additional condition also holds, the system is called a ring with identity.

(4) *The system contains an element 1 (not always to be identified with the integer 1), called the identity of the system, such that $1a = a1 = a$, for any a in the system.*

It is customary to designate indifferently either the *ring* or *its set of elements* by the same symbol—usually R. It should then be clear that the assertion made prior to the definition—that the set of all linear operators on a vector space forms a ring with identity—is true, the identity of this ring being I.

In addition to the *ring* operations of addition and multiplication, it is also possible to define an operation of *multiplication by a scalar* in a set of transformations. If T is a transformation of a vector space V and $r \in \mathbf{R}$, we may define the mapping rT as follows:

$$(r\mathsf{T})\alpha = r(\mathsf{T}\alpha) \qquad \text{for any } \alpha \in V$$

The domain of rT is clearly the same as the domain of T, and it is an elementary exercise (Prob. 9) to verify that rT is linear whenever T is linear.

EXAMPLE 3

Let $T(x_1,x_2,x_3) = (x_1 + x_2, x_2 - x_3, x_1 + x_2 + x_3)$ define T as a linear operator on V_3. By the preceding definition, 3T is also an operator on V_3 and $(3T)(x_1,x_2,x_3) = 3(x_1 + x_2, x_2 - x_3, x_1 + x_2 + x_3) = (3x_1 + 3x_2, 3x_2 - 3x_3, 3x_1 + 3x_2 + 3x_3)$. In particular, we see that $(3T)(1,2,3) = 3(3,-1,6) = (9,-3,18)$.

The observant reader will have noticed that, with the definition of multiplication by scalars, we have made *the set of linear transformations of a vector space V into a vector space W* into a vector space, because it is easy to check the requirements as listed in Sec. 1.1. In this new vector space, the "vectors" are the linear transformations of V into W, where W may possibly be identical with V. A mapping of a vector space V into a vector space W, such that the requirements (*i*) and (*ii*) of an isomorphic mapping (see Sec. 1.3) are satisfied but which is possibly not one-to-one or *onto* W, is said to be a *homomorphism* of V into W. Such a mapping is simply a linear transformation of V into W, and for this reason the vector space of all these linear transformations is often conveniently designated Hom (V,W). The vector space of operators on V is designated Hom (V,V).

If we recall that the real number system **R** can be regarded as a vector space of dimension 1 (over itself), an important special case of the preceding discussion occurs when we let $W = $ **R**. In this case, the linear transformations of V into **R** are called *linear functionals;* the space of these linear functionals, which make up Hom $(V,$**R**$)$, is the *conjugate* or *dual space* of V and is designated V^*. It is customary to write $f(\alpha)$ instead of $f\alpha$, for $f \in V^*$ and $\alpha \in V$. The study of linear functionals is of great current interest to mathematicians, and we devote the following section to it. We close this section with the summarizing remark that the set of linear operators on a vector space V can be considered to comprise any of the following algebraic systems, as desired: an *additive abelian group;* a *ring with identity;* a *vector space*.

Problems 2.3

1. If T_1 and T_2 are linear operators defined on V_3 by $T_1(x_1,x_2,x_3) = (2x_1 - x_2, x_1, x_2 - x_3)$ and $T_2(x_1,x_2,x_3) = (x_1 - x_3, x_2 + x_3, 0)$, determine (a) $T_1 + T_2$; (b) T_1T_2; (c) T_2T_1; (d) $2T_1$; (e) $2T_1 + 3T_2$.
2. If linear operators T_1, T_2 are defined on V_3 by $T_1E_1 = (1,2,-2)$, $T_1E_2 = (1,0,-2)$, $T_1E_3 = (0,-2,0)$, and $T_2E_1 =$

$(2,-1,1)$, $T_2E_2 = (2,1,1)$, $T_2E_3 = (1,1,1)$, determine (a) $(T_1 + T_2)X$; (b) $(T_1T_2)X$; (c) $(T_2T_1)X$, for an arbitrary $X = (x_1,x_2,x_3) \in V_3$.

3. Let f and g be linear functionals defined on V_2 so that $f(x_1,x_2) = 2x_1 + x_2$ and $g(x_1,x_2) = 3x_1 - 2x_2$. Then describe each of the following linear functionals: (a) $f + g$; (b) $f - g$; (c) $2f + g$; (d) $g - 3f$.

4. If T_1 and T_2 are linear transformations such that $T_1: V_3 \to V_3$ and $T_2: V_3 \to V_2$, explain why T_2T_1 is defined but T_1T_2 is not.

5. If T_1 is a dilation and T_2 is a shear (see Sec. 2.1) on V_2, check whether $T_1 + T_2$ and $T_2 + T_1$ are identical.

6. With T_1 and T_2 as in Prob. 5, check whether T_1T_2 is the same as T_2T_1, and interpret the result geometrically.

7. Explain why $(T_1 + T_2) + T_3 = T_1 + (T_2 + T_3)$ for mappings T_1, T_2, T_3 of a vector space V into a vector space W. Is it necessary that the mappings be linear for this result to hold?

8. If T_1, T_2, T_3 are mappings of a vector space V, explain why $(T_1T_2)T_3 = T_1(T_2T_3)$ whenever these products are defined. Is it necessary that the mappings be linear for this equality to hold?

9. Verify that, if T is a linear transformation of a vector space, so is (a) $-T$; (b) rT, for any $r \in \mathbf{R}$.

10. If T_1, T_2, T_3 are mappings of a vector space, explain why $T_1(T_2 + T_3) = T_1T_2 + T_1T_3$, provided all terms in this expressed equality are defined.

11. If D is the linear operator that transforms a polynomial p into its derivative p', explain why D is *nilpotent* (that is, $D^n = 0$, for some positive integer n) on the space P_k of polynomials of degree $k - 1$ or less.

12. With reference to Prob. 11, explain why D is not nilpotent on the space P of polynomials of arbitrary degree.

13. Let D be the derivative operator (see Prob. 11) on the space P of all polynomials in a real variable x. Then, if T is defined on P so that $Tp(t) = tp(t)$, for any $p(t) \in P$, verify that (a) $DT = TD + I$; (b) $(TD)^2 = T^2D^2 + TD$.

14. For arbitrary $\alpha (\neq 0), \beta \in V$, verify the existence of $f \in \text{Hom}(V,V)$ such that $f(\alpha) = \beta$.

15. For $\alpha (\neq 0), \beta \in V$ such that $\beta \neq k\alpha$, for any $k \in \mathbf{R}$, verify the existence of $g \in \text{Hom}(V,V)$ such that $g(\alpha) = 0$ and $g(\beta) \neq 0$.

16. Let V be the vector space of polynomial functions on **R**, with T_a defined on V so that $T_a(p) = p(a)$, for any $p \in V$ and $a \in \mathbf{R}$. Then verify that $T_a \in V^*$.
17. Let f be a linear functional on a vector space V; that is, $f \in V^*$. Then verify that T_α, defined on V^* by $T_\alpha(f) = f(\alpha)$, for any $\alpha \in V$, is a linear functional.
18. If T_1 and T_2 are elements of Hom (V,W), for vector spaces V and W, prove that rank $(T_1 + T_2) \leq$ rank $T_1 +$ rank T_2.
19. If T is a linear operator defined on V_3 by $T(x_1, x_2, x_3) = (2x_1 - x_2, 0, x_1 - x_3)$, describe the operator $T^2 + 2T$.
20. With T the operator defined in Prob. 19, describe $T^2 - 2T + 3I$. [Caution: I is the identity operator.]
21. If T is an *idempotent* ($T^2 = T$) linear operator on a finite-dimensional vector space V, prove that $V = W_1 \oplus W_2$, where $W_1 = \{\alpha \in V | T\alpha = \alpha\}$ and $W_2 = \{\alpha \in V | T\alpha = 0\}$. Observe that W_1 is the range and W_2 is the null space of T, and use this result to characterize any idempotent operator as a projection (see Sec. 1.9).

2.4 Linear Functionals

The idea of a linear functional, as a linear transformation of a vector space V into the space **R** of real numbers, was introduced near the end of Sec. 2.3, and several problems were included to expand somewhat on the idea. In this section we take another more serious look at the dual space V^* of linear functionals on V and examine—in part through the problems—the relationship between these two spaces. But first let us become more familiar with the basic concept through the medium of some examples.

EXAMPLE 1

If $V = \mathbf{R}$, the concept of a linear functional becomes identical with that of a linear function. Thus, a mapping f, defined on **R** so that $f(x) = rx$, for any $r \in \mathbf{R}$, is a linear functional on $V = \mathbf{R}$.

EXAMPLE 2

Let $\{\alpha_1, \alpha_2, \ldots, \alpha_n\}$ be a basis for a vector space V, with $\alpha = r_1\alpha_1 + r_2\alpha_2 + \cdots + r_n\alpha_n$ an arbitrary element of V. Then the mapping f_i, defined on V so that $f_i(\alpha) = r_i$, is a linear functional on V. This functional may be described as the function on V that maps each vector onto its ith coordinate, relative to the basis selected.

EXAMPLE 3

Let V be the space of all polynomial functions on **R**. Then $f_x(p) = p(x)$, for any fixed $x \in$ **R** and arbitrary $p \in V$, defines a linear functional f_x on V. This example illustrates the multiplicity of linear functionals, since there is a linear functional f_x on V associated with each real number x.

Let $\{\alpha_1, \alpha_2, \ldots, \alpha_n\}$ be a basis for a vector space V, and $\alpha = r_1\alpha_1 + r_2\alpha_2 + \cdots + r_n\alpha_n$ be an arbitrary element of V. Then, if f is a linear functional on V (that is, $f \in V^*$), $f(\alpha) = r_1 f(\alpha_1) + r_2 f(\alpha_2) + \cdots + r_n f(\alpha_n) = a_1 r_1 + a_2 r_2 + \cdots + a_n r_n$, where $a_1 = f(\alpha_1)$, $a_2 = f(\alpha_2)$, $\ldots$, $a_n = f(\alpha_n)$. It then follows from Theorem 2.1, for a given basis of V, that every n-tuple $(a_1, a_2, \ldots, a_n)$ of real numbers uniquely determines a linear functional f on V. Indeed, not only is it easy to see the one-to-one correspondence between these n-tuples and linear functionals on V, but it is not very difficult to verify that the vector spaces V_n and V^* are isomorphic. We leave this verification, however, to the reader (Prob. 10) and merely remark that this will imply that dim $V^* =$ dim $V_n =$ dim V. This important result also follows from Theorem 4.1 below, and for the theorem we give a proof that is quite independent of the above comments.

If the reader is not already familiar with the symbol δ_{ij} for the *Kronecker delta*, it should be noted that *by definition* $\delta_{ij} = 0$, if $i \neq j$, and $\delta_{ii} = 1$, for any i. In the Kronecker delta symbol, the indices i, j are understood to be positive integers that assume any of a prescribed set of values.

Theorem 4.1 *For any basis* $\{\alpha_1, \alpha_2, \ldots, \alpha_n\}$ *of V there exists a unique basis* $\{f_1, f_2, \ldots, f_n\}$ *of V^* such that* $f_i(\alpha_j) = \delta_{ij}$, $i, j = 1, 2, \ldots, n$. *Moreover, for each linear functional f on V, we have* $f = \sum_{i=1}^{n} f(\alpha_i) f_i$. *The basis* $\{f_1, f_2, \ldots, f_n\}$ *of V^* is said to be dual to the given basis* $\{\alpha_1, \alpha_2, \ldots, \alpha_n\}$ *of V.*

Proof
If we accept the isomorphism between the spaces V^* and V_n, as asserted above, the proof of the theorem is almost immediate. For the equations $f_i(\alpha_j) = \delta_{ij}$, $j = 1, 2, \ldots, n$, determine the unique n-tuple E_i of V_n and so, by Theorem 2.1, a unique linear functional f_i on V, $i = 1, 2, \ldots, n$. Moreover, the isomorphic map of V^* onto V_n maps f_i onto E_i, $i = 1, 2, \ldots, n$, and the independence of the E_i implies the independence of the f_i. It follows that $\{f_i\}$ is a basis for V^*, and the major portion of the theorem

is complete. However, it is instructive to give an independent proof of this important result, and we do so without reference to the isomorphism.

Theorem 2.1 implies the existence (for each $i = 1, 2, \ldots, n$) of a unique linear functional f_i on V such that

$$f_i(\alpha_j) = \delta_{ij} \qquad j = 1, 2, \ldots, n$$

These n linear functionals are linearly independent, for suppose $\sum_{i=1}^{n} c_i f_i = 0$, for $c_1, c_2, \ldots, c_n \in \mathbf{R}$, with 0 the zero functional. Then, in view of the defining operations in V^*, it follows that $0 = \left(\sum_{i=1}^{n} c_i f_i\right)(\alpha_j) = \sum_{i=1}^{n} c_i f_i(\alpha_j) = \sum_{i=1}^{n} c_i \delta_{ij} = c_j$, $j = 1, 2, \ldots, n$. That is, $c_1 = c_2 = \cdots = c_n = 0$, and the linear independence is established. To show that the functionals $f_1, f_2, \ldots, f_n$ span V^*, we take an arbitrary $f \in V^*$ and let $f(\alpha_i) = a_i$, $i = 1, 2, \ldots, n$. But now the functional $\sum_{i=1}^{n} a_i f_i$ has the same value as f, for each α_j, since $\left(\sum_{i=1}^{n} a_i f_i\right)(\alpha_j) = \sum_{i=1}^{n} a_i f_i(\alpha_j) = a_j = f(\alpha_j)$. Hence, again by Theorem 2.1, it follows that $f = \sum_{i=1}^{n} a_i f_i = \sum_{i=1}^{n} f(\alpha_i) f_i$, as asserted. The elements $f_1, f_2, \ldots, f_n$ then form a basis of V^*, and our proof has also shown that $f = \sum_{i=1}^{n} f(\alpha_i) f_i$. The proof of the theorem is complete. ∎

Corollary 1 $\quad Dim \, V^* = dim \, V.$

This result was noted earlier, but it may be seen to follow from the equal numbers of basis elements for the spaces V and V^*.

Corollary 2 $\quad$ *If $\{f_i\}$ is the dual basis for V^*, associated with the basis $\{\alpha_i\}$ for V, any vector $\alpha \in V$ can be expressed in the form*

$$\alpha = \sum_{i=1}^{n} f_i(\alpha) \alpha_i$$

Proof

If $\alpha = \sum_{i=1}^{n} r_i \alpha_i$ is an arbitrary vector of V, our definition of the dual basis $\{f_i\}$ implies that $f_j(\alpha) = \sum_{i=1}^{n} r_i f_j(\alpha_i) = \sum_{i=1}^{n} r_i \delta_{ij} = r_j$, $j = 1, 2, \ldots, n$. Thus

82 LINEAR TRANSFORMATIONS AND MATRICES

the unique representation of α, in terms of the $\{\alpha_i\}$ basis, is

$$\alpha = \sum_{i=1}^{n} f_i(\alpha)\alpha_i$$

As a result of Corollary 2, we are able to give a very concrete description of the members of the basis $\{f_i\}$ of V^* which is dual to a basis $\{\alpha_i\}$ of V. For f_i is precisely the function on V that assigns to each $\alpha \in V$ its ith coordinate, relative to the given basis $\{\alpha_i\}$. This is the linear functional described in Example 2, and it would be appropriate to call the functionals f_i of that example the *coordinate functions* for $\{\alpha_i\}$.

It may very well be that the reader is somewhat puzzled as to why the space V^* is called the *dual* of V, because we have given no real justification for this name. The general notion of *duality* between two algebraic systems usually implies, in some general sense, that a statement or theorem involving elements of the two systems remains true after certain basic elements of the systems have been interchanged. A comparison of the expression $f = \sum_{i=1}^{n} f(\alpha_i)f_i$, for an arbitrary $f \in V^*$, with the expression $\alpha = \sum_{i=1}^{n} f_i(\alpha)\alpha_i$, for an arbitrary $\alpha \in V$, is suggestive of some sort of symmetrical relationship between the spaces V and V^*, but it is *only* suggestive. It is a fact, however, that the space V is related to V^* in quite the same way that V^* is related to V, and it is not difficult to prove that the dual of V^*, usually designated V^{**}, is actually isomorphic to V. Although these results would be of essential interest in a more detailed study of the algebraic theory of vector spaces, it would take us too far from our central theme to prove them here. For the interested reader, however, we give an exact definition of the notion of *dual vector spaces* but leave further results for his attention in the problems.

In order to make the *duality* of two spaces more transparent, it is customary to introduce a notation for the value of a functional that is symmetric with respect to the two spaces. For this purpose, the symbol $[\alpha, f]$ might be used to designate the real number $f(\alpha)$, with $\alpha \in V$ and $f \in V^*$. However, in the general definition, even the "function" symbol f is discarded in favor of one that is more neutral.

Definition Two vector spaces V_1 and V_2 are said to be dual if there exists a real-valued function defined on $V_1 \times V_2$, whose value for $\alpha_1 \in V_1$ and $\alpha_2 \in V_2$ will be denoted by $[\alpha_1, \alpha_2]$ such that the following conditions are satisfied:

(i) $[\alpha_1, \alpha_2] = 0$, for every $\alpha_1 \in V_1$, implies that $\alpha_2 = 0$; and $[\alpha_1, \alpha_2] = 0$, for every $\alpha_2 \in V_2$, implies that $\alpha_1 = 0$.

(ii) *The function defined by $[\alpha_1,\alpha_2]$ is bilinear; that is, $[\alpha_1,\alpha_2]$ is linear in α_1, for a fixed α_2, and also linear in α_2, for a fixed α_1.*

It would be a good preliminary exercise for the reader to "translate" the above requirements into the usual symbolism of functions and then to prove that the spaces V and V^* are dual according to the definition. These and other interesting results will be suggested in the following set of problems.

Problems 2.4

1. Explain why the mapping $f\colon \mathbf{R} \to \mathbf{R}$, defined by $f(x) = x + 1$, is not a linear functional on $\mathbf{R}$.
2. Decide whether the mapping $g\colon V_2 \to \mathbf{R}$, defined by $g(x,y) = xy$, is a linear functional on V_2.
3. Let $x = 1$ in Example 3 of a linear functional, and determine $f_1(p)$ if (a) $p(x) = x^2 - x + 1$; (b) $p(x) = 2x^3 - x + 2$; (c) $p(x) = x^4 + x^2 + 1$.
4. Decide whether the mapping f, as defined below on the space of all real polynomials p in some indeterminate, is a linear functional: (a) $f(p) = \int_{-1}^{1} p(x)\, dx$; (b) $f(p) = \int_{0}^{1} [p(x)]^2\, dx$; (c) $f(p) = p'$, p' the derivative of p; (d) $f(p) = p'(1)$, with p' as in (c).
5. If f is a nonzero linear functional on a vector space V, decide whether there exists $\alpha \in V$ such that $f(\alpha) = a$, for an arbitrary $a \in \mathbf{R}$.
6. Define a linear functional f on V_3 such that $f(E_1) = f(E_2) = 0$, but $f(E_3) \neq 0$.
7. Write conditions (i) and (ii), as given in the definition of dual vector spaces, in the more usual functional notation.
8. If $X_1 = (1,1)$ and $X_2 = (1,-2)$, find the basis in V_2^* that is dual to $\{X_1, X_2\}$ in V_2.
9. If V is an n-dimensional vector space verify that there is a one-to-one correspondence between the elements of the dual space V^* and V_n.
10. Refer to Prob. 9 and verify that the spaces V^* and V_n are isomorphic.
11. With reference to the first paragraph in the proof of Theorem 4.1, explain why the linear independence of the $\{E_i\}$ guarantees that $\{f_i\}$ is a basis of V^*.

12. Verify that condition (i), in the definition of dual spaces, holds for any space V and its associated space V^* of linear functionals.
13. Verify that condition (ii), in the definition of dual spaces, holds for any space V and its associated space V^* of linear functionals.
14. Find the basis of V_3^* that is dual to the basis $\{(1,0,-1), (1,1,1),(2,2,0)\}$ of V_3.
15. Use the result in Prob. 14 to express the following vectors in V_3 in terms of the given basis: (a) $(2,-2,1)$; (b) $(0,-2,-3)$; (c) $(2,1,5)$.
16. A linear functional f on V_3 is defined by $f(x_1,x_2,x_3) = 2x_1 - x_2 + x_3$. Express f in terms of the basis V_3^*, as determined in Prob. 14.
17. If g is the linear functional defined on V_3 so that $g(E_1) = 1$, $g(E_2) = -1$, $g(E_3) = 2$, use Theorem 4.1 to determine $g(x_1,x_2,x_3)$ for an arbitrary $(x_1,x_2,x_3) \in V_3$.
18. If T_α is the linear functional defined on the dual V^* of a space V by $T_\alpha(f) = f(\alpha)$, for any $\alpha \in V$ (see Prob. 17 of Sec. 2.3), prove that the mapping $\alpha \to T_\alpha$ is an isomorphism of V onto V^{**}. [*Hint:* Use the result in Prob. 12 to see that $T_\alpha = 0$ if and only if $\alpha = 0$.]
19. If F is a linear functional on the space V^* dual to a given space V, use the result in Prob. 18 to verify the existence of a unique element $\alpha \in V$ such that $F(f) = f(\alpha)$, for every $f \in V^*$.
20. Use the result in Prob. 19 to prove that every basis of the dual V^* of a space V is dual to some basis of V.
21. Let $\{\alpha_1,\alpha_2,\ldots,\alpha_n\}$ be a basis for a vector space V, with $\{f_1,f_2,\ldots,f_n\}$ its dual basis in V^*. Then, for an arbitrary $\alpha \in V$ and an arbitrary $f \in V^*$, show that $f(\alpha) = a_1 b_1 + a_2 b_2 + \cdots + a_n b_n$, where a_i is the ith coordinate of f (relative to the basis $\{f_i\}$) and b_i is the ith coordinate of α (relative to the basis $\{\alpha_i\}$), $i = 1, 2, \ldots, n$.
22. Let W be a proper subspace of an n-dimensional vector space V, with $\alpha_0 \in V$, $\alpha_0 \notin W$. Then show that there exists $f \in V^*$ such that $f(\alpha_0) = 1$, but $f(\alpha) = 0$ for all $\alpha \in W$.
23. With $[\alpha_1,\alpha_2]$ defined as in the definition of dual spaces V_1 and V_2, show why this symbol is very suitable to define both a linear functional on V_1 and a linear functional on V_2, so as to emphasize the symmetrical role played by the two spaces.
24. Let V_1 and V_2 be dual vector spaces. Then show, using the symbolism of our definition, that the subset of all vectors

$\alpha_1 \in V_1$ such that $[\alpha_1, \alpha_2] = 0$, for any fixed $\alpha_2 \in V_2$, is a subspace of V_1.

25. Give what you would regard as a suitable definition of a *self-dual* vector space, and try to produce an example of such a space.
26. Let N be the null space of a linear functional $f \in V^*$, for some vector space V, and fix a vector $\alpha_0 \in V$, $\alpha_0 \notin N$. Then prove that every $\alpha \in V$ can be expressed in the form $c\alpha_0 + \beta$, for a unique $c \in \mathbf{R}$ and $\beta \in N$.

2.5 Annihilators

In this section we examine the interesting relationship that exists between subspaces and linear functionals on a vector space and show, in fact, that any subspace is actually determined by the set of linear functionals that vanish on it. To be more precise; if W is a subspace of a vector space V, where $\dim V = n$ and $\dim W = r$, there exist linear functionals $f_1, f_2, \ldots, f_{n-r} \in V^*$ such that $f_i(\alpha) = 0$, for any $\alpha \in W$, $i = 1, 2, \ldots, n - r$; and, moreover, W is precisely the subset of vectors in V which are mapped onto 0 by each of $f_1, f_2, \ldots, f_{m-n}$.

Definition   *The annihilator of a subset S of a vector space V is the subset S^0 of V^*, consisting of those linear functionals f such that $f(\alpha) = 0$, for every $\alpha \in S$.*

It is not difficult to see that S^0 is a subspace of V^*, regardless of whether S is a subspace or merely a subset of V (Prob. 7). Moreover, $V^0 = \mathcal{O}$ and $\mathcal{O}^0 = V^*$ (Probs. 2 and 3).

Theorem 5.1   *If W is a subspace of a finite-dimensional vector space V, then $\dim W + \dim W^0 = \dim V$.*

Proof
Let $\{\alpha_1, \alpha_2, \ldots, \alpha_k\}$ be a basis of W which we extend to a basis $\{\alpha_1, \alpha_2, \ldots, \alpha_k, \alpha_{k+1}, \ldots, \alpha_n\}$ of V. If $\{f_1, f_2, \ldots, f_n\}$ is the associated dual basis of V^*, we shall show that $\{f_{k+1}, f_{k+2}, \ldots, f_n\}$ is a basis for the annihilator W^0. By the definition of a dual basis, $f_i(\alpha_j) = \delta_{ij}$ $(i, j = 1, 2, \ldots, n)$; but $\delta_{ij} = 0$ if $j \leq k$ and $i \geq k + 1$, and so $f_i(\alpha_j) = 0$ for $i \geq k + 1$ and $j = 1, 2, \ldots, k$. Since any vector in W can be expressed as a linear combination of $\alpha_1, \alpha_2, \ldots, \alpha_k$, it follows that $f_i(\alpha) = 0$, for $\alpha \in W$ and $i \geq k + 1$, and so the subspace spanned by the functionals $f_{k+1}, f_{k+2}, \ldots, f_n$ annihilates W and is then a subspace of W^0. Since these functionals are already known to be linearly independent, all that remains is to verify that

they span W^0. But, if $f \in V^*$, we know from Theorem 4.1 that $f = \sum_{i=1}^{n} f(\alpha_i) f_i$ and, if $f \in W^0$, $f = \sum_{i=k+1}^{n} f(\alpha_i) f_i$. Hence W^0 is spanned by $\{f_{k+1}, f_{k+2}, \ldots, f_n\}$ and so dim $W^0 = n - k$. It follows that dim W + dim W^0 = dim V, as asserted. ∎

From this result, it is not difficult to establish the one-to-one correspondence implied earlier between the subspaces of a vector space V and the subspaces of V^* that annihilate them. This is the essence of our next theorem.

Theorem 5.2 *If W_1^0 and W_2^0 are the annihilators, respectively, of subspaces W_1 and W_2 of a vector space V, then $W_1 = W_2$ if and only if $W_1^0 = W_2^0$.*

Proof

If $W_1 = W_2$, it follows from the definition of an annihilator that $W_1^0 = W_2^0$, and so all that remains is the converse. It is conceivable that, although the annihilator of a subspace annihilates this subspace, it might very well annihilate other subsets of vectors in the space, so that distinct subsets would have the same annihilator. To see that this cannot be so for distinct *subspaces*, let us suppose that $W_1^0 = W_2^0$ for the respective annihilators W_1^0 and W_2^0 of subspaces W_1 and W_2 of a vector space V. If we designate by W the complete subspace of V that is annihilated by $W_1^0 (= W_2^0)$, it is clear that $W_1 \subset W$ and $W_2 \subset W$. But since W_1^0 is the annihilator of both W_1 and W, it follows from Theorem 5.1 that dim W = dim W_1. A similar argument shows that dim W = dim W_2, and this, along with the above inclusions, requires that $W = W_1 = W_2$, as asserted. An alternative, but somewhat more sophisticated, proof can be based on Prob. 18 of Sec. 2.4 by identifying V^{**} with V and considering $W^{00} = (W^0)^0$ as the "annihilator of the annihilator" of W. This proof is suggested to the reader in Prob. 25. The reader is also directed to Prob. 26 for an extension of the idea of an annihilator to any pair of dual vector spaces. ∎

A nice application of linear functionals and annihilators can be obtained by considering a system of homogeneous linear equations, as shown in system (1) of Sec. 1.5 with $b_1 = b_2 = \cdots = b_m = 0$. If we let f_i ($i = 1, 2, \ldots, m$) be the linear functional defined on V_n by $f_i(x_1, x_2, \ldots, x_n) = a_{i1}x_1 + a_{i2}x_2 + \cdots + a_{in}x_n$, the system of equations can be expressed in the form $f_i(x_1, x_2, \ldots, x_n) = 0$, $i = 1, 2, \ldots, m$. Moreover, if we agree to regard the n-tuple $(c_1, c_2, \ldots, c_n)$ as a solution of this system, provided $f_i(c_1, c_2, \ldots, c_n) = 0$, for each i, the solutions comprise a subset of V_n. In fact, it is easy to verify (Prob. 9) that this set of solutions or *solution set* is a subspace of V_n, under the usual operations in this coordinate space. To use language appropriate to the present context, we may state that, in solving

SEC. 2.5 ANNIHILATORS

the system of homogeneous equations, we are seeking the subspace of V_n that is annihilated by the functionals $f_1, f_2, \ldots, f_m$. The information supplied here by Theorem 5.1 is that the sum of the dimensions of the solution space and the space $\mathcal{L}\{f_1, f_2, \ldots, f_m\}$ is equal to the dimension n of V_n. However, instead of proceeding further with abstractions, let us illustrate these ideas with examples.

EXAMPLE 1

We consider Example 3 of Sec. 1.5 from the present point of view. The three equations in the system may be expressed in terms of three functionals f_1, f_2, f_3 on V_3 as follows:

$$f_1(x,y,z) = 2x + 3y - z = 0$$
$$f_2(x,y,z) = x - y + 2z = 0$$
$$f_3(x,y,z) = x + 2y - z = 0$$

The solution space, as discovered in the earlier example but in the present notation, consists of all n-tuples of the form $(-c,c,c)$, with $c \in \mathbf{R}$. Since $(-c,c,c) = c(-1,1,1)$, the solution space is a one-dimensional subspace of V_3; it follows from Theorem 5.1 that the space of linear functionals on V_3 generated by f_1, f_2, f_3 must have dimension 2. There then must exist a linear relationship of the form $a_1 f_1 + a_2 f_2 + a_3 f_3 = 0$, for certain real numbers a_1, a_2, a_3 (not all 0). (See Prob. 12.)

It is also possible to look at the system of equations $f_i(x_1, x_2, \ldots, x_n) = 0$ ($i = 1, 2, \ldots, m$) from the *dual* point of view. That is, given a set of vectors $\{(c_1, c_2, \ldots, c_n)\}$ in V_n, we may be interested in a determination of the annihilator of the subspace spanned by these vectors. Since a typical linear functional f on V_n has the form $f(x_1, x_2, \ldots, x_n) = a_1 x_1 + a_2 x_2 + \cdots + a_n x_n$ (Prob. 8), the essential condition being imposed on f is that $f(c_1, c_2, \ldots, c_n) = a_1 c_1 + a_2 c_2 + \cdots + a_n c_n = 0$, for each $(c_1, c_2, \ldots, c_n)$ in a generating set for the solution space. It is now possible to use the Gauss reduction method to obtain a concrete description of the space of annihilators. Again, we illustrate with an example.

EXAMPLE 2

If $W = \mathcal{L}\{\alpha_1, \alpha_2\}$, where $\alpha_1 = (1, 2, -1, 1)$ and $\alpha_2 = (0, -1, 1, 0)$, determine W^0.

Solution

The problem is to determine all linear functionals $f(x_1, x_2, x_3, x_4) = a_1 x_1 + a_2 x_2 + a_3 x_3 + a_4 x_4$ such that $f(\alpha_1) = f(\alpha_2) = 0$. These conditions are equivalent to the following system of equations:

$$a_1 + 2a_2 - a_3 + a_4 = 0$$
$$- a_2 + a_3 = 0$$

Ordinarily, at this stage, a Gauss reduction would be needed to solve this system of equations, but this particular system is already essentially in reduced form. We see immediately that the complete solution is obtained by letting a_3 and a_4 be arbitrary real numbers, say $a_3 = a$ and $a_4 = b$, from which we obtain the general solution: $a_1 = -a - b$, $a_2 = a$, $a_3 = a$, $a_4 = b$. The form of a general $f \in W^0$ may now be described as $f(x_1,x_2,x_3,x_4) = (-a-b)x_1 + ax_2 + ax_3 + bx_4$, for arbitrary $a,b \in \mathbf{R}$. The dimension of W^0 is known, by Theorem 5.1, to be 2, and so we may obtain a basis for this space of linear functionals by letting $a = -1$, $b = 0$, and then $a = 0$, $b = -1$. On doing this, we obtain $\{f_1, f_2\}$ as a basis for W^0, where $f_1(x_1,x_2,x_3,x_4) = x_1 - x_2 - x_3$ and $f_2(x_1,x_2,x_3,x_4) = x_1 - x_4$. It may be included as a final note of some possible interest that the equality $W^{00} = W$, referred to earlier, may be interpreted in this case to mean that W consists *exactly* of those vectors $(x_1,x_2,x_3,x_4) \in V_4$ such that $x_1 - x_2 - x_3 = 0$, $x_1 - x_4 = 0$.

There are elementary connecting links between the contents of this section and the analytic geometry of 3-space. The subspaces (proper and improper) of this geometric space are the origin, the lines and planes that pass through the origin, and the whole 3-space itself. The asserted connection with analytic geometry is then obtained by the association of the "equation" of a subspace with the linear functional that annihilates it. For example, the plane whose equation is $2x - 3y + z = 0$ is annihilated by the linear functional f defined on V_3 by $f(x,y,z) = 2x - 3y + z$. The linear functional f, in turn, is the annihilator of the subspace consisting of the points of the plane. It is immediate that any line through the origin can be regarded as the subspace of V_3 that is annihilated by two properly chosen linear functionals, and the origin itself has the complete dual space V^* as its annihilator. The geometric illustrations of subspaces and their annihilators are quite limited in 3-space, but these ideas can be extended by analogy to general n-space where a greater variety of situations may be conceived intuitively in geometric terms.

Problems 2.5

1. Check that the relationship between a linear functional and its null space is a special case of the relationship that exists between a subspace and its annihilator.

2. Explain why $\mathcal{O}^0 = V^*$, for any vector space V.
3. Explain why $V^0 = \mathcal{O}$, for any vector space V.
4. If S is any nonzero subset of a vector space V, show, without reference to Prob. 3, that $S^0 \neq V^*$.
5. If f_1, f_2 are linearly independent functionals on V_5, determine the dimension of the subspace of vectors $\alpha \in V_5$ such that $f_1(\alpha) = f_2(\alpha) = 0$.
6. If S_1 and S_2 are subsets of a vector space V, with $S_1 \subset S_2$, explain why $S_2^0 \subset S_1^0$.

7. Verify that S^0 is a subspace of V^* for any *subset* S of a vector space V.
8. Explain why any linear functional f on V_n must have the form $f(x_1, x_2, \ldots, x_n) = a_1 x_1 + a_2 x_2 + \cdots + a_n x_n$, for $a_1, a_2, \ldots, a_n \in \mathbf{R}$.
9. Verify that the solutions of a system of homogeneous linear equations in n unknowns, regarded as elements of V_n, make up a subspace W of V_n. Describe the annihilator W^0.
10. Explain why the result in Prob. 9 fails if the equations are not homogeneous.
11. If the dimension of the solution space (see Prob. 9) of a system of m homogeneous linear equations in n unknowns ($m \leq n$) is greater than $n - m$, what can be concluded about the functionals that define the equations?
12. Find the linear relationship asserted to exist in Example 1.

13. Determine the annihilator of $\mathcal{L}\{(1,0,0),(0,1,0),(1,0,1)\} \subset V_3$.
14. If $W = \mathcal{L}\{(0,1,1),(1,0,0)\}$, determine W^0.
15. Explain why the annihilator of a subset of a vector space V must be identical with the annihilator of the smallest subspace that contains the subset.
16. If $W = \mathcal{L}\{(1,0,1)\}$, determine a basis for W^0.
17. If W is the subspace of V_4, consisting of all vectors (x_1, x_2, x_3, x_4) such that $x_1 - x_2 + x_3 = 0$, determine W^0. Interpret this result in geometric terms.
18. Find the subspace of V_4 that is annihilated by $\mathcal{L}\{f_1, f_2, f_3\}$, where $f_1(X) = 2x_1 - 2x_2 + x_3 + x_4, f_2(X) = x_1 - 2x_3 + 2x_4, f_3(X) = x_1 - 4x_2 + 8x_3 - 4x_4$, for arbitrary $X = (x_1, x_2, x_3, x_4) \in V_4$.
19. If W_1 and W_2 are subspaces of a vector space V, prove that $(W_1 + W_2)^0 = W_1^0 \cap W_2^0$.

20. If W_1 and W_2 are subspaces of a vector space V, prove that $(W_1 \cap W_2)^0 = W_1^0 + W_2^0$.
21. Let W be the subspace of V_n, consisting of all vectors $(x_1, x_2, \ldots, x_n)$ such that $x_1 + x_2 + \cdots + x_n = 0$. Prove that W^0 consists of the set of all linear functionals f such that $f(x_1, x_2, \ldots, x_n) = k(x_1 + x_2 + \cdots + x_n)$, for $k \in \mathbf{R}$.
22. With W defined as in Prob. 21, prove that W^* may be identified with the subset of functions f such that $f(x_1, x_2, \ldots, x_n) = a_1 x_1 + a_2 x_2 + \cdots + a_n x_n$, where $a_1 + a_2 + \cdots + a_n = 0$.
23. A *hyperplane* in a vector space V may be defined as the null space of a nonzero linear functional in V^*. Explain why every subspace of V may be regarded as the intersection of hyperplanes.
24. A straight line in 3-space is defined by the equations $2x - 3y + z = 0$, $x + y - 2z = 0$. Consider the line as a subspace of V_3, and describe its annihilator.
25. If we agree to identify V^{**} with V (see Prob. 18 of Sec. 2.4), show how this may be used to conclude that $W^{00} = W$, for a subspace W of V. [*Hint:* Apply Theorem 5.1 to the space V^*, and use $W \subset W^{00} \subset V^{**}$.]
26. If V_1 and V_2 are dual spaces, as defined in Sec. 2.4, use the bracket symbol to define the concept of an annihilator and then prove that $V_1^{00} = V_1$ and $V_2^{00} = V_2$.

2.6 Nonsingular Linear Transformations

We have seen that the set of all linear transformations of a vector space V into a space W forms an additive abelian group, and, in case $W = V$, even a ring with identity. However, it is easy to see that such a system of transformations does not share its additive properties with its underlying multiplicative system, even when one is defined. We have been referring to conditions A1 to A4 in the definition of a vector space (see Sec. 1.1) as the defining properties of a group, but no special significance should be attached to the fact that the operation indicated there is addition. However, if the operation is denoted as multiplication, it is customary to make certain appropriate changes in the form of the group postulates. In brief, a *group under multiplication*, or a *multiplicative group*, is a set G in which a *product* is defined, subject to the following conditions: $ab \in G$, for any $a, b \in G$; $a(bc) = (ab)c$, for any $a, b, c \in G$; there exists a unique element $1 \in G$, called the *identity element*, such that $a1 = 1a = a$, for any $a \in G$; with each $a \in G$, there is associated a unique element $a^{-1} \in G$, called the *inverse* of a, such that

$aa^{-1} = a^{-1}a = 1$. If, in addition, $ab = ba$, for any $a,b \in G$, the group is a multiplicative *abelian* group.

If we attempt to consider the set of linear transformations of V into W as a multiplicative system, we are confronted with difficulties at every turn. In fact, a product is not even defined unless we assume $W = V$. However, even in this subset of linear *operators* on V, although an associative product is always defined and the identity operator is available as the identity element of the multiplicative system, this system still falls short of being a group: for there is no guarantee of the existence of *inverse* operators, as required by the definition of a group. One of the objectives of this section is to discover a subset of linear operators on a vector space V that does comprise a multiplicative group. However, before doing this, we examine more thoroughly the whole idea of an inverse of a linear transformation in a general context.

As suggested in the definition of a multiplicative group above, the effect of an inverse mapping must be to reverse some original mapping. Thus, if T: $V \to W$ is a mapping of a vector space V into a vector space W, the *inverse mapping* T^{-1}, if it exists, must be such that the following requirements are satisfied: T^{-1}T *is the identity map on* V; TT^{-1} *is the identity map on* W. It is almost immediate that two conditions are both necessary and sufficient for the existence of T^{-1}.

(1) Since T^{-1} is to be a mapping of W, the domain of T^{-1} must be W; and since TT^{-1} is the identity on W, the range of T must be the whole of W (Prob. 9). Hence T must map V *onto* W.

(2) If T^{-1} is to reverse the mapping of T, it is clear that distinct elements of V must be mapped onto distinct elements of W. That is, Tα = Tβ, for $\alpha, \beta \in V$, if and only if $\alpha = \beta$, a condition described by stating that the mapping T is *one-to-one*. (See Prob. 2.)

We are now able to state the pertinent theorem.

Theorem 6.1 *If* T *is a mapping of a vector space* V *into a vector space* W, *the (unique) inverse* T^{-1} *of* T *exists if and only if* T *is one-to-one and onto* W. *Moreover,* T^{-1} *is defined so that* T$^{-1}\beta = \alpha$ *if* T$\alpha = \beta$, *with* $\alpha \in V$ *and* $\beta \in W$.

Proof
The remarks preceding the statement of the theorem show that it is *necessary* for the existence of T^{-1} that T be a one-to-one mapping onto W. On the other hand, *if* T is this type of mapping, it is possible to define T^{-1} as in the statement of the theorem, and it is not difficult (Prob. 11) to show that TT^{-1} is the identity on W and T^{-1}T is the identity on V. It is also easy to verify that there is only one mapping with this characteristic *inverse* property, so that T^{-1} is unique (Prob. 8). ∎

A mapping may be said to be *invertible* if its inverse exists. It is then possible to paraphrase Theorem 6.1 in part by asserting that *a mapping of V into W is invertible if and only if it is one-to-one and onto W.* In most cases of interest to us, the mapping T is a linear transformation, and the following important definition is applicable.

Definition   *A linear transformation* $T: V \to W$, *for vector spaces V and W, is nonsingular if the null space N_T is the zero subspace of V. That is, $T\alpha = 0$, for a nonsingular T, implies that $\alpha = 0$.*

Since $T\alpha = T\beta$ if and only if $T(\alpha - \beta) = 0$, the statement "T is nonsingular" is equivalent to "T is one-to-one" for a *linear transformation* T. Hence we may state the following corollary to Theorem 6.1.

Corollary   *A linear transformation* $T: V \to W$ *is invertible if and only if* T *is nonsingular and its range is W. The mapping* T^{-1} *is defined as in the theorem.*

The next result is an important one in a study of systems of linear transformations.

Theorem 6.2   *The inverse T^{-1} of an invertible linear transformation T of a vector space V is a linear transformation of the range W of T. Moreover, T^{-1} is invertible and $(T^{-1})^{-1} = T$.*

Proof
Let $\beta_1, \beta_2 \in W$, where $T\alpha_1 = \beta_1$ and $T\alpha_2 = \beta_2$, with $\alpha_1, \alpha_2 \in V$. Then $T(r_1\alpha_1 + r_2\alpha_2) = r_1(T\alpha_1) + r_2(T\alpha_2) = r_1\beta_1 + r_2\beta_2$, with $r_1, r_2 \in \mathbf{R}$; and the basic property of T^{-1} implies that $T^{-1}(r_1\beta_1 + r_2\beta_2) = r_1\alpha_1 + r_2\alpha_2 = r_1(T^{-1}\beta_1) + r_2(T^{-1}\beta_2)$. Hence T^{-1} is a linear transformation of W into V. Let us designate the identity mappings on V and W, respectively, by I_V and I_W. Then, since T and T^{-1} play symmetrical roles in the conditions $TT^{-1} = I_W$ and $T^{-1}T = I_V$, it is clear that $(T^{-1})^{-1}$ exists if and only if T^{-1} exists and that $(T^{-1})^{-1} = T$. ∎

It is interesting and of basic importance that a nonsingular linear transformation does not disturb the linear independence of any set of vectors. The following theorem shows that this is a characteristic property of such transformations.

Theorem 6.3   *A linear transformation $T: V \to W$ is nonsingular if and only if linearly independent vectors of V are mapped onto vectors of W that are also linearly independent.*

Proof
First, let us suppose that T is nonsingular and that $r_1(T\alpha_1) + r_2(T\alpha_2) + \cdots + r_n(T\alpha_n) = 0$, for $r_1, r_2, \ldots, r_n \in \mathbf{R}$ and linearly independent vectors $\alpha_1, \alpha_2, \ldots, \alpha_n \in V$. Then $T(r_1\alpha_1 + r_2\alpha_2 + \cdots + r_n\alpha_n) = 0$ and the nonsingular nature of T requires that $r_1\alpha_1 + r_2\alpha_2 + \cdots + r_n\alpha_n = 0$. In view of the assumed linear independence of $\alpha_1, \alpha_2, \ldots, \alpha_n$, we conclude that $r_1 = r_2 = \cdots = r_n = 0$, whence $T\alpha_1, T\alpha_2, \ldots, T\alpha_n$ are also linearly independent. Conversely, let us suppose that T has the property that it maps any set of linearly independent vectors onto vectors that are also linearly independent. Since any nonzero vector α is linearly independent, whereas 0 is a linearly dependent vector, we conclude that $T\alpha \neq 0$ for any $\alpha \, (\neq 0) \in V$. But then $N_T = \mathfrak{O}$, and T is nonsingular, as desired. ∎

Corollary *A linear transformation* $T: V \to W$, *where* $\dim V = \dim W$, *is nonsingular if and only if the image under* T *of any basis of* V *is a basis of* W.

From this point on in the section, we shift our attention from linear transformations $T: V \to W$, where V and W are arbitrary vector spaces, to the very important special case where $W = V$. We shall then consider linear *operators* on a vector space V. Everything that has been said earlier concerning linear transformations remains valid for linear operators, but in most cases the results can be stated more simply. Let us itemize a few of these important simplified statements for a linear operator T on V.

(i) If T is invertible, its inverse T^{-1} can be characterized uniquely by the condition that $TT^{-1} = T^{-1}T = I$, where I is the identity operator on V.

(ii) As a result of Theorem 2.2, the notions of *one-to-one* and *onto* are redundant (if applicable) properties of T. That is, T *is one-to-one if and only if* T *is onto* (Prob. 13). Hence (Theorem 6.1), T is invertible if and only if *either* (and hence both) T is a *one-to-one* mapping *or* the range of T is V. The definition of T^{-1} is the same as in Theorem 6.1 but with both α and β in V.

(iii) The notions of *invertible* and *nonsingular* are equivalent for T. That is, a linear operator T is invertible if and only if it is nonsingular.

(iv) If the operator T is nonsingular, so is T^{-1} and $(T^{-1})^{-1} = T$.

(v) A linear operator T on V is nonsingular if and only if linearly independent vectors of V are mapped by T onto vectors that are also linearly independent.

(vi) A linear operator T on V is nonsingular if and only if the map under T of an arbitrary basis of V is also a basis of V.

It is now possible to sum up most of the important properties of nonsingular linear operators in the following theorem, which may be regarded as one of the important goals of this section.

94 LINEAR TRANSFORMATIONS AND MATRICES

Theorem 6.4 *The set of nonsingular linear operators on a vector space V comprises a multiplicative group. This group is called the full linear group of operators on V.*

Proof
The only part of the proof that remains is to check the closure of the system, and so let T_1, T_2 be arbitrary nonsingular linear operators on V. The pertinent discussion in Sec. 2.3 assures us that T_1T_2 is a linear operator, and so we have merely to verify that this composite operator is nonsingular. We know that T_1^{-1} and T_2^{-1} exist, and the associative property of mappings makes it possible to write $(T_1T_2)(T_2^{-1}T_1^{-1}) = T_1(T_2T_2^{-1})T_1^{-1} = T_1(I)T_1^{-1} = T_1T_1^{-1} = I$. We may then assert, using the uniqueness of an inverse operator, that $(T_1T_2)^{-1}$ exists and equals $T_2^{-1}T_1^{-1}$. Hence T_1T_2 is invertible or (equivalently) nonsingular, and so the system of nonsingular operators is closed. The proof that the system is a group is now completed with the following observations: associativity is a property of general mappings, whenever products are defined; the identity operator is nonsingular, because it is its own inverse; since any nonsingular operator is invertible, each such operator has a unique inverse, and this inverse is also nonsingular by statement (**iv**). ∎

EXAMPLE 1

Let T_1 and T_2 be linear operators on V_3, defined by $T_1(x_1,x_2,x_3) = (x_1 + x_3, x_2 - x_1, x_3)$ and $T_2(x_1,x_2,x_3) = (x_1 - x_3, x_1 + x_2 - x_3, x_3)$. Then $(T_2T_1)(x_1,x_2,x_3) = T_2(x_1 + x_3, x_2 - x_1, x_3) = (x_1,x_2,x_3)$, so that $T_2T_1 = I$. A similar computation shows that $T_1T_2 = I$, and so $T_2 = T_1^{-1}$ and $T_1 = T_2^{-1}$. In this connection, the reader is referred to Prob. 10.

Our final example illustrates the fact that it is possible for the conditions of Theorem 6.1 to be satisfied, so that a mapping has an inverse, without the mapping being a linear transformation.

EXAMPLE 2

Find the inverse of the mapping T, defined on V_3 so that $T(x_1,x_2,x_3) = (x_1 + x_2 - 1, x_2, x_3 - x_2 + 2)$, and verify that T is not a linear transformation.

Solution
If we let $x_1 + x_2 - 1 = y_1$, $x_2 = y_2$, $x_3 - x_2 + 2 = y_3$, elementary algebra yields $x_1 = y_1 - y_2 + 1$, $x_2 = y_2$, $x_3 = y_3 + y_2 - 2$. Hence T^{-1} may be defined by the mapping $T^{-1}(y_1,y_2,y_3) = (x_1,x_2,x_3) = (y_1 - y_2 + 1, y_2, y_3 +$

$y_2 - 2$); and, if desired, it is easy to verify that $TT^{-1} = T^{-1}T = I$ (Prob. 14). On the other hand, we may note that $T[2(x_1,x_2,x_3)] = T(2x_1,2x_2,2x_3) = (2x_1 + 2x_2 - 1, 2x_2, 2x_3 - 2x_2 + 2)$, and $2T(x_1,x_2,x_3) = 2(x_1 + x_2 - 1, x_2, x_3 - x_2 + 2) = (2x_1 + 2x_2 - 2, 2x_2, 2x_3 - 2x_2 + 4)$. Hence T cannot be linear.

Problems 2.6

1. If T is an invertible mapping of a vector space V, explain why it would *not* be appropriate to state the defining property of T^{-1} as $TT^{-1} = T^{-1}T = I$.

2. Explain the distinction between a one-to-one mapping and a one-to-one correspondence. [Cf. discussion in Sec. 1.3.] Give an example of a mapping of V_2 into V_3 that is one-to-one but does not set up a one-to-one correspondence between the two spaces.

3. Prove that a linear transformation $T: V \to W$, for vector spaces V and W, is one-to-one if and only if dim $R_T = $ dim V.

4. Give an example of a nonsingular linear transformation that is not invertible.

5. If T is the nonsingular linear operator on V_2, defined by $T(x_1,x_2) = (x_1 + x_2, x_2 - x_1)$, determine T^{-1}.

6. If $T_1, T_2, \ldots, T_n$ are nonsingular linear operators on a vector space, prove that $T_1T_2 \cdots T_n$ is nonsingular and that $(T_1T_2 \cdots T_n)^{-1} = T_n^{-1} \cdots T_2^{-1}T_1^{-1}$.

7. Prove that *invertible* and *nonsingular* are equivalent properties, when applicable to a linear transformation $T: V \to W$ such that dim $V = $ dim W.

8. Explain why the inverse of an invertible linear transformation is a unique mapping.

9. With $T: V \to W$, explain why the condition $TT^{-1} = I_W$ (identity on W) requires that $R_T = W$.

10. With T and T_1 linear operators defined so that $TT_1 = I$, show that $T_1T = I$ and hence $T_1 = T^{-1}$.

11. With T^{-1} defined as in Theorem 6.1, verify that $T^{-1}T = I_V$ and $TT^{-1} = I_W$, with I_V and I_W the respective identities on V and W.

12. Give the proof of the corollary following Theorem 6.3.

13. Explain why *one-to-one* and *onto* are equivalent properties of a linear operator on a vector space.

14. With reference to Example 2, verify that $TT^{-1} = T^{-1}T = I$.

15. If T is an invertible linear transformation, and $c \ (\neq 0) \in \mathbf{R}$, prove that $c\mathrm{T}$ is invertible and $(c\mathrm{T})^{-1} = c^{-1}\mathrm{T}^{-1}$.
16. If T is the nonsingular linear operator, defined on V_3 so that $\mathrm{T}(x_1,x_2,x_3) = (x_1,\ x_1 + x_2,\ x_1 + x_2 + x_3)$, determine T^{-1}.
17. Verify that the mapping T, defined on V_3 so that $\mathrm{T}(x_1,x_2,x_3) = (2x_1 + 1,\ x_2 + 2,\ x_1 + x_3)$, is invertible but not linear.
18. Let D be the differentiation operator, defined on the space P of all polynomials in an indeterminate so that $\mathrm{D}p = p'$, where p' is the derivative of the polynomial $p \in P$. Then decide whether D is a nonsingular operator, and either give a proof that it is or supply a counterexample to show that it is not.
19. If T is a linear transformation of a vector space V, prove that vectors $\alpha_1, \alpha_2, \ldots, \alpha_n \in V$ are linearly independent if $\mathrm{T}\alpha_1, \mathrm{T}\alpha_2, \ldots, \mathrm{T}\alpha_n$ are linearly independent, regardless of whether T is nonsingular.
20. If the product $\mathrm{T}_1\mathrm{T}_2$ of two linear operators is nonsingular, prove that both T_1 and T_2 are nonsingular.
21. If T is a linear operator such that $\mathrm{T}^2 - \mathrm{T} - \mathrm{I} = 0$, prove that T is nonsingular and that $\mathrm{T}^{-1} = \mathrm{T} - \mathrm{I}$.
22. Find two linear operators $\mathrm{T}_1, \mathrm{T}_2$ on V_2 such that $\mathrm{T}_1\mathrm{T}_2 = 0$ but $\mathrm{T}_2\mathrm{T}_1 \neq 0$.
23. Let T be a linear operator, defined on V_3 so that $\mathrm{T}(x_1,x_2,x_3) = (2x_1 + x_2,\ x_2 - x_3,\ x_1 + x_2 + x_3)$.
    (a) Prove that T is nonsingular, without actually finding T^{-1}.
    (b) Determine T^{-1}.
    (c) Find the basis onto which the $\{E_i\}$ basis of V_3 is mapped by T.
24. If T is a linear operator, defined on V_3 so that $\mathrm{T}(x_1,x_2,x_3) = (3x_1,\ x_1 - x_2,\ 2x_1 + x_2 + x_3)$, determine (a) T^2; (b) T^{-1}.
25. Prove that the operator T in Prob. 24 satisfies the equation $(x^2 - \mathrm{I})(x^2 - 3\mathrm{I}) = 0$. [Caution: I is the identity operator.]
26. Determine T^{-1}, if T is the *rotation* operator defined in the xy plane (as the space V_2) by $\mathrm{T}(x,y) = (x \cos \theta - y \sin \theta,\ x \sin \theta + y \cos \theta)$.
27. Determine T^{-1}, if T is the *shear* operator defined in the xy plane (as the space V_2) by $\mathrm{T}(x,y) = (x + cy,\ y)$, for $c \in \mathbf{R}$.

2.7 Matrices of Linear Transformations

The notion of a *matrix* arose in Sec. 1.5 in connection with the solution of a system of linear equations; in that context a matrix was merely a rectangular array of numbers, which had a convenient and unique association with the equations. In this section, we first show that it is possible to associate such an array of numbers, also in a unique manner, with any linear transformation of a (*finite-dimensional*) vector space. But here we go farther. In the present context we define *operations* on matrices, in such a way that there is generated an algebraic system that is isomorphic to the associated system of transformations, be it an additive group, a vector space, a multiplicative group, or a ring with identity element. In fact, the association between matrices and linear transformations or operators is so close that even the symbol T for a transformation is sometimes replaced by the symbol A of the associated matrix. The image $T\alpha$ of a vector α in the vector space transformed by T might then be designated $A\alpha$. This practice is probably justifiable in the work of an experienced scientist, who is interested only in his final results; in an introductory text, such as this one, it is best to keep linear transformations and their associated matrices quite distinct, in spite of their close connection. Accordingly, we shall adhere to a strictly conventional usage of the symbols for transformations and matrices.

We shall consider a linear transformation T: $V \to W$, where V and W are possibly distinct vector spaces with dim $V = n$ and dim $W = m$. If $\{\alpha_1, \alpha_2, \ldots, \alpha_n\}$ and $\{\alpha'_1, \alpha'_2, \ldots, \alpha'_m\}$ are respective index-ordered bases of V and W, it follows from the definition of a basis that

$$T\alpha_j = a_{1j}\alpha'_1 + a_{2j}\alpha'_2 + \cdots + a_{mj}\alpha'_m$$

for each $j = 1, 2, \ldots, n$ and unique real numbers $a_{1j}, a_{2j}, \ldots, a_{mj}$. In view of the nature of a linear transformation, as described in Theorem 2.1, the transformation T is completely determined by the numbers $\{a_{ij}\}$, *after the bases $\{\alpha_j\}$ and $\{\alpha'_j\}$ have been chosen.* It is then convenient and also quite appropriate to represent T by an array of these numbers, which have been suitably ordered in the form of a matrix. We recall that the horizontal subarrays of a matrix are called its *rows* and the vertical subarrays are its *columns*. We then elect to arrange the numbers $\{a_{ij}\}$ in matrix form so that the coefficients of $T\alpha_j$, in their natural order above, make up the jth column of the matrix. The number a_{ij} will then appear at the intersection of the ith row and jth column and, as in Sec. 1.5, it will often be convenient to designate this matrix A by $[a_{ij}]$. The matrix A, which *represents* the transformation T, will then be an $m \times n$ matrix, and

$$A = [a_{ij}] = \begin{bmatrix} a_{11} & a_{12} & \cdots & a_{1n} \\ a_{21} & a_{22} & \cdots & a_{2n} \\ \cdot & \cdot & \cdot & \cdot \\ a_{m1} & a_{m2} & \cdots & a_{mn} \end{bmatrix}$$

The matrix A will also be known as the *matrix of the linear transformation* T, relative to the $\{\alpha_i\}$ and $\{\alpha'_i\}$ bases, and we shall often indicate the association between T and A by T $\leftrightarrow A$. For our immediate purposes, then, the matrix of a transformation of a vector space is simply a convenient means of describing the transformation by storing, in an orderly fashion, coordinates of the images under the transformation of a given ordered set of basis elements of the space. It should be apparent that, *if we select different bases for the spaces, the matrix of the transformation (or the transformation matrix) will change*, and a large part of matrix theory is concerned with the effects on a transformation matrix of various changes of bases. It should be emphasized that *only* transformations on *finite-dimensional* vector spaces have matrix representations and so *in the context of matrices this type of space should be understood* without explicit statement to this effect.

EXAMPLE 1

Let us suppose that a linear transformation T: $V_3 \to V_2$ has been defined by means of the following equations, the $\{E_i\}$ bases being used in the case of both spaces:

$$T(1,0,0) = (2,1) = 2(1,0) + 1(0,1)$$
$$T(0,1,0) = (1,-1) = 1(1,0) - 1(0,1)$$
$$T(0,0,1) = (2,3) = 2(1,0) + 3(0,1)$$

It is now clear from the remarks immediately preceding this example that the matrix of T, relative to the $\{E_i\}$ bases, is

$$\begin{bmatrix} 2 & 1 & 2 \\ 1 & -1 & 3 \end{bmatrix}$$

Before proceeding further, we shall comment briefly on the form of the matrix that we shall use to represent a linear transformation. The reader who has referred to other books on the subject may have noticed that, although we have used the coefficients (or *coordinates*) of Tα_j to make up the *j*th *column* of the matrix A, in some books these same numbers may have been used to form the *j*th *row*. The reason for the difference in choice is not apparent at this time, but it stems from our decision to interpret a product T_1T_2 of two transformations as T_2 followed by T_1, rather than as T_1 followed by T_2. The effect of this interpretation on our *system* of representation matrices will be made clear in the following section.

The most common linear transformations are those which map a vector space into itself, or, in already familiar language, the *operators* on the space.

SEC. 2.7 MATRICES OF LINEAR TRANSFORMATIONS

It is immediate that a representation matrix of an operator on an n-dimensional vector space is an $n \times n$ matrix, otherwise called a *square matrix of order n*.

EXAMPLE 2

A rotation of the plane through an angle θ, in which the point (x,y) is mapped onto the point (x',y'), may be described by the following familiar equations:

$$x' = x \cos \theta - y \sin \theta \qquad y' = x \sin \theta + y \cos \theta$$

This rotation can be expressed as a linear (cf. Prob. 26 of Sec. 2.6) operator on V_2 by $T(x,y) = (x \cos \theta - y \sin \theta, x \sin \theta + y \cos \theta)$. In particular, we see from this that $T(1,0) = (\cos \theta, \sin \theta) = (\cos \theta)(1,0) + (\sin \theta)(0,1)$ and $T(0,1) = (-\sin \theta, \cos \theta) = (-\sin \theta)(1,0) + (\cos \theta)(0,1)$. Hence the matrix of T, relative to the $\{E_i\}$ basis of V_2, is

$$\begin{bmatrix} \cos \theta & -\sin \theta \\ \sin \theta & \cos \theta \end{bmatrix}$$

EXAMPLE 3

As a further illustration of a transformation matrix, let us examine the differentiation operator D on the space P_4 of polynomials of degree 3 or less in a real variable t. The most common basis for this space is $\{1, t, t^2, t^3\}$, and we know that $D1 = 0$, $Dt = 1$, $Dt^2 = 2t$, $Dt^3 = 3t^2$. It then follows that the matrix of D, relative to the chosen basis, is

$$\begin{bmatrix} 0 & 1 & 0 & 0 \\ 0 & 0 & 2 & 0 \\ 0 & 0 & 0 & 3 \\ 0 & 0 & 0 & 0 \end{bmatrix}$$

We have seen that, if vector spaces V and W have dimensions n and m, respectively, any linear transformation T: $V \to W$ determines a *unique* $m \times n$ matrix that represents it, relative to some selection of bases for the spaces. On the other hand, if we are given an arbitrary $m \times n$ matrix, it is clear that it can be regarded as the representing matrix of some linear transformation, and this transformation is *uniquely* determined by the matrix—again relative to the bases selected. Thus, *for a given choice of bases for the spaces*, there is a one-to-one correspondence between the linear transformations of V into W and the set of $m \times n$ matrices. We shall next show that it is possible to define matrix operations so that linear transformations and their associated transformation matrices comprise isomorphic systems. In making these definitions, we shall be guided by the operations on transformations and how these transformations and the matrices are associated. In this section, only *addition* and *multiplication by scalars* will be considered.

In any construction of an algebraic *system* from a mere set of elements, it is important that we understand what we mean by *equal* or *identical* elements. This is usually implicit in the concept of the elements under study, but let us be precise in the case of matrices:

Definition If $A = [a_{ij}]$ and $B = [b_{ij}]$ are two $m \times n$ matrices, we say that A and B are equal and write $A = B$ if and only if $a_{ij} = b_{ij}$, for $i = 1, 2, \ldots, m$ and $j = 1, 2, \ldots, n$. That is, two matrices are equal provided they are the same size and their elements in corresponding positions are equal.

In view of the association that we have already made between matrices and linear transformations, it is clear (Prob. 11) that this definition is inevitable because of the uniqueness of a transformation matrix. Since we wish the sum of two transformation matrices to be the transformation matrix of the sum of the associated transformations, it is only natural that we give the following definition of addition:

Definition If $A = [a_{ij}]$ and $B = [b_{ij}]$ are two $m \times n$ matrices, we define the sum $A + B$ to be the $m \times n$ matrix $C = [c_{ij}]$, where $c_{ij} = a_{ij} + b_{ij}$, $i = 1, 2, \ldots, m$ and $j = 1, 2, \ldots, n$. That is, we add two matrices of the same size by adding their elements in corresponding positions.

It is noteworthy that addition of matrices is defined only if they are of the same size. The following theorem now follows easily, after a quick review of the postulates of an additive abelian group as listed (A1 to A5) in Sec. 1.1.

Theorem 7.1 The system G of all $m \times n$ matrices comprises an additive abelian group.

Proof
Since the sum of two $m \times n$ matrices is also an $m \times n$ matrix, the system is closed under addition. If A, B, C are arbitrary matrices of G, the equality $A + (B + C) = (A + B) + C$ follows from the associativity of addition in **R**. The $m \times n$ matrix, all of whose entries are 0, is the *zero matrix* (to be denoted by O) of G, for $A + O = O + A = A$, for any $A \in G$. If $A = [a_{ij}]$ is any matrix in G, the *additive inverse* $-A$ can be defined so that $-A = [a'_{ij}]$, where

$a'_{ij} = -a_{ij}$; for, clearly, $A + (-A) = O$. The *uniqueness* of the zero and additive inverse matrices also follows immediately from the intrinsic properties of these matrices (Prob. 12). Hence G is an additive group. The abelian nature of G is a consequence of the commutative property of real numbers under addition, and the proof of the theorem is complete. ∎

EXAMPLE 4

Let us illustrate the group properties, as established in Theorem 7.1, with the use of the following 2 × 3 matrices:

$$A = \begin{bmatrix} 1 & 0 & 2 \\ 2 & -1 & 1 \end{bmatrix} \quad B = \begin{bmatrix} -2 & 3 & 4 \\ 0 & 1 & -2 \end{bmatrix} \quad C = \begin{bmatrix} 2 & 0 & -4 \\ 1 & 3 & 5 \end{bmatrix}$$

The sum $A + B = \begin{bmatrix} -1 & 3 & 6 \\ 2 & 0 & -1 \end{bmatrix}$ illustrates that the sum of two 2 × 3 matrices is also a 2 × 3 matrix, which is the rule of *closure*. Since

$$A + (B + C) = \begin{bmatrix} 1 & 0 & 2 \\ 2 & -1 & 1 \end{bmatrix} + \begin{bmatrix} 0 & 3 & 0 \\ 1 & 4 & 3 \end{bmatrix} = \begin{bmatrix} 1 & 3 & 2 \\ 3 & 3 & 4 \end{bmatrix}$$

and

$$(A + B) + C = \begin{bmatrix} -1 & 3 & 6 \\ 2 & 0 & -1 \end{bmatrix} + \begin{bmatrix} 2 & 0 & -4 \\ 1 & 3 & 5 \end{bmatrix} = \begin{bmatrix} 1 & 3 & 2 \\ 3 & 3 & 4 \end{bmatrix}$$

we have an illustration of the associative law of addition. The zero 2 × 3 matrix is $\begin{bmatrix} 0 & 0 & 0 \\ 0 & 0 & 0 \end{bmatrix}$, and the additive inverse $-A$ of A is $\begin{bmatrix} -1 & 0 & -2 \\ -2 & 1 & -1 \end{bmatrix}$, observing that $A + (-A) = \begin{bmatrix} 0 & 0 & 0 \\ 0 & 0 & 0 \end{bmatrix}$. Finally, to illustrate the abelian nature of the additive system of 2 × 3 matrices, we may note that

$$A + B = \begin{bmatrix} -1 & 3 & 6 \\ 2 & 0 & -1 \end{bmatrix} = B + A$$

Before stating our next result—the focal point in this section—we must make precise what we mean by the notion of *isomorphism* as applied to groups. In general, as was implied in Sec. 1.3, two systems that are indistinguishable *as systems* are isomorphic, but the following definition makes our meaning clear for isomorphic *groups*.

Definition   Two additive (multiplicative) groups G and G' are isomorphic if there exists a one-to-one correspondence $g \leftrightarrow g'$ between the elements $g \in G$ and $g' \in G$, such that

$$g_1 + g_2 \leftrightarrow g'_1 + g'_2 \qquad (g_1 g_2 \leftrightarrow g'_1 g'_2)$$

for arbitrary $g_1, g_2 \in G$ and their correspondents $g'_1, g'_2 \in G'$.

We are now ready for the theorem to which we referred above.

Theorem 7.2   If V and W are vector spaces of respective dimensions n and m, the additive group of linear transformations of V into W is isomorphic to the additive group of $m \times n$ matrices.

Proof

There is really nothing left to prove, because our definition of addition of matrices was motivated with this result in mind. However, let us outline the details of a proof, in the light of the preceding definition. If we designate the group of linear transformations by G_T and the group of matrices by G_A, we have already emphasized the one-to-one correspondence between the members of the two systems, in which a matrix A is the transformation matrix of a linear transformation T. Let $T_1 \leftrightarrow A_1 = [a_{ij}]$ and $T_2 \leftrightarrow A_2 = [b_{ij}]$, in this correspondence, for arbitrary $T_1, T_2 \in G_T$ and their associated transformation matrices A_1 and A_2. The identification of a transformation matrix assumes a choice of ordered bases for V and W, and let us suppose that $\{\alpha_j\}$ and $\{\beta_j\}$ are the basis choices for V and W, respectively. The definition of addition in G_T requires that $(T_1 + T_2)\alpha = T_1\alpha + T_2\alpha$, for any $\alpha \in V$; and so, in particular, $(T_1 + T_2)\alpha_j = T_1\alpha_j + T_2\alpha_j = (a_{1j}\beta_1 + a_{2j}\beta_2 + \cdots + a_{mj}\beta_m) + (b_{1j}\beta_1 + b_{2j}\beta_2 + \cdots + b_{mj}\beta_m) = (a_{1j} + b_{1j})\beta_1 + (a_{2j} + b_{2j})\beta_2 + \cdots + (a_{mj} + b_{mj})\beta_m$, $j = 1, 2, \ldots, n$. But then if $C = [c_{ij}]$ is the transformation matrix of $T_1 + T_2$, this result shows that $c_{ij} = a_{ij} + b_{ij}$, for any $i = 1, 2, \ldots, m$, and so the jth column of C is made up by adding corresponding entries of A_1 and A_2. That is, in our adopted notation, if $T_1 \leftrightarrow A_1$ and $T_2 \leftrightarrow A_2$, then $T_1 + T_2 \leftrightarrow A_1 + A_2$. This completes the proof that the two additive groups are isomorphic. ∎

It was observed in Sec. 2.3 that, with the definition of rT, for any $r \in \mathbf{R}$ and a linear transformation T of a vector space V, the set of all such mappings of V can be considered to constitute a vector space in its own right. Indeed, the dual space V^* of V arose as a special case of this, when the linear transformations T are real-valued functions and so map V into $\mathbf{R}$. With the definition of rT and the association between linear transformations and matrices in mind, we now define the operation of *multiplication by scalars* on matrices as follows:

Definition For any matrix $A = [a_{ij}]$ and any $r \in \mathbf{R}$, we define rA so that $rA = [b_{ij}]$, where $b_{ij} = ra_{ij}$, for all i and j involved in the definition of A.

The definition of *isomorphism*, as it is applied to vector spaces, was detailed in Sec. 1.3, and so we leave the proof of the following theorem to the reader (Prob. 14).

Theorem 7.3 If V and W are vector spaces of respective dimensions n and m, the vector space of linear transformations of V into W is isomorphic to the vector space of all $m \times n$ matrices.

Our final result is merely a special case of Theorems 7.2 and 7.3.

Theorem 7.4 The additive group (or vector space) of all linear operators on an n-dimensional vector space is isomorphic to the additive group (or vector space) of square matrices of order n.

Problems 2.7

1. If
$$A = \begin{bmatrix} 1 & 3 & -2 \\ 2 & 0 & 2 \\ -1 & 4 & 1 \end{bmatrix} \quad \text{and} \quad B = \begin{bmatrix} -1 & 4 & 6 \\ 2 & 5 & -2 \\ 1 & 3 & 9 \end{bmatrix}$$
determine (a) $A + B$; (b) $A - B$; (c) $2A + 3B$; (d) $2A - 3B$.

2. If
$$\begin{bmatrix} 2 & 1 & -1 \\ 1 & 2 & 1 \\ 1 & -2 & 2 \end{bmatrix}$$
is the matrix of a linear transformation T, determine the matrix (relative to the same bases) of (a) 2T; (b) $-$T.

3. If
$$\begin{bmatrix} 1 & 2 & -1 \\ 1 & 3 & 2 \\ -2 & 0 & 1 \end{bmatrix}$$

is the matrix of a linear transformation T, determine the matrix (relative to the same bases) of $3T + 2I$. [I is the identity.]

4. Explain why the associative law of addition for linear transformations implies the associative law of addition for matrices.

5. Discuss the abelian property of addition just as was done with associativity in Prob. 4.

6. Find the matrix representation, relative to the $\{E_i\}$ bases, of each of the following indicated transformations: (a) $T(x,y,z) = (2x - y, \; y + z)$; (b) $T(x,y,z) = (x, \; y - z, \; y + 2z)$; (c) $T(x,y,z) = (x + y, \; y + 2z, \; x + 3z)$.

7. Repeat the directions in Prob. 6 for each of the following: (a) $T(x,y) = (x,y)$; (b) $T(x,y,z) = (0,0,0)$; (c) $T(x,y,z,w) = (x + y, \; z, \; y + w)$.

8. If $T(x,y,z) = (x - 2y, \; x + 3z, \; y - z)$, for a linear operator T on V_3, find the matrix of T relative to the $\{E_i\}$ basis.

9. Find the matrix of the transformation T in Prob. 8, but relative to the basis $\{(1,1,0),(1,0,1),(0,1,1)\}$.

10. Find two matrix representations (relative to different bases) of the linear operator T defined on V_2 by (a) $T(1,1) = (0,1)$, $T(-1,1) = (3,4)$; (b) $T(4,1) = (1,1)$, $T(2,-1) = (2,3)$; (c) $T(1,2) = (3,-2)$, $T(1,0) = (2,-4)$.

11. Explain why the definition given for the equality of matrices is demanded by the uniqueness of a transformation matrix.

12. Explain why the zero matrix and the additive inverse of any matrix are uniquely determined by their intrinsic properties as group elements.

13. Explain why the transformation matrix of the zero transformation is the zero matrix, regardless of the choice of bases for the spaces involved.

14. Give the proof for Theorem 7.3.

15. Let $A = [a_{ij}]$, where $a_{ij} = 0$ for $i \neq j$, and $a_{ii} = c \; (\neq 0)$ for all indices i. If A is the matrix of a linear transformation $T: V \to W$, describe T and decide whether A is invariant under any change of basis for either V or W.

16. Describe the linear transformation T, if its transformation matrix (relative to some given bases) is

$$\begin{bmatrix} 2 & 0 & 0 \\ 0 & 3 & 0 \\ 0 & 0 & 1 \end{bmatrix}$$

17. If
$$\begin{bmatrix} 1 & 2 & 1 \\ -1 & 1 & 2 \\ 1 & 0 & -2 \end{bmatrix}$$
is the matrix of a linear operator T on V_3, relative to the $\{E_i\}$ basis, find $T(x,y,z)$, where (x,y,z) is (a) $(1,1,0)$; (b) $(-1,1,0)$; (c) $(2,-2,3)$.

18. Describe $T(x,y,z)$, if
$$\begin{bmatrix} 2 & 1 & -1 \\ 1 & 2 & 3 \\ -1 & 1 & 1 \end{bmatrix}$$
is the matrix of T relative to (a) the $\{E_i\}$ basis of V_3; (b) the basis $\{(1,1,0),(1,0,1),(0,1,1)\}$ of V_3.

19. Apply the directions in Prob. 18 to the matrix
$$\begin{bmatrix} 1 & -2 & 3 \\ 0 & 2 & 4 \\ 3 & 4 & 1 \end{bmatrix}$$

20. If
$$\begin{bmatrix} 2 & 1 & 0 \\ -1 & 2 & 0 \\ 1 & 0 & 1 \end{bmatrix}$$
is the matrix of the linear operator T on V_3, relative to the $\{E_i\}$ basis, find the matrix representation of T relative to the basis $\{(1,1,0),(1,0,1),(0,0,1)\}$.

21. Let the operators T_1 and T_2 be defined on V_3 so that $T_1E_1 = (1,0,1)$, $T_1E_2 = (0,1,1)$, $T_1E_3 = (1,1,2)$, and $T_2E_1 = (2,2,1)$, $T_2E_2 = (0,1,2)$, $T_2E_3 = (-1,2,1)$. Find the matrix, relative to the $\{E_i\}$ basis of V_3, of (a) $T_1 + T_2$; (b) T_1T_2; (c) T_2T_1; (d) $T_1 + T_1T_2$.

22. If $\begin{bmatrix} a & b \\ c & d \end{bmatrix}$ is the matrix of a linear transformation T, prove that $T^2 - (a+d)T + (ad-bc)I = 0$.

23. If T_a is the linear operator on V_n defined by $T(x_1,x_2,\ldots,x_n) = (ax_1,ax_2,\ldots,ax_n)$, find the matrix of T_aT_b, relative to the $\{E_i\}$ basis of V_n, for real numbers a, b. Is this matrix the same as the matrix of T_{ab}?

24. Find the matrix of T on V_2, relative to the $\{E_i\}$ basis, if (a) T is a shear (see Example 4 of Sec. 2.1) in the x direction: $T(x,y) = (x + ay, y)$, a ($\neq 0$) $\in$ **R**; (b) T is a rotation (see Example 2 of Sec. 2.7) of the plane through an angle of $\pi/2$; (c) T is a reflection defined by $T(x,y) = (y,x)$.

25. Verify that $\{1, t - 1, t^2 + t, 2t^3 - 3t\}$ is a basis of the space P_4 in Example 3, and find the matrix of D relative to this basis.

26. Verify that the matrices $\begin{bmatrix} 1 & 0 \\ 0 & 0 \end{bmatrix}, \begin{bmatrix} 0 & 0 \\ 1 & 0 \end{bmatrix}, \begin{bmatrix} 0 & 1 \\ 0 & 0 \end{bmatrix}, \begin{bmatrix} 0 & 0 \\ 0 & 1 \end{bmatrix}$ form a basis of the vector space of 2×2 matrices, and thence conclude that it has dimension 4. [See Prob. 27.]

27. Show that the vector space of $n \times n$ matrices has dimension n^2, and find a basis for this space. Illustrate with $n = 3$.

2.8 Matrices as Multiplicative Systems

In this section we shall expand the algebraic systems of representation matrices, defined in Sec. 2.7, to rings and multiplicative groups. We have already seen how to associate matrices with linear transformations and how to define operations of *addition* and *multiplication by scalars* in sets of matrices so that those associated with either an additive group or vector space of linear transformations form an isomorphic matrix system. We emphasize that it was with isomorphism in mind that we were led from the operations on transformations to the corresponding operations on matrices. If a set of linear transformations forms a ring or multiplicative group, we again wish the associated matrices to make up an *isomorphic* system, and this desire guides us to the rule for the multiplication of matrices. Before getting to this rule, however, we should make precise what is to be meant by *isomorphic rings*.

Definition Two rings R and R' are isomorphic if there exists a one-to-one correspondence $r \leftrightarrow r'$ between the elements $r \in R$ and $r' \in R'$ such that (i) $r_1 + r_2 \leftrightarrow r_1' + r_2'$ and (ii) $r_1 r_2 \leftrightarrow r_1' r_2'$, for arbitrary $r_1, r_2 \in R$ and their correspondents $r_1', r_2' \in R'$.

As is the case with any isomorphic systems, rings that are isomorphic are indistinguishable *as rings*, except for symbolism. It is an important consequence of the definition that, if 0, 0' and 1, 1' are the respective additive and multiplicative (if any) identities of R and R', $0 \leftrightarrow 0'$ and $1 \leftrightarrow 1'$. If a ring or multiplicative group of linear transformations is to be associated with a system of matrices, as in Sec. 2.7, with the transformation and matrix systems isomorphic, it is a result of condition (ii) above that the matrix associated with a product of linear transformations must be the product, *in the same order*, of the matrices of the transformations. An illustration may help to clarify this point.

If T_1 and T_2 are linear operators on V_3 represented, respectively, by matrices A and B, let us find the matrix representation of $T_2 T_1$, relative to the same basis of the space, where

$$A = \begin{bmatrix} 1 & 2 & 1 \\ 0 & -1 & 2 \\ -1 & 2 & 3 \end{bmatrix} \quad \text{and} \quad B = \begin{bmatrix} 2 & 4 & -1 \\ 0 & 2 & 6 \\ -5 & -2 & 1 \end{bmatrix}$$

If we denote the chosen basis of V_3 by $\{\alpha_1, \alpha_2, \alpha_3\}$, the problem is equivalent to a determination of the images under $T_2 T_1$ of α_1, α_2, and α_3. Our interpretation of the matrices A and B leads to the following equations: $T_1 \alpha_1 = \alpha_1 - \alpha_3$, $T_1 \alpha_2 = 2\alpha_1 - \alpha_2 + 2\alpha_3$, $T_1 \alpha_3 = \alpha_1 + 2\alpha_2 + 3\alpha_3$; $T_2 \alpha_1 = 2\alpha_1 - 5\alpha_3$, $T_2 \alpha_2 = 4\alpha_1 + 2\alpha_2 - 2\alpha_3$, $T_2 \alpha_3 = -\alpha_1 + 6\alpha_2 + \alpha_3$. From these equalities, it follows that $(T_2 T_1)\alpha_1 = T_2(T_1 \alpha_1) = T_2(\alpha_1 - \alpha_3) = T_2 \alpha_1 - T_2 \alpha_3 = (2\alpha_1 - 5\alpha_3) - (-\alpha_1 + 6\alpha_2 + \alpha_3) = 3\alpha_1 - 6\alpha_2 - 6\alpha_3$; $(T_2 T_1)\alpha_2 = T_2(T_1 \alpha_2) = T_2(2\alpha_1 - \alpha_2 + 2\alpha_3) = 2(T_2 \alpha_1) - T_2 \alpha_2 + 2(T_2 \alpha_3) = 2(2\alpha_1 - 5\alpha_3) - (4\alpha_1 + 2\alpha_2 - 2\alpha_3) + 2(-\alpha_1 + 6\alpha_2 + \alpha_3) = -2\alpha_1 + 10\alpha_2 - 6\alpha_3$; $(T_2 T_1)\alpha_3 = T_2(T_1 \alpha_3) = T_2(\alpha_1 + 2\alpha_2 + 3\alpha_3) = T_2 \alpha_1 + 2(T_2 \alpha_2) + 3(T_2 \alpha_3) = 2\alpha_1 - 5\alpha_3 + 2(4\alpha_1 + 2\alpha_2 - 2\alpha_3) + 3(-\alpha_1 + 6\alpha_2 + \alpha_3) = 7\alpha_1 + 22\alpha_2 - 6\alpha_3$. Hence the product $T_2 T_1$ is represented by the matrix C, where

$$C = \begin{bmatrix} 3 & -2 & 7 \\ -6 & 10 & 22 \\ -6 & -6 & -6 \end{bmatrix}$$

It is now our intention to define matrix multiplication in such a way that, in particular, $BA = C$:

$$\begin{bmatrix} 2 & 4 & -1 \\ 0 & 2 & 6 \\ -5 & -2 & 1 \end{bmatrix} \begin{bmatrix} 1 & 2 & 1 \\ 0 & -1 & 2 \\ -1 & 2 & 3 \end{bmatrix} = \begin{bmatrix} 3 & -2 & 7 \\ -6 & 10 & 22 \\ -6 & -6 & -6 \end{bmatrix}$$

In so doing, we shall note that condition (ii) in the definition of a ring isomorphism is satisfied. We now consider the most general situation with regard to the product of two matrices.

Let $T_1: U \to V$ and $T_2: V \to W$ define linear transformations for vector spaces U, V, W, where $\dim U = n$, $\dim V = m$, and $\dim W = r$. If we assume that T_1 and T_2 are represented by the $m \times n$ and $r \times m$ matrices $A = [a_{ij}]$ and $B = [b_{ij}]$, respectively, we wish to determine the matrix C that will represent $T_2 T_1$, relative to the same bases. Let us designate the respec-

tive ordered bases for the spaces U, V, and W by $\{\alpha_1, \alpha_2, \ldots, \alpha_n\}$, $\{\beta_1, \beta_2, \ldots, \beta_m\}$, and $\{\gamma_1, \gamma_2, \ldots, \gamma_r\}$. Our interpretation of the matrices A and B then leads to the following: $(T_2 T_1)\alpha_j = T_2(T_1 \alpha_j) = T_2 \left(\sum_{k=1}^{m} a_{kj}\beta_k \right) =$
$\sum_{k=1}^{m} a_{kj}(T_2 \beta_k) = \sum_{k=1}^{m} a_{kj} \left(\sum_{i=1}^{r} b_{ik}\gamma_i \right) = \sum_{i=1}^{r} \left(\sum_{k=1}^{m} b_{ik} a_{kj} \right) \gamma_i = \sum_{i=1}^{r} c_{ij}\gamma_i$, where $c_{ij} = \sum_{k=1}^{m} b_{ik}a_{kj}$, and j assumes any subscript (from 1 to n) of the basis vectors of U. It follows that $T_2 T_1$ is represented by the matrix $C = [c_{ij}]$, where $c_{ij} = \sum_{k=1}^{m} b_{ik} a_{kj}$, and this result provides the motivation for the following definition, in which the roles of A and B have been reversed from the foregoing.

Definition   If $A = [a_{ij}]$ is an $m \times r$ and $B = [b_{ij}]$ is an $r \times n$ matrix, the product AB is the $m \times n$ matrix $C = [c_{ij}]$, where $c_{ij} = \sum_{k=1}^{r} a_{ik} b_{kj}$.

It is important to notice that the product AB is defined *only if the number of columns of A is equal to the number of rows of B*. It should also be observed that *it is a consequence of our definition of matrix multiplication* that the transformation matrix of the product $T_2 T_1$ of the linear transformations T_2 and T_1 is the product BA of the respective transformation matrices B and A of T_2 and T_1. The rule for the multiplication of two matrices, as described in the definition, is usually known as the *row-by-column* rule, for the entry at the intersection of the ith row and jth column of a product AB is obtained by adding the products of corresponding entries in the ith row of A and the jth column of B.

EXAMPLE 1

Find the product BA, where

$$B = \begin{bmatrix} 2 & 1 & 1 \\ 1 & -1 & 0 \\ 0 & 2 & 5 \end{bmatrix} \quad \text{and} \quad A = \begin{bmatrix} 0 & 2 & 5 \\ -2 & 1 & 6 \\ 0 & -4 & 3 \end{bmatrix}$$

Solution
If $BA = C = [c_{ij}]$, we use the rule for matrix multiplication to see that $c_{11} = 2(0) + 1(-2) + 1(0) = -2$; $c_{12} = 2(2) + 1(1) + 1(-4) = 1$; $c_{21} = 1(0) +$

$(-1)(-2) + 0(0) = 2$; etc. Hence

$$BA = \begin{bmatrix} -2 & 1 & 19 \\ 2 & 1 & -1 \\ -4 & -18 & 27 \end{bmatrix}$$

If T is a linear transformation of a vector space V, the rule just given for the product of two matrices can be used to very great advantage in determining the transform $T\alpha$ of any $\alpha \in V$. Our first observation in this connection is that a vector $(a_1, a_2, \ldots, a_n)$, *as an ordered n-tuple*, is indistinguishable in any essential detail from either the $1 \times n$ matrix $[a_1 \, a_2 \, \cdots \, a_n]$ or the $n \times 1$ matrix

$$\begin{bmatrix} a_1 \\ a_2 \\ \cdot \\ a_n \end{bmatrix}$$

If $A = [a_{ij}]$ is the $m \times n$ matrix of T: $V \to W$, relative to bases $\{\alpha_i\}$ and $\{\beta_i\}$ of V and W, respectively, it follows from the definition of A that

$$T\alpha_j = \sum_{i=1}^{m} a_{ij}\beta_i \qquad j = 1, 2, \ldots, n$$

Hence

$$T\left(\sum_{j=1}^{n} x_j \alpha_j\right) = \sum_{j=1}^{n} x_j(T\alpha_j) = \sum_{j=1}^{n} x_j \left(\sum_{i=1}^{m} a_{ij}\beta_i\right) = \sum_{i=1}^{m} \left(\sum_{j=1}^{n} a_{ij} x_j\right) \beta_i$$

so that T induces a mapping of the coordinate vector $(x_1, x_2, \ldots, x_n)$ onto the coordinate vector $(y_1, y_2, \ldots, y_m)$, where $y_i = \sum_{j=1}^{n} a_{ij} x_j$. We now make the very important observation that *it is a consequence of matrix multiplication* that

$$\begin{bmatrix} a_{11} & a_{12} & \cdots & a_{1n} \\ a_{21} & a_{22} & \cdots & a_{2n} \\ \cdots & \cdots & \cdots & \cdots \\ a_{m1} & a_{m2} & \cdots & a_{mn} \end{bmatrix} \begin{bmatrix} x_1 \\ x_2 \\ \cdot \\ x_n \end{bmatrix} = \begin{bmatrix} y_1 \\ y_2 \\ \cdot \\ y_m \end{bmatrix}$$

where $y_i = \sum_{j=1}^{n} a_{ij}x_j$. This leads to the following very practical rule for determining the image under T of any vector $\alpha \in V$:

> Let A be the matrix of a linear transformation $T\colon V \to W$, relative to given bases of V and W, and suppose $T\alpha = \beta$, for $\alpha \in V$ and $\beta \in W$. Then, if $(x_1, x_2, \ldots, x_n)$ is the coordinate vector of α, the coordinate vector $(y_1, y_2, \ldots, y_m)$ of β can be determined from the matrix equation
>
> $$AX = Y$$
>
> where X and Y are the coordinate vectors of α and β, respectively, written as column matrices.

We shall see later that, if X and Y are written as row matrices, there is only a superficial change in the form of the equation $AX = Y$.

EXAMPLE 2

If

$$A = \begin{bmatrix} 1 & -2 & 3 & 0 \\ 2 & 3 & 0 & 1 \\ 1 & -1 & 1 & 1 \end{bmatrix}$$

is the matrix of a linear transformation $T\colon V \to W$, relative to bases $\{\alpha_i\}$ and $\{\beta_i\}$ of V and W, respectively, determine $T\alpha$, where $\alpha = 3\alpha_1 + \alpha_2 - 2\alpha_3$.

Solution

The coordinate vector of α is $(3, 1, -2, 0)$, and so we merely compute

$$\begin{bmatrix} 1 & -2 & 3 & 0 \\ 2 & 3 & 0 & 1 \\ 1 & -1 & 1 & 1 \end{bmatrix} \begin{bmatrix} 3 \\ 1 \\ -2 \\ 0 \end{bmatrix} = \begin{bmatrix} -5 \\ 9 \\ 0 \end{bmatrix}$$

from which we conclude that $T\alpha = -5\beta_1 + 9\beta_2 + 0\beta_3 = -5\beta_1 + 9\beta_2$.

We now return to the subject of systems of linear transformations and consider the special case of the set of all linear operators on a vector space V of dimension n. The following results have already been obtained for such a system of operators, when regarded as a ring.

(1) The correspondence T ↔ A is one-to-one between the linear operators T on V and the $n \times n$ matrices A, where A is the transformation matrix relative to some selected basis of V.

(2) The correspondence is *preserved under addition:* if T$_1$ ↔ A and T$_2$ ↔ B, then T$_1$ + T$_2$ ↔ $A + B$.

(3) The correspondence is *preserved under multiplication:* if T$_1$ ↔ A and T$_2$ ↔ B, then T$_2$T$_1$ ↔ BA.

If T$_1$, T$_2$, T$_3$ are any three linear operators on V that are associated (relative to some fixed basis) with transformation matrices A_1, A_2, A_3, respectively, the following results are corollary to the above:

$$T_3(T_2T_1) \leftrightarrow A_3(A_2A_1) \qquad (T_3T_2)T_1 \leftrightarrow (A_3A_2)A_1$$
$$T_3(T_1 + T_2) \leftrightarrow A_3(A_1 + A_2) \qquad T_3T_1 + T_3T_2 \leftrightarrow A_3A_1 + A_3A_2$$
$$(T_1 + T_2)T_3 \leftrightarrow (A_1 + A_2)A_3 \qquad T_1T_3 + T_2T_3 \leftrightarrow A_1A_3 + A_2A_3$$

Since the operation of multiplication is associative and also distributive with respect to addition within the ring of linear operators, the uniqueness of the operator-matrix correspondence implies that the multiplication of matrices has these same characteristics. We are then able to state the following important fact:

> *The set of all $n \times n$ matrices forms a ring, and the three results itemized above show that this ring is actually isomorphic to the ring of linear operators on V.*

The $n \times n$ matrix

$$I_n = \begin{bmatrix} 1 & 0 & 0 & \cdots & 0 \\ 0 & 1 & 0 & \cdots & 0 \\ 0 & 0 & 1 & \cdots & 0 \\ \cdot & \cdot & \cdot & \cdot & \cdot \\ 0 & 0 & 0 & \cdots & 1 \end{bmatrix}$$

in which all entries are 0 except for 1 in each position along the main diagonal (i.e., elements a_{kk} in $[a_{ij}]$) is the *identity* in the ring of matrices and is the transformation matrix of the identity operator I on V. When there is no danger of confusing orders, the subscript n will be dropped and an identity matrix will be designated simply as I. The symbol I has been used to denote an *identity operator*, while the symbol I denotes an *identity matrix*. It should be pointed out that the associative and distributive laws, which we have just deduced from the corresponding laws in the ring of operators, can also be verified directly. However, this method of proof is very tedious even for matrices of low orders (Prob. 17), whereas the approach we have taken avoids computation completely.

It is shown in Example 2 of Sec. 2.6 that the property of linearity is not necessary for a mapping T of a vector space to be invertible. However, if T: $V \to W$ is a *linear* transformation for vector spaces V and W, it is a consequence of Theorem 2.2 and the basic idea of an inverse mapping (see Sec. 2.6) that dim V = dim W is a necessary condition if T^{-1} is to exist as a linear transformation of W. Our main interest is in the case where $W = V$, and here the condition is automatically satisfied. If T is an invertible (or nonsingular) linear transformation with matrix A of an n-dimensional vector space V, it follows from the matrix ring isomorphism that there exists a unique $n \times n$ matrix A^{-1} such that $AA^{-1} = A^{-1}A = I$. This matrix A^{-1} is called the *inverse* of A, and A is said to be a *nonsingular* matrix. We note, in passing, that, although the matrix of a nonsingular linear operator must be nonsingular, the matrix of a nonsingular linear transformation is not necessarily nonsingular, nor is it even of necessity a square matrix (Prob. 13).

It is of some interest to note that, if A is a nonsingular matrix, either one of $AA_1 = I$ or $A_1A = I$ will imply the other, and it may be inferred further from the uniqueness of A^{-1} that $A_1 = A^{-1}$. This may be seen from Prob. 10 of Sec. 2.6, but a simple direct proof may also be given. For, if $AA_1 = I$ and A^{-1} exists, then $A^{-1} = A^{-1}I = A^{-1}(AA_1) = (A^{-1}A)A_1 = IA_1 = A_1$. It follows that *a determination of the inverse of a nonsingular matrix then requires merely a determination of a "one-sided" inverse*. It should be emphasized that a nonsingular matrix must of necessity be square. In addition to the *identity*, there are two other special types of matrices that should be mentioned: *diagonal* and *scalar*. The matrix $A = [a_{ij}]$ is *diagonal* if all its entries off the main diagonal are 0; that is, $a_{ij} = 0$, for $i \neq j$. If $a_1, a_2, \ldots, a_n$ are the diagonal entries, in order, of a diagonal matrix, it is convenient to write A = diag $\{a_1, a_2, \ldots, a_n\}$. If all diagonal entries of a diagonal matrix are equal, the matrix is said to be a *scalar matrix;* and, if the common diagonal entry is c, the matrix is written cI. It is clear that I itself is a scalar—and diagonal—matrix.

Since the nonsingular $n \times n$ matrices form a subset of the ring of all $n \times n$ matrices, the identity matrix $I_n = I$ being a member of this subset, the following result may be stated as a theorem.

Theorem 8.1 *The set of all $n \times n$ nonsingular matrices forms a multiplicative group. This group is isomorphic to the multiplicative group of nonsingular linear operators on an n-dimensional vector space and may be referred to as the full linear group of $n \times n$ matrices.*

Proof
The theorem is a direct consequence of the unique correspondence (relative to a fixed basis) between linear operators on an n-dimensional vector space

(Theorem 6.4) and the definition of isomorphism as it applies to groups. We note that the full linear group of $n \times n$ matrices is then isomorphic to the full linear group of linear operators on an n-dimensional vector space. ∎

A discussion of the most efficient methods of finding the inverse of a nonsingular matrix will be postponed until Chap. 3, but we shall give two illustrations of how a direct approach may be used in simple cases.

EXAMPLE 3

If we wish to determine A^{-1} (provided it exists) from $A = \begin{bmatrix} 2 & 3 \\ 3 & 5 \end{bmatrix}$, we must find a matrix $\begin{bmatrix} a & b \\ c & d \end{bmatrix}$ such that

$$\begin{bmatrix} 2 & 3 \\ 3 & 5 \end{bmatrix}\begin{bmatrix} a & b \\ c & d \end{bmatrix} = \begin{bmatrix} a & b \\ c & d \end{bmatrix}\begin{bmatrix} 2 & 3 \\ 3 & 5 \end{bmatrix} = \begin{bmatrix} 1 & 0 \\ 0 & 1 \end{bmatrix}$$

As a consequence of the "one-sided" inverse result, to which reference was made in this section, it will be sufficient to use only the first product in the system of equalities. Thus,

$$\begin{bmatrix} 2a + 3c & 2b + 3d \\ 3a + 5c & 3b + 5d \end{bmatrix} = \begin{bmatrix} 1 & 0 \\ 0 & 1 \end{bmatrix}$$

and from this we obtain the following equations to be satisfied:

$$\begin{array}{ll} 2a + 3c = 1 & 2b + 3d = 0 \\ 3a + 5c = 0 & 3b + 5d = 1 \end{array}$$

The first pair of equations yields quite easily that $a = 5$ and $c = -3$, and the second pair gives $b = -3$ and $d = 2$. Hence $A^{-1} = \begin{bmatrix} 5 & -3 \\ -3 & 2 \end{bmatrix}$, and a simple check may be used to verify that $AA^{-1} = A^{-1}A = I$. This method reduces a matrix-inversion problem to one of solving a system of equations, and it may be applied to a square matrix of any order.

In some simple cases, it is easy to invert a matrix by considering the associated linear operator on some suitable vector space.

EXAMPLE 4

Find the inverse (if it exists) of the matrix $A = \begin{bmatrix} 2 & 1 \\ 3 & 4 \end{bmatrix}$.

Solution

We may consider A to be the matrix of a linear operator T on V_2, relative to the basis $\{E_1, E_2\}$. Then, $TE_1 = 2E_1 + 3E_2$ and $TE_2 = E_1 + 4E_2$ and, more generally, $T(x,y) = T(xE_1 + yE_2) = x(TE_1) + y(TE_2) = x(2E_1 + 3E_2) + y(E_1 + 4E_2) = (2x + y)E_1 + (3x + 4y)E_2 = (2x + y, 3x + 4y)$. A little computation shows that

$$\frac{4(2x+y) - (3x+4y)}{5} = x \quad \text{and} \quad \frac{2(3x+4y) - 3(2x+y)}{5} = y$$

It is clear from this that the inverse operator T^{-1} exists, and it may be defined on V_2 by $T^{-1}(x,y) = [(4x - y)/5, (2y - 3x)/5]$. In particular, $T^{-1}E_1 = T^{-1}(1,0) = (\frac{4}{5}, -\frac{3}{5}) = \frac{4}{5}E_1 - \frac{3}{5}E_2$ and $T^{-1}E_2 = T^{-1}(0,1) = (-\frac{1}{5}, \frac{2}{5}) = (-\frac{1}{5})E_1 + \frac{2}{5}E_2$, whence

$$A^{-1} = \begin{bmatrix} \frac{4}{5} & -\frac{1}{5} \\ -\frac{3}{5} & \frac{2}{5} \end{bmatrix}$$

It is probable that the method of Example 4 will not be viewed very favorably, but it nonetheless illustrates an available method; and, more important, it reemphasizes the close relationship that exists between matrices and linear transformations.

It is often an aid in matrix computation to *partition* the matrices into suitable blocks of submatrices. For example, a matrix A may be expressed in the form $A = \begin{bmatrix} A_1 & A_2 \\ A_3 & A_4 \end{bmatrix}$, where the submatrices A_1, A_2 and A_3, A_4 have the same numbers of rows and A_1, A_3 and A_2, A_4 have the same numbers of columns. It is easy to see that, if $B = \begin{bmatrix} B_1 & B_2 \\ B_3 & B_4 \end{bmatrix}$ is another partitioned matrix, then the product AB can be computed by the rule

$$AB = \begin{bmatrix} A_1B_1 + A_2B_3 & A_1B_2 + A_2B_4 \\ A_3B_1 + A_4B_3 & A_3B_2 + A_4B_4 \end{bmatrix}$$

provided the submatrices have sizes that are compatible with the indicated multiplications. It may be observed that this method of multiplication is especially attractive if each of A_2, A_3, B_2, B_3 is a zero submatrix, for then $AB = \begin{bmatrix} A_1B_1 & 0 \\ 0 & A_4B_4 \end{bmatrix}$. In this simplification, we have found the product of two matrices, possibly of large size, by multiplying two pairs of smaller submatrices. If A is a nonsingular matrix, which can be partitioned as $A =$

$\begin{bmatrix} A_1 & O \\ O & A_4 \end{bmatrix}$, where A_1 and A_4 are square, it is relatively easy to find A^{-1}, for it is clear that $A^{-1} = \begin{bmatrix} A_1^{-1} & O \\ O & A_4^{-1} \end{bmatrix}$. In this case, we have inverted a matrix by finding the inverses of two submatrices whose orders are less than that of the original. In most cases, it is much easier to work with several matrices of small orders than with one matrix of large order. If a matrix A is partitioned in such a way that all the partitioning submatrices are zero except for $A_1, A_2, \ldots, A_n$ along the main diagonal, it is convenient to denote A by diag $\{A_1, A_2, \ldots, A_n\}$. This notation will be used, in particular, in Chap. 6.

EXAMPLE 5

If

$$A = \begin{bmatrix} 2 & 3 & 0 & 0 \\ 3 & 5 & 0 & 0 \\ 0 & 0 & 2 & 1 \\ 0 & 0 & 3 & 4 \end{bmatrix} \quad \text{and} \quad B = \begin{bmatrix} 1 & 2 & 0 & 0 \\ 2 & -1 & 0 & 0 \\ 0 & 0 & 3 & -2 \\ 0 & 0 & 1 & 3 \end{bmatrix}$$

determine AB and A^{-1} by partitioning A and B into submatrices of order 2.

Solution

We may denote A as diag $\{A_1, A_2\}$ and B as diag $\{B_1, B_2\}$, where

$$A_1 = \begin{bmatrix} 2 & 3 \\ 3 & 5 \end{bmatrix} \quad A_2 = \begin{bmatrix} 2 & 1 \\ 3 & 4 \end{bmatrix} \quad B_1 = \begin{bmatrix} 1 & 2 \\ 2 & -1 \end{bmatrix} \quad B_2 = \begin{bmatrix} 3 & -2 \\ 1 & 3 \end{bmatrix}$$

But then $AB = $ diag $\{A_1 B_1, A_2 B_2\}$, and we easily compute that $A_1 B_1 = \begin{bmatrix} 8 & 1 \\ 13 & 1 \end{bmatrix}$ and $A_2 B_2 = \begin{bmatrix} 7 & -1 \\ 13 & 6 \end{bmatrix}$. Hence

$$AB = \begin{bmatrix} 8 & 1 & 0 & 0 \\ 13 & 1 & 0 & 0 \\ 0 & 0 & 7 & -1 \\ 0 & 0 & 13 & 6 \end{bmatrix}$$

We know from Example 3 that $A_1^{-1} = \begin{bmatrix} 5 & -3 \\ -3 & 2 \end{bmatrix}$, and, from Example 4, that $A_2^{-1} = \begin{bmatrix} \frac{4}{5} & -\frac{1}{5} \\ -\frac{3}{5} & \frac{2}{5} \end{bmatrix}$. It then follows from the discussion preceding this example that

$$A^{-1} = \begin{bmatrix} 5 & -3 & 0 & 0 \\ -3 & 2 & 0 & 0 \\ 0 & 0 & \frac{4}{5} & -\frac{1}{5} \\ 0 & 0 & -\frac{3}{5} & \frac{2}{5} \end{bmatrix}$$

Problems 2.8

1. Verify that $BA = C$, for the matrices occurring in the preliminary discussion of this section.
2. If

$$A = \begin{bmatrix} 2 & 1 & 1 \\ 1 & -2 & 1 \end{bmatrix} \quad \text{and} \quad B = \begin{bmatrix} 1 & 4 & -6 \\ 2 & 0 & 1 \\ 3 & 2 & -3 \end{bmatrix}$$

find (a) AB; (b) B^2.

3. Determine AB and BA, where

$$A = \begin{bmatrix} 3 & 4 & 5 & 1 \\ 2 & 4 & 3 & 6 \end{bmatrix} \quad \text{and} \quad B = \begin{bmatrix} 2 & 5 \\ 1 & 2 \\ 4 & 0 \\ 1 & 4 \end{bmatrix}$$

4. If matrix A in Prob. 2 represents a linear transformation T, find the coordinate vector into which each of the following is mapped by T: (a) $(2,-1,4)$; (b) $(-3,6,2)$; (c) $(5,0,-3)$.
5. If B in Prob. 3 is the matrix of a linear transformation T, relative to the basis $\{\alpha_1, \alpha_2\}$ of a vector space V, use B to determine Tα, where (a) $\alpha = 2\alpha_1 - 3\alpha_2$; (b) $\alpha = 5\alpha_1 + \alpha_2$; (c) $-3\alpha_1 + 2\alpha_2$.
6. If matrices A and B in Prob. 3 are respective matrices of the transformations T_1 and T_2, use the matrices to determine the coordinate vector into which each of the following is mapped by T_2T_1: (a) $(1,-2,4,3)$; (b) $(2,-3,-4,0)$; (c) $(1,1,-2,-2)$.
7. If A and B are $n \times n$ matrices, express $(A + B)^3$ as a polynomial in A and B.
8. Use the method of Example 3 to find the inverse of each of the following matrices: (a) $\begin{bmatrix} 1 & 3 \\ 4 & 1 \end{bmatrix}$; (b) $\begin{bmatrix} 5 & 2 \\ 8 & 3 \end{bmatrix}$; (c) $\begin{bmatrix} 1 & 2 \\ 5 & 9 \end{bmatrix}$; (d) $\begin{bmatrix} 2 & -4 \\ 6 & -2 \end{bmatrix}$.
9. Use the method of Example 4 to find the inverse of each of the matrices listed in Prob. 8.

10. If

$$A = \begin{bmatrix} 1 & 0 & 1 \\ 1 & 1 & 0 \end{bmatrix} \quad \text{and} \quad B = \begin{bmatrix} 1 & -1 \\ -1 & 2 \\ 0 & 1 \end{bmatrix}$$

show that $AB = I$ but $BA \neq I$. Explain why neither A nor B can represent invertible linear transformations.

11. Criticize the statement that "a $1 \times n$ matrix is a vector."
12. Fill in the details of the proof of Theorem 8.1.
13. Give an example of a nonsingular linear transformation whose associated transformation matrix is not square.
14. Explain how the expressions for x and y in terms of $2x + y$ and $3x + 4y$ were obtained in Example 4.
15. Explain why the inverse of a nonsingular matrix is unique.
16. If A and B are nonsingular matrices of the same order, show that AB is also nonsingular and that $(AB)^{-1} = B^{-1}A^{-1}$. Generalize this result to the case of n matrices (cf. Prob. 6 of Sec. 2.6).
17. Use general 3×3 matrices to verify directly that the multiplication of matrices is associative, and distributive with respect to addition.
18. If A and B are respective matrices associated with the transformations T_1 and T_2, explain why it follows from the appropriate theory of isomorphic systems that (a) $AB = BA$ if and only if $T_1T_2 = T_2T_1$; (b) $AB = I$ implies $BA = I$, for nonsingular matrices A, B (cf. Prob. 10); (c) A^{-1} exists if T^{-1} exists.

19. Partition each of the matrices A and B, and use the partitioned form to compute AB and BA, where

$$A = \begin{bmatrix} 2 & 1 & 2 & 1 \\ 1 & 1 & -1 & 1 \\ 0 & 2 & 1 & 2 \\ 1 & 1 & 2 & 3 \end{bmatrix} \quad \text{and} \quad B = \begin{bmatrix} 1 & 2 & 1 & 0 \\ -1 & 1 & 2 & 1 \\ 3 & 0 & 0 & 2 \\ 2 & 4 & -1 & 1 \end{bmatrix}$$

20. Find AB and BA, where $A = \text{diag } \{A_1, A_2\}$, $B = \text{diag } \{B_1, B_2\}$ and $A_1 = \begin{bmatrix} 2 & 1 \\ 1 & 1 \end{bmatrix}$, $A_2 = \begin{bmatrix} 1 & 2 \\ 1 & 1 \end{bmatrix}$, $B_1 = \begin{bmatrix} 2 & 1 \\ -1 & 1 \end{bmatrix}$, $B_2 = \begin{bmatrix} 1 & -1 \\ 3 & 1 \end{bmatrix}$.

21. Find the inverse of each of the following matrices: (a) A in Prob. 20; (b) B in Prob. 20; (c) diag $\{A_1, A_2\}$, where $A_1 = \begin{bmatrix} 1 & 2 \\ 3 & 5 \end{bmatrix}$ and $A_2 = \begin{bmatrix} 2 & 1 \\ 1 & 1 \end{bmatrix}$.

22. Use the method of Example 3 to find the inverse of each of the following matrices, solving the system of equations by Gauss reduction:

    (a) $\begin{bmatrix} 1 & 0 & -1 \\ 2 & 1 & 1 \\ 0 & 2 & 1 \end{bmatrix}$ (b) $\begin{bmatrix} 2 & 1 & 1 \\ 1 & 2 & -1 \\ 1 & 0 & 3 \end{bmatrix}$ (c) $\begin{bmatrix} 1 & -2 & 3 \\ 2 & 4 & -2 \\ -1 & 1 & 2 \end{bmatrix}$

23. Use the method of Example 3 to find A^{-1}, where

    $$A = \begin{bmatrix} 0 & 1 & 2 & 3 \\ 1 & 1 & 2 & 3 \\ 2 & 2 & 2 & 3 \\ 3 & 3 & 3 & 3 \end{bmatrix}$$

    solving the system of equations by Gauss reduction.

24. If AB and BA are identity matrices, is it necessary that A and B be nonsingular? Explain, and illustrate with linear transformations.

25. If $T(x,y,z) = (x + 1, y - 2, z)$, find T^{-1}. Note that T is not a linear transformation and so has no associated transformation matrix.

26. Let T be defined on V_3 so that $T(x,y,z) = (x + y, y + z, z)$. Then (a) find the transformation matrices of T and T^{-1}, relative to the $\{E_i\}$ basis of V_3; (b) use the matrices found in (a) to determine $T(1, -1, 2)$ and $T^{-1}(1, 0, -3)$.

27. Let D be the differentiation operator in the space P_4 of polynomials of degree 3 or less in a real variable t. Then find the transformation matrix of D^2, relative to the basis $\{1, t, t^2, t^3\}$ of P_4, and use this matrix to determine $D^2(1 + 2t^2 - 3t^3)$ and $D^2(2 + 4t^3)$.

28. If A is an $m \times n$ matrix and B is an $n \times m$ matrix such that $AB = I_m$ and $BA = I_n$, prove that $m = n$.

29. Verify that

    $$A = \begin{bmatrix} 1 & 0 & 1 \\ 0 & 1 & 1 \\ 0 & 0 & 1 \end{bmatrix}$$

    is a "zero" of the polynomial $x^3 - 3x^2 + 3x - 1$, in the sense that $A^3 - 3A^2 + 3A - I = 0$.

30. A Lorentz matrix $L(v)$ is defined as one of the form $b \begin{bmatrix} 1 & -1 \\ -v/c^2 & 1 \end{bmatrix}$, where v is the velocity of a moving body, c is the velocity of light, and $b = c/\sqrt{c^2 - v^2}$. Show that $L(v)$ is nonsingular if $|v| < c$.

31. Let V be a vector space with basis $\{\alpha_1, \alpha_2, \ldots, \alpha_n\}$, and T a linear operator defined on V so that $T\alpha_i = \alpha_{i+1}$, $i = 1, 2, \ldots, n-1$, $T\alpha_n = 0$.
    (a) Find the matrix A of T relative to the basis $\{\alpha_i\}$.
    (b) Show that $T^n = 0$.
    (c) Explain why the result in (b) implies that $A^n = 0$.

2.9 Change of Basis

It will have been noted, in the discussions of the preceding two sections, that, whenever the matrix of a linear transformation was mentioned, we had to refer it to some choice of (ordered) bases for the spaces involved. In fact, the multiplicity of choices for a basis of a vector space is responsible for much of the complication—and also some of the simplicity—of the results of linear algebra. Although the great number of possible bases makes it difficult to discover "invariants" of a space, which are independent of a choice of basis (the *dimension* of the vector space is one), it is also true that the existence of so many bases can often be used to advantage in the selection of one that will greatly simplify a particular problem. We shall elaborate on this remark in subsequent discussions, but in this section we have two specific problems: to study the effect of a change of basis for a space V on (1) the coordinates of an arbitrary vector in V and (2) the matrix of a given linear transformation of V. It should be recalled that the *coordinate vector* or n-tuple of coefficients of a vector α in an n-dimensional vector space V is the image of α in the isomorphism of V onto V_n, as described in Theorem 9.1 (Chap. 1). It would have been possible to have discussed the first problem in Sec. 1.8, but it is easier to obtain results in the context of matrices, with which we are now more familiar. First, it is necessary to make precise what is to be meant by a *matrix of transition* from one basis of a vector space to another. It will also be useful to introduce the *transpose* A' of a matrix $A = [a_{ij}]$ as the matrix $[a'_{ij}]$ in which $a'_{ij} = a_{ji}$. That is, the *transpose* of a matrix is obtained by interchanging, in order, its rows and columns.

If $\{\alpha_1, \alpha_2, \ldots, \alpha_n\}$ and $\{\beta_1, \beta_2, \ldots, \beta_n\}$ are two bases of the vector space V, an essential property of a basis permits us to write the following equations, for real numbers p_{ij}:

$$\beta_j = \sum_{i=1}^{n} p_{ij} \alpha_i \qquad j = 1, 2, \ldots, n$$

120 LINEAR TRANSFORMATIONS AND MATRICES

The matrix $P = [p_{ij}]$ is called the *matrix of transition* from the "old" basis $\{\alpha_i\}$ to the "new" basis $\{\beta_i\}$. It should be noted that, by the definition just given, P is the transpose of the matrix of coefficients that would be exhibited in the right members, if the above n equations were written in detail. In other words, *the columns of P are n-tuples that are the ordered coefficients of the new basis vectors when they are expressed in terms of the old;* or, expressed otherwise, *the columns of P may be identified with the coordinates of the new basis vectors relative to the old basis.* It follows from the discussions in Sec. 2.7 that P may be regarded as the matrix of a linear operator on V that maps α_i on β_i, $i = 1, 2, \ldots, n$. As a result of Theorem 6.3 [in particular item (vi) following the theorem] and the isomorphism between systems of linear transformations and the associated systems of transformation matrices, *the matrix of transition P is nonsingular and so P^{-1} exists.*

Now let $\sum_{i=1}^{n} x_i \alpha_i$ and $\sum_{i=1}^{n} y_i \beta_i$ be the representations of an arbitrary vector of V relative to the $\{\alpha_i\}$ and $\{\beta_i\}$ bases, respectively. It follows by some elementary substitutions that $\sum_{j=1}^{n} y_j \beta_j = \sum_{j=1}^{n} y_j \left(\sum_{i=1}^{n} p_{ij} \alpha_i \right) = \sum_{i=1}^{n} \left(\sum_{j=1}^{n} p_{ij} y_j \right) \alpha_i,$ and the uniqueness of the representation of a vector in terms of a basis yields the following equations:

$$x_i = \sum_{j=1}^{n} p_{ij} y_j \qquad i = 1, 2, \ldots, n$$

It will be noted that the matrix of transition P is involved in expressing the *old coordinates x_i* in terms of the *new coordinates y_j* of a vector in much the same way as it is involved in expressing the *new basis vectors β_j* in terms of the *old basis vectors α_i*. But there is one notable difference, observable from the connecting equations: whereas it is the *columns* of P that may be identified with *the coordinates of the new basis vectors in terms of the old*, it is the *rows* of P that designate *the appropriate coefficients for expressing the old coordinates of the vector in terms of the new coordinates.*

EXAMPLE 1

Let us suppose that

$$P = \begin{bmatrix} 2 & 1 & -1 \\ 1 & 0 & 1 \\ -1 & 1 & 1 \end{bmatrix}$$

is the matrix of transition from a basis $\{\alpha_1,\alpha_2,\alpha_3\}$ to a basis $\{\beta_1,\beta_2,\beta_3\}$ of a three-dimensional vector space V. Then, as a result of what we have said, the columns of P give the information that $\beta_1 = 2\alpha_1 + \alpha_2 - \alpha_3$, $\beta_2 = \alpha_1 + \alpha_3$, $\beta_3 = -\alpha_1 + \alpha_2 + \alpha_3$. At the same time, we observe from the rows of P that if $\alpha = x_1\alpha_1 + x_2\alpha_2 + x_3\alpha_3$, with coordinate vector (x_1,x_2,x_3) relative to the $\{\alpha_i\}$ basis, the coordinate vector of α, relative to the $\{\beta_i\}$ basis, is (y_1,y_2,y_3), where $x_1 = 2y_1 + y_2 - y_3$, $x_2 = y_1 + y_3$, $x_3 = -y_1 + y_2 + y_3$.

It will not have gone unnoticed that, although the rows of P give interesting information concerning the old and new coordinates of a vector, after a change of basis for the space, it would be much more useful to have a convenient means of expressing the *new* in terms of the *old* coordinates. It is easy to obtain the desired result, after we have available the following lemma.

Lemma   *If* $AB = C$, *for* $n \times n$ *matrices* A, B, C, *then* $(AB)' = B'A'$.

Proof
Let $A = [a_{ij}]$, $B = [b_{ij}]$, and $C = [c_{ij}]$, so that $A' = [a'_{ij}]$, $B' = [b'_{ij}]$, and $C' = [c'_{ij}]$, where $a'_{ij} = a_{ji}$, $b'_{ij} = b_{ji}$, and $c'_{ij} = c_{ji}$. It then follows directly from the rule for matrix multiplication that $(AB)' = C' = [c'_{ij}]$, where $c'_{ij} = c_{ji} = \sum_{k=1}^{n} a_{jk}b_{ki} = \sum_{k=1}^{n} b_{ki}a_{jk} = \sum_{k=1}^{n} b'_{ik}a'_{kj}$. But this means that $C' = B'A'$, as asserted. ∎

It now appears that, if P is the matrix of transition from one basis to another of an n-dimensional vector space, the relationship between the old coordinates $(x_1,x_2,\ldots,x_n)$ and the new coordinates $(y_1,y_2,\ldots,y_n)$ of the same vector in V may be expressed by the matrix equation

$$X = PY$$

in which, of course, we identify the old and new *coordinate vectors* as the *column matrices* X and Y, respectively. Since P is a nonsingular matrix, the equation $X = PY$ is equivalent to $Y = P^{-1}X$, and it is of interest to compare this with the *transformation* equation $Y = AX$ of Sec. 2.8. It should be noted that, although a change of basis *can* be interpreted as a linear operator on V, we are concerned in this section with a comparison of coordinates of the *same vector* relative to *two different bases*; only one basis for a space is involved in the discussion of Sec. 2.8. It is clear (Prob. 12), from the definition of matrix multiplication and the above lemma, that the equation $Y = P^{-1}X$ is equivalent to $Y' = X'(P^{-1})'$ or, if we agree that X and Y are to be regarded as row or column matrices according as which makes sense in the *indicated matrix product*, to $Y = X(P^{-1})'$. We have established the following theorem.

Theorem 9.1   *If P is the matrix of transition from one basis to another of a vector space, either $Y = P^{-1}X$ or $Y = X(P^{-1})'$ describes the induced change in coordinates of a vector, according as the coordinate vectors (old and new) are identified with column or row matrices (X and Y).*

The equation $Y = X(P^{-1})'$ permits a better comparison of the change in coordinates with the change in basis than could be achieved with the equation $Y = P^{-1}X$ (see Sec. 2.11): for the columns of $(P^{-1})'$ can be regarded as "coefficients" of the new coordinates if they are expressed as a linear combination of the old, just as the columns of P are the coefficients of the new basis vectors in terms of the old. In fact, it may be useful to paraphrase the result described in Theorem 9.1 as follows:

> *If a basis of a vector space is transformed by a matrix P, the coordinates of any vector in the space are transformed by a matrix that is the transpose of the inverse of P.* (See Prob. 11.)

Although this result is quite satisfactory from a theoretical point of view, the fact that we have not discussed *in general* the problem of matrix inversion detracts from the *practical* value of the result. This general problem of inverting a nonsingular matrix will be studied in Chap. 3 but until then we must be content with inverting only very simple matrices, by the methods given in Sec. 2.8. We now turn to the second matter under study in this section: *the effect of a change of basis on a representing matrix of a linear transformation of a space.* It is easy to obtain this result, in view of the way in which matrix multiplication was defined.

We consider the general problem in which T is a linear transformation of an n-dimensional vector space V into an m-dimensional space W. Let $\{\alpha_1, \alpha_2, \ldots, \alpha_n\}$, $\{\alpha_1', \alpha_2', \ldots, \alpha_n'\}$ and $\{\beta_1, \beta_2, \ldots, \beta_m\}$, $\{\beta_1', \beta_2', \ldots, \beta_m'\}$ be alternative bases of the spaces V and W, respectively. Then, if $A = [a_{ij}]$ and $B = [b_{ij}]$ are the transformation matrices of T, relative to the bases $\{\alpha_i\}$, $\{\beta_i\}$ and $\{\alpha_i'\}$, $\{\beta_i'\}$, respectively, we know, for $k = 1, 2, \ldots, n$, that

$$T\alpha_k = \sum_{i=1}^{m} a_{ik}\beta_i \quad \text{and also} \quad T\alpha_k' = \sum_{i=1}^{m} b_{ik}\beta_i'$$

If P is the matrix of transition from the basis $\{\alpha_i\}$ to $\{\alpha_i'\}$ in V, and if Q plays a similar role in W for the bases $\{\beta_i\}$ and $\{\beta_i'\}$, we may regard both P

and Q as the matrices of linear operators S: $V \to V$ and U: $W \to W$, respectively, defined by

$$S\alpha_k = \alpha'_k \quad k = 1, 2, \ldots, n \quad \text{and} \quad U\beta_i = \beta'_i \quad i = 1, 2, \ldots, m$$

But then $T(S\alpha_k) = \sum_{i=1}^{m} b_{ik}(U\beta_i)$ and, since U is invertible, we may write

$$U^{-1}[T(S\alpha_k)] = (U^{-1}TS)\alpha_k = U^{-1}\Big[\sum_{i=1}^{m} b_{ik}(U\beta_i)\Big] = (U^{-1}U)\Big(\sum_{i=1}^{m} b_{ik}\beta_i\Big)$$

and it is easy to see that this is equivalent to the following equality:

$$(U^{-1}TS)\alpha_k = \sum_{i=1}^{m} b_{ik}\beta_i \quad k = 1, 2, \ldots, n$$

This states, in effect, that B is the representing matrix of the transformation $U^{-1}TS: V \to W$, relative to the bases $\{\alpha_i\}$ and $\{\beta_i\}$. The uniqueness of such a transformation matrix, along with the product-preserving correspondence between linear transformations and their transformation matrices for fixed bases, now (see Prob. 13) shows that $B = Q^{-1}AP$, an important result which we restate as a theorem.

Theorem 9.2   *If A is the matrix of a linear transformation* $T: V \to W$, *relative to given bases of V and W, and if P and Q are the respective matrices of transition to other bases of V and W, then the matrix of* T, *relative to the new bases, is $Q^{-1}AP$.*

A very important special case arises when V and W are the same space.

Corollary   *Let A be the matrix of a linear operator on a vector space V, relative to some given basis. Then, if P is the matrix of transition from this to another basis of V, the matrix of the operator, relative to the new basis, is $P^{-1}AP$.*

Definition   *Two square matrices A and B are said to be similar if there exists a nonsingular matrix P of the same order such that $B = P^{-1}AP$.*

The preceding corollary can now be phrased in terms of similar matrices, and, in view of its importance, we do this in the form of a theorem.

Theorem 9.3   *Any two matrix representations of a linear operator on a vector space are similar.*

EXAMPLE 2

If $A = \begin{bmatrix} 2 & 1 \\ 1 & 2 \end{bmatrix}$ is the matrix of a linear operator T on V_2, relative to the basis $\{E_1, E_2\}$, find the matrix of T, relative to the basis $\{\alpha_1, \alpha_2\}$, where $\alpha_1 = (1,1)$, $\alpha_2 = (1,2)$.

Solution

Since $\alpha_1 = E_1 + E_2$ and $\alpha_2 = E_1 + 2E_2$, the matrix of transition from the $\{E_i\}$ basis to the $\{\alpha_i\}$ basis is $P = \begin{bmatrix} 1 & 1 \\ 1 & 2 \end{bmatrix}$. It follows from the corollary to Theorem 9.2 that the desired matrix is $P^{-1}AP$. But a short computation, by the method of either Example 3 or Example 4 of Sec. 2.8, shows that $P^{-1} = \begin{bmatrix} 2 & -1 \\ -1 & 1 \end{bmatrix}$, and hence

$$P^{-1}AP = \begin{bmatrix} 2 & -1 \\ -1 & 1 \end{bmatrix}\begin{bmatrix} 2 & 1 \\ 1 & 2 \end{bmatrix}\begin{bmatrix} 1 & 1 \\ 1 & 2 \end{bmatrix} = \begin{bmatrix} 3 & 0 \\ -1 & 1 \end{bmatrix}\begin{bmatrix} 1 & 1 \\ 1 & 2 \end{bmatrix} = \begin{bmatrix} 3 & 3 \\ 0 & 1 \end{bmatrix}$$

This final matrix shows that $T\alpha_1 = (3,0)$ and $T\alpha_2 = (3,1)$ (relative, of course, to the basis $\{\alpha_1, \alpha_2\}$), and these may be verified by checking with TE_1 and TE_2.

Problems 2.9

1. If

$$A = \begin{bmatrix} 2 & 1 & 1 \\ -1 & 1 & 2 \\ 0 & 1 & 1 \end{bmatrix} \quad \text{and} \quad B = \begin{bmatrix} 1 & 1 & 2 \\ -2 & 1 & 1 \\ 1 & 0 & -1 \end{bmatrix}$$

verify that $(AB)' = B'A'$.

2. State at least one important difference between the *vector* $(x_1, x_2, \ldots, x_n)$ and the *matrix* $[x_1 \ x_2 \ \cdots \ x_n]$.
3. Determine the matrix of transition from the $\{E_i\}$ basis of V_3 to the basis $\{(1,1,0),(0,1,1),(2,1,-2)\}$.
4. Determine the matrix of transition from the $\{E_i\}$ basis of V_3 to the basis $\{(1,2,-1),(2,0,5),(0,-1,2)\}$.
5. If

$$P = \begin{bmatrix} 2 & 1 & -2 \\ 3 & 1 & 4 \\ 1 & 2 & 1 \end{bmatrix}$$

is given as the matrix of transition from a basis $\{\alpha_i\}$ to a basis $\{\beta_i\}$ of a vector space V, express the new basis vectors in terms of the old.

6. If
$$\begin{bmatrix} 1 & -1 & 1 \\ 2 & 2 & 1 \\ 0 & -1 & 2 \end{bmatrix}$$
is the inverse of the matrix of transition from one basis of a vector space to another, find the new coordinate vector that corresponds to the old vector (a) $(1,-1,1)$; (b) $(2,1,1)$; (c) $(-2,1,3)$.

7. If $A = \begin{bmatrix} 1 & 2 \\ -1 & 3 \end{bmatrix}$, verify that $(A^{-1})' = (A')^{-1}$. [Cf. Prob. 11.]

8. Verify that *similarity* ($\sim$) is an *equivalence relation* in the set of all $n \times n$ matrices. That is, for arbitrary matrices A, B, C in the set, show that (a) $A \sim A$; (b) if $A \sim B$, then $B \sim A$; (c) if $A \sim B$ and $B \sim C$, then $A \sim C$.

9. Make the check suggested at the end of Example 2.

10. Explain why it is permissible to place U and U^{-1} in juxtaposition in the system of equalities in the proof preceding the statement of Theorem 9.2.

11. Prove that $(A^{-1})' = (A')^{-1}$ for any nonsingular matrix A.

12. Explain why the equations $Y = P^{-1}X$ and $Y' = X'(P^{-1})'$ are equivalent, for column matrices X and Y.

13. In the discussion just prior to the statement of Theorem 9.2, it is shown that the matrix of U^{-1}TS is B, relative to the $\{\alpha_i\}$ and $\{\beta_i\}$ bases. In justifying the subsequent matrix substitution $Q^{-1}AP$ for U^{-1}TS, a reader may argue that, since S maps α_i onto α_i', the matrix of T should be B, since this is the matrix of T with respect to the $\{\alpha_i'\}$ basis. How would you answer this argument?

14. If T: $V_3 \to V_2$ is a linear transformation with matrix $\begin{bmatrix} 2 & -1 & 3 \\ 1 & 1 & -2 \end{bmatrix}$, relative to the $\{E_i\}$ bases of both spaces, find the matrix of T, relative to the bases $\{\alpha_i\}$ and $\{\beta_i\}$, where $\alpha_1 = (1,1,1)$, $\alpha_2 = (0,-2,3)$, $\alpha_3 = (2,0,1)$ and $\beta_1 = (2,1)$, $\beta_2 = (3,-2)$.

15. If the matrix of a linear operator T on a two-dimensional space is $\begin{bmatrix} 2 & 1 \\ 1 & 4 \end{bmatrix}$, relative to a basis $\{\alpha_1,\alpha_2\}$, determine the matrix of T, relative to the basis $\{\beta_1,\beta_2\}$, where $\beta_1 = 3\alpha_1 + \alpha_2$, $\beta_2 = 5\alpha_1 + 2\alpha_2$.

16. Use the directions given in Prob. 15, but with $\beta_1 = 2\alpha_1 - 3\alpha_2$, $\beta_2 = \alpha_1 + \alpha_2$.

17. With reference to the transformation T, defined in Prob. 15, verify that $T(\alpha_1 + 2\alpha_2)$ is independent of which matrix representation of T is used to describe it.

18. Explain what change results in the matrix of a linear transformation if the vectors of the given (ordered) basis are permuted. If two spaces are involved, give consideration to permutations of the members of both bases.

19. Use associated transformation matrices with the space P_3 of polynomials of degree 2 or less in an indeterminate t, to express each of the following polynomials as a polynomial in 1, $1 - t$, $t + t^2$: (a) $2 - 2t + t^2$; (b) $1 - 3t^2$; (c) $3 - t + 2t^2$.

20. In the space P_3 of Prob. 19, find the matrix of transition from the basis $\{1,t,t^2\}$ to the basis $\{2, 1 - t, 2 + t^2\}$.

21. Let α be a nonzero vector of an n-dimensional vector space V, with T a linear operator on V such that $T^n = 0$ and $\{\alpha,T\alpha,T^2\alpha, \ldots, T^{n-1}\alpha\}$ is a basis of V. Determine the matrix of T, relative to this basis. Is it clear that a vector such as α will always exist in V?

22. Let T be a linear operator defined on V_2 so that $T(x,y) = (-y,x)$.
    (a) Find the matrix of T, relative to the $\{E_i\}$ basis.
    (b) Find the matrix of T, relative to the basis $\{(1,2), (1,-1)\}$.
    (c) Verify that the two representing matrices are similar.

23. If T is the linear operator defined in Prob. 22, verify that the operator $T - cI$ is nonsingular for any real number c.

24. If $\{\alpha_i\}$, $\{\beta_i\}$, and $\{\gamma_i\}$ are bases of a vector space V, with P and Q the matrices of transition from $\{\alpha_i\}$ to $\{\beta_i\}$ and $\{\beta_i\}$ to $\{\gamma_i\}$, respectively, decide whether PQ or QP is the matrix of transition from $\{\alpha_i\}$ to $\{\gamma_i\}$.

25. Prove that AB and BA are similar for $n \times n$ matrices A and B, provided at least one of A and B is nonsingular. Does the same conclusion hold if neither A nor B is nonsingular?

26. Let T be the linear transformation defined in Prob. 14. Then, if P in Prob. 5 and A in Prob. 7 are the transition matrices to new bases in V_3 and V_2, respectively, determine the new coordinate vector associated with (a) $(1,1,-1)$; (b) $(2,1,2)$; (c) $(-1,1,1)$.

2.10 Rank

The notion of *rank* occurs in the theory of linear transformations and also (in several different connections) within the theory of matrices. However, even though the various ideas are conceptually distinct, it will ultimately be shown that all are equivalent. We shall obtain our most important results on *rank* in this section, but our discussion will be completed in Chap. 3.

The *rank* of a *linear transformation* has already been defined in Sec. 2.2. Thus, if T is a linear transformation of a finite-dimensional vector space, the *rank* of T is the *dimension of the range space* R_T of T. We now focus our attention on matrices.

Let A be any $m \times n$ matrix. The columns of A are m-tuples which may be regarded as elements of the coordinate vector space V_m. The subspace of V_m, which is generated by these m-tuples, is known as the *column space* of A, and the dimension of this column space is called the *column rank* of A. In the same way, the rows of A may be considered elements of V_n and will generate a subspace of V_n, known as the *row space* of A, the dimension of this subspace being called the *row rank* of A. For example, consider the matrix $A = \begin{bmatrix} 2 & 1 & 4 \\ 1 & 2 & 3 \end{bmatrix}$. The columns of A may be identified with the vectors $\{(2,1),(1,2),(4,3)\}$, and these will generate the column space of A. In this case, it is clear that this subspace is the whole space V_2, and so the column rank of A is 2. The row space is the subspace of V_3 which is generated by the vectors $\{(2,1,4),(1,2,3)\}$, and, since these vectors are not multiples of each other, they are linearly independent and generate a two-dimensional subspace. Since the row rank of A is then also 2, we have illustrated a result (to be established later) that the row rank and the column rank of a matrix are equal. But first we connect the column rank of a matrix with the rank of a linear transformation.

Theorem 10.1 *The rank of a linear transformation on a finite-dimensional vector space is equal to the column rank of any matrix that represents it.*

Proof
If T is a linear transformation of a vector space V into a vector space W, where $\dim V = n$ and $\dim W = m$, it is well known that any matrix that represents T is $m \times n$. It is also known, and indeed was emphasized in Sec. 2.9, that the columns of the matrix are the ordered coordinates, relative to the basis selected for W, of the images under T of the basis selected for V. Hence, by Theorem 9.1 (Chap. 1), the subspace generated by the columns of the matrix is isomorphic to the range space R_T of T. The theorem now follows from the definition of rank, as applied to the transformation T. ∎

Inasmuch as any matrix can be regarded as the matrix of a linear transformation, relative to spaces of the proper dimensions, it is apparent that the notions of *rank of a linear transformation* and *column rank of a matrix* are essentially equivalent concepts. In order to study the row rank of a matrix and its connection with the column rank, it is convenient to relate the matrix with another transformation which we now define.

Definition Let V^* and W^* be the dual spaces of linear functionals, associated, respectively, with vector spaces V and W. Then, if T is a linear transformation of V into W, we define the transpose transformation T' of W^* into V^* so that $T'g = gT$, for any $g \in W^*$.

Thus T' maps $g \in W^*$ onto the functional $T'g \in V^*$, so that $(T'g)\alpha = g(T\alpha)$, for any $\alpha \in V$. It is easy to see that T' is a linear mapping. Suppose $g_1, g_2 \in W^*$, and c_1, c_2 are real numbers. Then $[T'(c_1g_1 + c_2g_2)]\alpha = (c_1g_1 + c_2g_2)(T\alpha) = c_1g_1(T\alpha) + c_2g_2(T\alpha) = c_1(T'g_1)\alpha + c_2(T'g_2)\alpha = [c_1(T'g_1) + c_2(T'g_2)]\alpha$, from which we conclude that $T'(c_1g_1 + c_2g_2) = c_1(T'g_1) + c_2(T'g_2)$. Hence T' is linear as asserted.

The next theorem shows that the *rows* of a matrix bear a relationship to the *transpose* T' of a linear transformation T, which is quite similar to that which we have already seen existing between the *columns* of the matrix and T.

Theorem 10.2 Let $T: V \to W$, where $\dim V = n$ and $\dim W = m$, and let $A = [a_{ij}]$ be the matrix of T, relative to bases $\{\alpha_i\}$ and $\{\beta_i\}$ of V and W, respectively. If $B = [b_{ij}]$ is the matrix of T', relative to the respective dual bases $\{f_i\}$ and $\{g_i\}$ of V^* and W^*, then $B = A'$.

Proof

By definition of A, $T\alpha_i = \sum_{k=1}^{m} a_{ki}\beta_k$, $i = 1, 2, \ldots, n$; and, by definition of B, $T'g_j = \sum_{i=1}^{n} b_{ij}f_i$, $j = 1, 2, \ldots, m$. However, $(T'g_j)\alpha_i = g_j(T\alpha_i) = g_j\left(\sum_{k=1}^{m} a_{ki}\beta_k\right) = \sum_{k=1}^{m} a_{ki}g_j(\beta_k) = \sum_{k=1}^{m} a_{ki}\delta_{jk} = a_{ji}$, where δ_{jk} is the familiar Kronecker delta (see Sec. 2.4). If f is an arbitrary linear functional on V, we know from Theorem 4.1 that $f = \sum_{i=1}^{n} f(\alpha_i)f_i$; and so, in particular for the functional $T'g_j$, we find that

$$T'g_j = \sum_{i=1}^{n} [(T'g_j)\alpha_i]f_i = \sum_{i=1}^{n} a_{ji}f_i$$

But this result implies that the matrix of T', relative to the given dual bases, is $[b_{ij}]$ where $b_{ij} = a_{ji}$, and so $B = A'$, as asserted in the theorem. ∎

Corollary *The rank of the transformation* T', *which is the transpose of a transformation* T, *is equal to the row rank of any matrix that represents* T.

Proof
The rows of any matrix that represents T may be regarded as coordinate vectors that generate a space isomorphic to the range space of T'. ∎

After the proof of a very simple lemma, we shall be able to obtain the important result that the three notions of rank, as introduced in this section, are essentially equivalent.

Lemma *For any linear transformation* T, *the null space of* T' *is the annihilator of the range of* T.

Proof
Let T: $V \to W$, with V^* and W^* the associated dual spaces of linear functionals on V and W, respectively. Then, for any $g \in W^*$, the definition of T' requires that $(T'g)\alpha = g(T\alpha)$, for each $\alpha \in V$. But this equality implies that g is in the null space of T' if and only if $T\alpha$ is mapped onto 0 by g. Hence the null space of T' is exactly the same as the annihilator of the range of T. ∎

Theorem 10.3 *If* T *is any linear transformation on a finite-dimensional vector space,* rank T = rank T'.

Proof
As in the lemma, let T: $V \to W$, where dim $V = n$ and dim $W = m$. If r is the rank of T, it follows from Theorem 5.1 that the annihilator of the range of T has dimension $m - r$. By the lemma, $m - r$ must also be the nullity of T'. But now, since T' is a linear transformation of an m-dimensional vector space, it is a consequence of Theorem 2.2 that the rank of T' is $m - (m - r) = r$. Hence T and T' have the same rank. ∎

Corollary *If* A *is any matrix and* T *is a linear transformation represented by* A, *then the following sequence of equalities is valid:*

Rank of T = row rank of A = column rank of A

Definition *The common value of the row rank and column rank of a matrix* A *is called the rank of* A.

It follows from this definition that the rank of any matrix is the same as its row rank or its column rank, and this common value is the rank of any linear transformation that may be represented by the matrix.

In this section we have been discussing the notion of *rank* as it applies to a linear transformation and also as it applies to an arbitrary matrix. At the same time, we have emphasized the fact that any matrix can be regarded as representing some linear transformation. Since the rank of a linear transformation cannot depend on any particular choice of bases for the spaces involved, the results in this and the preceding sections then lead to the following matrix-theoretic result.

Theorem 10.4 *If P and Q are nonsingular matrices such that the product $Q^{-1}AP$ is defined, the rank of this matrix is the same as that of the matrix A.*

Corollary *The ranks of similar matrices are equal.*

We have phrased the discussions in this section for the general situation where T is a linear transformation of an n-dimensional vector space V into an m-dimensional vector space W. It is clear, of course, that everything remains valid if V and W are the same space. In this very important special case, T is a linear operator on a finite-dimensional space V, and the transpose T′ becomes identified with a linear operator on the dual space V^*. The characteristic property of T′ is then that it maps an arbitrary functional $f \in V^*$ onto the functional $T'f \in V^*$ so that $(T'f)\alpha = f(T\alpha)$, for any $\alpha \in V$.

Although there is one further characterization of *rank* (as applied to a matrix) which we shall mention in Chap. 3, we shall close this section with a comprehensive example that will illustrate many of the notions discussed here.

EXAMPLE

We let D designate the differentiation operator on the space P_4 of polynomials of degree 3 or less in a real variable x. The operator D will be represented, relative to the basis $\{1, x, x^2, x^3|$ of P_4, by the matrix A, where

$$A = \begin{bmatrix} 0 & 1 & 0 & 0 \\ 0 & 0 & 2 & 0 \\ 0 & 0 & 0 & 3 \\ 0 & 0 & 0 & 0 \end{bmatrix}$$

It is known from elementary calculus that the range space of D is the subspace of all polynomials of degree 2 or less in P_4; and it is immediate that the dimension of this subspace, or the *rank of* D, is 3. The column space of A, to which the range of D is isomorphic, is generated by the vectors (1,0,0,0), (0,2,0,0),(0,0,3,0) [the vector (0,0,0,0) having been discarded] and, since these vectors are clearly independent, the column rank of A is 3. The row space of A is generated by the vectors (0,1,0,0),(0,0,2,0),(0,0,0,3) and, again, it is clear that the space generated by these three vectors has dimension 3. The fact that the rank of D, the column rank of A, and the row rank of A are all equal is in accord with the results established in this section. We are quite familiar with the fact that the effect of D is to map any polynomial $p \in P_4$ onto its derivative polynomial p', but the transpose operator D' may be somewhat less familiar. However, the definition of a transpose operator requires that D' map any linear functional $f \in P_4^*$ onto the linear functional $fD \in P_4^*$. In particular, if f is the linear functional defined on P_4 by $f(p) = \int_0^1 p(x)\,dx$, then D'f is the functional such that $D'f(p) = \int_0^1 p'(x)\,dx$. To be precise, if p designates an arbitrary polynomial $a + bx + cx^2 + dx^3$ in P_4, so that $f(p) = \int_0^1 (a + bx + cx^2 + dx^3)\,dx = a + b/2 + c/3 + d/4$, then $Dp = p' = b + 2cx + 3dx^2$, and $D'f(p) = \int_0^1 p'(x)\,dx = \int_0^1 (b + 2cx + 3dx^2)\,dx = b + c + d$. It is instructive to obtain these same results with the use of the matrix A and its transpose A'. Thus, since

$$\begin{bmatrix} 0 & 1 & 0 & 0 \\ 0 & 0 & 2 & 0 \\ 0 & 0 & 0 & 3 \\ 0 & 0 & 0 & 0 \end{bmatrix} \begin{bmatrix} a \\ b \\ c \\ d \end{bmatrix} = \begin{bmatrix} b \\ 2c \\ 3d \\ 0 \end{bmatrix}$$

we see that D maps the coordinate vector (a,b,c,d) onto $(b,2c,3d,0)$; interpreted in terms of the polynomial p, this means that $Dp = b + 2cx + 3dx^2 = p'$, as above. Similarly, if the linear functional f defined above is expressed in terms of the basis $\{f_1,f_2,f_3,f_4\}$, which is dual to $\{1,x,x^2,x^3\}$, it may be seen (Prob. 11) from Theorem 4.1 that $f = f_1 + \frac{1}{2}f_2 + \frac{1}{3}f_3 + \frac{1}{4}f_4$. But now, since

$$\begin{bmatrix} 0 & 0 & 0 & 0 \\ 1 & 0 & 0 & 0 \\ 0 & 2 & 0 & 0 \\ 0 & 0 & 3 & 0 \end{bmatrix} \begin{bmatrix} 1 \\ \frac{1}{2} \\ \frac{1}{3} \\ \frac{1}{4} \end{bmatrix} = \begin{bmatrix} 0 \\ 1 \\ 1 \\ 1 \end{bmatrix}$$

it follows that the effect of D' is to map f onto the functional $f_2 + f_3 + f_4$. As before, we observe that $D'f(p) = f_2(p) + f_3(p) + f_4(p) = b + c + d$.

Problems 2.10

1. Write the transpose of each of the following matrices:

   (a) $\begin{bmatrix} 2 & 1 \\ 4 & 1 \end{bmatrix}$
   (b) $\begin{bmatrix} 1 & 2 & -1 \\ 1 & 2 & 3 \\ 3 & 1 & 5 \end{bmatrix}$
   (c) $\begin{bmatrix} 2 & -1 & 3 \\ -1 & 4 & 1 \\ 3 & 1 & 5 \end{bmatrix}$

2. Show that the row rank and column rank of the matrix

   $$\begin{bmatrix} 1 & 3 & -1 & 2 & 4 \\ 7 & -2 & 1 & 5 & -1 \end{bmatrix}$$

   are equal to 2.

3. Show that the row rank and column rank of the matrix

   $$\begin{bmatrix} 2 & 1 & 1 \\ 1 & -1 & 2 \\ 0 & 0 & 0 \end{bmatrix}$$

   are equal to 2.

4. Let T be the linear operator on V_2 where $T(x,y) = (x - y, y + 2x)$. If g is the linear functional on V_2, defined so that $g(x,y) = 2x - y$, use the definition of a transpose to determine T'g.

5. With T the operator in Prob. 4, determine T'g, where $g(x,y) = x + 2y$.

6. With T the operator in Prob. 4, determine T'g, where $g(x,y) = 3x$.

7. Describe the row space and the column space of the matrix A, where

   $$A = \begin{bmatrix} 1 & -1 & 2 \\ 2 & 0 & 1 \\ 1 & -1 & 2 \end{bmatrix}$$

8. If

   $$A = \begin{bmatrix} 1 & 4 & -1 & 1 \\ 1 & 2 & 1 & 0 \\ 2 & -1 & 3 & 1 \end{bmatrix}$$

   is the representation of a linear transformation $T: V \to W$, relative to bases $\{\alpha_i\}$ and $\{\beta_i\}$ of V and W, respectively, use A' to describe the transpose T' on W^*.

9. Prove Theorem 10.4 and its corollary.

10. Verify that the vectors, said to generate the row (column) space of matrix A in the example of this section, are linearly independent.

11. Verify that $f = f_1 + \frac{1}{2}f_2 + \frac{1}{3}f_3 + \frac{1}{4}f_4$, for the functional f in the example of this section.

12. With f and D defined as in the example in this section, use the basis $\{1, 1 + t, t + t^2, t^3\}$ of P_4 and the associated matrix of D' to determine D'f. Check that D'$f(p) = f($D$p)$, for a general $p \in P_4$.

13. If S and T are linear operators on a vector space, prove the following:
    (a) Rank $(S + T) \leq$ rank S + rank T.
    (b) Rank ST $\leq$ min {rank S, rank T}.
    (c) Rank ST = rank TS = rank S, if T is invertible.

14. If
$$\begin{bmatrix} 3 & 1 & 4 \\ 2 & -1 & 1 \\ 1 & 2 & 1 \end{bmatrix}$$
is the matrix of a linear operator T on a vector space, relative to a given basis, express T'g_1, T'g_2, T'g_3, where $\{g_1, g_2, g_3\}$ is the associated dual basis of the space of functionals.

15. If $T(x,y,z) = (2x, x - y, z)$ defines the linear operator T on V_3, use a matrix representation of T' to determine T'f, where $f(x,y,z) = x + y - z$. Use the definition of a transpose to check your result.

16. With the operator T defined on V_2 as follows, use a matrix representation of T' to determine T'f, where $f(x,y) = 2x + 3y$.
    (a) $T(x,y) = (x,0)$; (b) $T(x,y) = (-x,-y)$; (c) $T(x,y) = (x - y, 2x + y)$.

17. Let P_5 be the vector space of polynomials of degree 4 or less in a real variable x, and let D^2 designate the "second derivative" operator on P_5. If $f(p) = p(1)$, for any $p \in P_5$, use a matrix representation to describe the transpose of D^2 on P_5^*, and use the definition of a transpose to check your result.

18. If T: $V \to W$, for vector spaces V and W, prove that $R_{T'}$ is the annihilator of N_T.

19. Illustrate the result in Prob. 18, using the operator T in the example of this section.

20. If $A = [a_{ij}]$ is an $n \times n$ matrix, we define the *trace* (tr) of A by tr $(A) = a_{11} + a_{22} + \cdots + a_{nn}$. Prove that the trace function is a linear functional on the space V of $n \times n$ matrices.

21. Refer to Prob. 20, and prove that tr $(A'A) = 0$ if and only if $A = 0$.
22. With tr and V defined as in Prob. 20, let T be the linear operator defined on V by $TA = KA - AK$, for some fixed matrix $K \in V$, and describe T'(tr).
23. With the notation of Prob. 20, and f a linear functional defined on V so that $f(AB) = f(BA)$, for any matrices $A, B \in V$, prove that $f = c(\text{tr})$, for some $c \in \mathbf{R}$.
24. In Prob. 23, if we include $f(I) = n$, prove that $c = 1$.
25. A square matrix $A = [a_{ij}]$ is *symmetric* if $a_{ij} = a_{ji}$, for all i, j. If the matrix of a linear transformation T is symmetric, what significance, if any, does this have for the nature of the transpose T'?

2.11 An Important Note of Clarification

If the reader of this book consults other books on linear algebra or vector analysis, he may become confused over slight changes in the form of some of the basic equations that he encounters. The corresponding formulas in the various books will be very similar, but there will be differences which, to say the least, will be disturbing to the neophyte. The confusion seems to arise from two sources:

(1) Some people express coordinate vectors as *row* n-tuples, whereas others prefer to use columns.

(2) In many books, two kinds of vectors, *covariant* and *contravariant*, are introduced, and these types are distinguished from each other in an essential way. The very fact that they are not even mentioned in most books on linear algebra is in itself somewhat of a disturbing factor, since one will encounter them in courses in vector and tensor analysis.

In this section, we shall attempt to clarify some of the confusion related to these two points.

With regard to the first item above, in modern usage a vector is an abstract entity that is *neither* a row *nor* a column n-tuple: it is an element of an abstract algebraic system called a *vector space*. However, in the case of finite-dimensional spaces, it is often a great convenience to represent a vector by an ordered n-tuple, the components of which are the coordinates of the vector relative to some basis that is assumed given. [This practice is justified by Theorem 9.1 (Chap. 1)]. It should be clear that it does not matter whether the n-tuple is written as a horizontal or vertical array, or even in some other orderly form, because only *the identification of its components and how these change with a change of basis* for the space are of importance. However, for the reason of compactness of certain formulas, it has become the custom to associate a coordinate vector with a "thin" matrix and thereby take advantage of the rules for matrix multiplication. But it now makes a difference whether this is a row or column *matrix*. For example, if X is a $1 \times n$ or "row" matrix, and A is an $n \times n$ matrix, the product XA is defined whereas the indicated "product" AX is not. On the other hand, the situation is reversed if Y is a $n \times 1$ or "column" matrix, for in this case AY is defined but the "product" YA is without meaning. Moreover, even if X and Y are matrix forms of the *same* n-tuple, the products AY and XA are usually different n-tuples, unless A is a *symmetric* matrix with $A = A'$ (see Prob. 25 of Sec. 2.10). If A is a symmetric matrix, it is indeed true that AX and XA are the same, *as n-tuples*, where it is understood that X is a column in the first occurrence and a row in the second. This may then explain the reason for some of the variations found in writing the matrix equivalent even of such a simple equation as $T\alpha = \beta$, for a linear transformation T.

If X and Y are matrices representing coordinate vectors of α and β, respectively, and A is the matrix of T as we have defined it, it seems only natural to regard X and Y as *column* matrices and write $AX = Y$. On the other hand, if the matrix A of T is defined (as is sometimes done) as the transpose of what we have called the matrix of T, it would be natural to express the same equality in the form $XA = Y$, with X and Y regarded as *row* matrices. It would be appropriate to make similar comments on the form of the change-of-basis equation which we have expressed earlier as $Y = X(P^{-1})'$, with X and Y row matrices that designate, respectively, the old and new coordinates of a vector in the space. It would then appear that whether a coordinate vector is represented as a row or column matrix depends on how it is used in a matrix equation and on other pertinent definitions.

The second item, mentioned at the beginning of this section, is a much more subtle one and requires a more elaborate discussion, because the point of confusion lies in the very nature of a vector. A vector in an n-dimensional vector space is an abstract object that has n components in each coordinate system of the space, but it is important to understand that the description of a vector is incomplete unless it is known how to obtain these components in an arbitrary coordinate system. In practice, this amounts to requiring that one know the rule by which the coordinates of a vector are changed as the bases of the space are changed.

The classical discussion of vectors introduces two kinds of vectors, one called *covariant* and the other *contravariant*. Although there is no uniformity about this, the *contravariant* vectors are usually the members of some initial space V, whereas the *covariant* ones belong to the space V^* which is dual to V. The very fact that V and V^* are dual makes this "distinguishing" feature between covariant and contravariant vectors a rather superficial one! On the other hand, if we consider how the coordinate vectors of V and V^* change (when related to dual bases), with a change of basis of V, we see that the two rules of transformation are different. And so, from the point of view of how they transform, there are indeed two different "kinds" of vectors associated with a space V. However, the modern point of view is simply to observe that V and V^* are *different* (albeit related) vector spaces and, in this sense, there are as many different "kinds" of vectors as there are vector spaces. At the same time, there are certain computational advantages in considering a vector as "covariant" or "contravariant" in the classical way; this is particularly true in connection with a more general sort of vector called a *tensor*. For this reason, and in order to make a point of contact with books on vector and tensor analysis, we shall investigate the nature of covariant and contravariant vectors somewhat more carefully. We shall freely use the notations introduced earlier, without further elaborate explanations.

If $P = [p_{ij}]$ is the matrix of transition from an $\{\alpha_i\}$ basis to a $\{\beta_i\}$ basis, we know that

$$\beta_j = \sum_{i=1}^{n} p_{ij} \alpha_i \qquad j = 1, 2, \ldots, n \qquad (1)$$

This change of basis results in a change of coordinates for each vector of the space; and, if $X = [x_1 \ x_2 \ \cdots \ x_n]'$ and $Y = [y_1 \ y_2 \ \cdots \ y_n]'$ are the respective old and new coordinates, written as column matrices, of a vector in V, we know $X = PY$. In terms of the components of X and Y, this matrix equation is equivalent to

$$x_i = \sum_{j=1}^{n} p_{ij} y_j \qquad i = 1, 2, \ldots, n \qquad (2)$$

We note that, whereas Eq. (1) expresses the new basis vectors in terms of the old, Eq. (2) expresses the old components of a coordinate vector in terms of the new. Hence, in order that we may compare the laws of transformation for basis vectors and coordinates, we must solve Eq. (2) for y_j. We have done this in an earlier section, and the matrix equation $X = PY$ is then replaced by the matrix equation $Y = P^{-1}X$. It is clear that X and Y are column matrices in the latter equation and so, in order to obtain an equation that will relate the components of X and Y in a form similar to Eq. (1), we take the transpose of both sides and obtain $Y' = X'(P^{-1})' = X'Q$, where $Q = (P^{-1})'$. Hence, if $Q = [q_{ij}]$, we may express Eq. (2) in an equivalent form as

$$y_j = \sum_{i=1}^{n} q_{ij} x_i \qquad j = 1, 2, \ldots, n \qquad (3)$$

We now take a look at the dual space V^*, and recall that the representation of a linear functional $f \in V^*$, relative to the basis $\{f_i\}$ which is dual to $\{\alpha_i\}$, is $f = f(\alpha_1)f_1 + f(\alpha_2)f_2 + \cdots + f(\alpha_n)f_n$. The coordinate vector of f is $[f(\alpha_1), f(\alpha_2), \ldots, f(\alpha_n)]$ which, as a row matrix, takes the form $[f(\alpha_1) \; f(\alpha_2) \; \cdots \; f(\alpha_n)]$. It is known, however, that if we designate this matrix by Z_1, a change of basis in V, with matrix of transition P, transforms the coordinate vectors in V^* so that Z_1 is mapped onto Z_2, where $Z_2 = Z_1 P$. Of course, $Z_2 = [g(\alpha_1) \; g(\alpha_2) \; \cdots \; g(\alpha_n)]$, for some $g \in V^*$, and so the coordinates of vectors in the dual space V^* may be seen to transform, when so induced by a matrix P of basis transition in V, according to the law

$$g(\alpha_j) = \sum_{i=1}^{n} p_{ij} f(\alpha_i) \qquad j = 1, 2, \ldots, n \tag{4}$$

If we now look at the three laws of transformation, exhibited in Eqs. (1), (3), and (4), we note that laws (1) and (4) are identical whereas (3) is different. That is, the coordinates of vectors in V^* may be said to *co-vary* or transform according to the *same* law as the basis vectors of V, whereas the coordinates of vectors in V transform according to a *different* or *contra* law. Although the origin of the words *covariant* and *contravariant* is not made clear in most books, the above discussion should provide at least a plausible explanation for their use. We again emphasize that covariant and contravariant vectors are not elements of the same space, for if P is the matrix of transition for a basis in V, the covariant vectors in V^* transform with P whereas the contravariant vectors in V transform with $(P^{-1})'$.

Finally, a word about notation. We have already said that the idea of a vector may be generalized to that of a *tensor*. In an n-dimensional space, a tensor is an object that has n^r components in any coordinate system, the nonnegative integer r being the *order* of the tensor. Each of its sets of n components will in general lie in a distinct coordinate space, each with its own law of transformation. Some of these laws will be according to the covariant rule and some according to the contravariant rule, and, in order to lessen the confusion due to this, a certain practice is followed in designating indices: a covariant index is written as a subscript, while a superscript indicates a contravariant index. For example, a_k^{ij} would be a component of a tensor of order 3, which is covariant in the one index k and contravariant in the two indices i, j. If we associate a scalar with all coordinate systems without discrimination, we may regard a scalar as a tensor of order 0. We have seen that the vectors of an "initial" space V are contravariant of order 1, and in the notation of tensors it would be appropriate to express the law of transformation of their coordinates in the form $y^i = \sum_{i=1}^{n} q_j^i x^j$, in which the superscript i indicates the contravariant nature of the coordinates of vectors in V. On the other hand, the law of transformation of a linear functional in V^* would be expressed in the form $g(\alpha_i) = \sum_{j=1}^{n} p_i^j f(\alpha_j)$; and in this case the subscript i is indicative of the covariant nature of the coordinates of vectors in V^*. As a final passing remark, we state that it is customary in treatises on tensor analysis to omit all summation signs and simply to understand that any repeated index is to be summed over its entire range of values.

EXAMPLE

A familiar example of the two kinds of vectors that we have been discussing occurs in the context of differentials and gradients of functions of n variables $x_1, x_2, \ldots, x_n$. The *initial* vector space V is, in this case, the space of differentials $(dx_1, dx_2, \ldots, dx_n)$ to be associated with any point $(x_1, x_2, \ldots, x_n)$ of V_n, and the dual space V^* is the space of linear functionals on the space of differentials. Since a typical element of V has the form $a_1(dx_1) + a_2(dx_2) + \cdots + a_n(dx_n)$, and this can be expressed as dw, where $a_i = \partial w/\partial x_i$, $i = 1, 2, \ldots, n$, we see that the coordinate vectors of V^* have the form $(\partial w/\partial x_1, \partial w/\partial x_2, \ldots, \partial w/\partial x_n)$ for differentiable functions $w = w(x_1, x_2, \ldots, x_n)$. Such a vector is called the *gradient* of w and is often abbreviated to ∇w.

We are interested in the manner in which the differential or *displacement* vectors and the gradient vectors transform with a change of variable. Thus, suppose $y_1, y_2, \ldots, y_n$ are the members of a new set of variables such that $y_i = y_i(x_1, x_2, \ldots, x_n)$ is a differentiable function of x_j, and $x_i = x_i(y_1, y_2, \ldots, y_n)$ is a differentiable function of y_j, for $i, j = 1, 2, \ldots, n$. We know, from multivariate calculus, that the law of transformation for the displacement vectors is given by

$$dy_j = \sum_{i=1}^{n} \frac{\partial y_j}{\partial x_i} dx_i \qquad j = 1, 2, \ldots, n$$

Furthermore, it is also well known from calculus that the associated law of transformation for the gradient vectors is given by

$$\frac{\partial w}{\partial y_j} = \sum_{i=1}^{n} \frac{\partial w}{\partial x_i} \frac{\partial x_i}{\partial y_j} = \sum_{i=1}^{n} \frac{\partial x_i}{\partial y_j} \frac{\partial w}{\partial x_i}$$

We have observed that if $X = (dx_1, dx_2, \ldots, dx_n)$ and $Y = (dy_1, dy_2, \ldots, dy_n)$, the matrix that transforms X onto Y is $Q = [q_{ij}] = \left[\dfrac{\partial y_j}{\partial x_i}\right]$, so that $Y = QX$ where X and Y are written as column matrices. By the general theory in Sec. 2.9 (or the result can be obtained independently), the transformation of the coordinate vector X can be induced by a transformation of the $\{E_i\}$ basis onto an $\{F_i\}$ basis of V, the matrix of transition being $P = [p_{ij}]$, where $P = (Q^{-1})'$. It is now an elementary computation to verify that

$$F_j = \sum_{i=1}^{n} \frac{\partial x_i}{\partial y_j} E_i \qquad j = 1, 2, \ldots, n$$

Hence, the coordinate (gradient) vectors of V^* transform like the basis vectors of V (that is, covariantly), whereas the coordinate (displacement) vectors of V transform in a contravariant manner. This is in accord with what we have noted earlier in a general context.

Selected Readings

Bing, Kurt: A Construction of the Null Space of a Linear Transformation, *Amer. Math. Monthly*, **67**(1): 34–38 (January, 1960).

May, K. O.: The Impossibility of a Division Algebra of Vectors in Three Dimensional Space, *Amer. Math. Monthly*, **73**(3): pp. 289–291 (March, 1966).

Parker, F. D.: Inverses of Vandermonde Matrices, *Amer. Math. Monthly*, **71**(4): 410–411 (April, 1964).

3

DETERMINANTS AND SYSTEMS OF LINEAR EQUATIONS

3.1 Matrices and Linear Systems

In the preceding chapters we occasionally encountered two difficulties of a practical nature, with which we chose not to cope at that time: Our discussion of linear independence of vectors was left in an incomplete state, because it was seen to be intimately connected with the solvability of systems of linear equations, a matter that we investigated only to meet our immediate needs in Chap. 1; and in Chap. 2 we often had occasion to refer to and use nonsingular matrices, but we did not present a general method for determining the inverse that is characteristic of such a matrix. One of the objectives of this chapter is to remedy both these deficiencies.

In most of the general investigations of this chapter, we shall consider a system of equations of the form (1) given in Sec. 1.5 but which we rewrite for easy reference:

$$\begin{aligned} a_{11}x_1 + a_{12}x_2 + \cdots + a_{1n}x_n &= b_1 \\ a_{21}x_1 + a_{22}x_2 + \cdots + a_{2n}x_n &= b_2 \\ &\cdots \\ a_{m1}x_1 + a_{m2}x_2 + \cdots + a_{mn}x_n &= b_m \end{aligned} \quad (1)$$

SEC. 3.1 MATRICES AND LINEAR SYSTEMS

With matrix multiplication now available, it is easily seen that this system of equations is equivalent to the single matrix equation

$$AX = B$$

in which $A = [a_{ij}]$ is the matrix of coefficients in the left members of Eqs. (1), $B = [b_1 \; b_2 \; \cdots \; b_m]'$ is the column matrix of right members, and $X = [x_1 \; x_2 \; \cdots \; x_n]'$ is the column matrix of unknowns. It is appropriate to consider X as *the* unknown of the matrix equation $AX = B$, and the solution of system (1) for $x_1, x_2, \ldots, x_n$ is equivalent to the solution of $AX = B$ for X.

EXAMPLE 1

Let us consider the following system of equations:

$$\begin{aligned} 2x - 3y + z &= -1 \\ x + 2y - z &= 3 \\ 4x + y + 2z &= 5 \end{aligned}$$

It is clear that the matrix form of this system is $AX = B$, where

$$A = \begin{bmatrix} 2 & -3 & 1 \\ 1 & 2 & -1 \\ 4 & 1 & 2 \end{bmatrix} \quad B = \begin{bmatrix} -1 \\ 3 \\ 5 \end{bmatrix} \quad X = \begin{bmatrix} x \\ y \\ z \end{bmatrix}$$

The coefficient matrix of the system is A, and the *unknown* in the matrix equation is X, the solution for which is equivalent to the solution of the given system.

It is already apparent that, if the general matrix equation $AX = B$ is solved, any natural solution for X will be in the form of a column matrix. Up to this stage in the book, we have used $(x_1, x_2, \ldots, x_n)$ to designate a typical element of the coordinate space V_n but, as was pointed out in Sec. 2.8, it is unimportant how this ordered n-tuple is expressed. Either of the notations $[x_1 \; x_2 \; \cdots \; x_n]$ or $[x_1 \; x_2 \; \cdots \; x_n]'$ would be acceptable as an alternative way of writing the coordinate vector $(x_1, x_2, \ldots, x_n)$; but, now that matrices have been introduced, we shall often find it convenient to consider a coordinate vector as a row or column matrix and take advantage of matrix multiplication. We have already done this essentially in the latter sections of Chap. 2. In the context of this chapter, we have noted that any solutions of $AX = B$ will appear naturally as column matrices and, if we agree to identify V_n with the vector space of all $n \times 1$ column matrices, any solution of the equation will be an element of this familiar space. We shall do this here and, whenever convenient, at any point in the sequel. We offer no apology for this variation in notation, for we feel that there may be some real advantage to the reader in adjusting to such a change. In this way, he may avoid some of the pitfalls inherent in the constant use of an inflexible symbolism.

Our first result is one of theoretical interest in connection with the solutions of the system of equations (1). Throughout this chapter, unless stated otherwise, it should be understood that the equation $AX = B$ is the matrix form of this general system.

Theorem 1.1 *The solutions of $AX = 0$ constitute a subspace of V_n.*

Proof
If we observe that the solutions comprise the *null space* of any linear transformation on V_n represented by A, there is nothing further to prove. However, it is also easy to give a direct and independent proof. All that is necessary is to prove *closure* of the solution set under addition and multiplication by scalars. If X_1 and X_2 are solutions of $AX = 0$, then $A(X_1 + X_2) = AX_1 + AX_2 = 0 + 0 = 0$, so that $X_1 + X_2$ is a solution. Moreover, for any real number c, we see (Prob. 7) that $A(cX_1) = c(AX_1) = 0$, whence cX_1 is in the solution set of $AX = 0$. The proof is then complete. ∎

We know that V_n has dimension n, and so the space of solutions or the *solution space* of $AX = 0$ has dimension $k \leq n$. If $\{X_1, X_2, \ldots, X_k\}$ is a basis for the solution space, *every* solution X can be expressed in the form $X = c_1 X_1 + c_2 X_2 + \cdots + c_k X_k$, for real numbers $c_1, c_2, \ldots, c_k$. Inasmuch as every solution has this form, this general expression is sometimes called the *complete* solution of the homogeneous equation $AX = 0$. The next theorem shows the relationship between the solutions of $AX = 0$ and those of a nonhomogeneous system $AX = B$.

Theorem 1.2 *Let $\{X_1, X_2, \ldots, X_k\}$ be a basis for the solution space of $AX = 0$. Then, if X_p is any particular solution of $AX = B$, every solution of $AX = B$ can be expressed in the form $c_1 X_1 + c_2 X_2 + \cdots + c_k X_k + X_p$, for suitable real numbers $c_1, c_2, \ldots, c_k$.*

Proof
Let us suppose that $\tilde{X}$ is an arbitrary solution of $AX = B$. Then, since $AX_p = B$, it follows that $A\tilde{X} - AX_p = 0 = A(\tilde{X} - X_p)$, so that $\tilde{X} - X_p$ is a solution of $AX = 0$. But then $\tilde{X} - X_p = c_1 X_1 + c_2 X_2 + \cdots + c_k X_k$,

for suitable real numbers $c_1, c_2, \ldots, c_k$, whence $\tilde{X} = c_1 X_1 + c_2 X_2 + \cdots + c_k X_k + X_p$, as asserted. ∎

We should point out that a given system of equations $AX = B$ may have an empty solution set, in which case the formal equations are said to be *inconsistent*. Any homogeneous system $AX = 0$ always has at least the *trivial* solution $X = 0 = [0 \ 0 \ \cdots \ 0]'$, and this may or may not be the only solution. If 0 is the only solution, the solution space is the zero subspace. In subsequent sections we shall continue the investigation (begun in Chap. 1) of conditions under which $AX = B$ has solutions, or is *consistent*, as well as new methods of finding them.

Since the equation $AX = 0$ always has at least the trivial solution, the solvability of $AX = B$ must rest ultimately on the nature of B. This dependence is clearly described in the following theorem.

Theorem 1.3 The system of equations $AX = B$ is consistent if and only if $B \in \mathcal{L}\{A_1, A_2, \ldots, A_n\}$, where

$$A_1 = [a_{11} \ a_{21} \ \cdots \ a_{m1}]'$$
$$A_2 = [a_{12} \ a_{22} \ \cdots \ a_{m2}]$$
$$\cdots \cdots \cdots \cdots \cdots \cdots \cdots$$
$$A_n = [a_{1n} \ a_{2n} \ \cdots \ a_{mn}]'$$

Proof

The proof of the theorem depends on the single observation that the system of equations is equivalent to the single matrix equation: $x_1 A_1 + x_2 A_2 + \cdots + x_n A_n = B$. For, now, the solvability of this equation allows us to assert that $B \in \mathcal{L}\{A_1, A_2, \ldots, A_n\}$; and, if $B \in \mathcal{L}\{A_1, A_2, \ldots, A_n\}$, this equation must be solvable. ∎

The following interesting (and sometimes useful) corollary is a consequence of this theorem and the equality of column rank with rank of a matrix.

Corollary The system of equations $AX = B$ is solvable if and only if the ranks of the coefficient matrix A and its augmented matrix are equal. This common rank of A and its augmented matrix is sometimes called the rank of the system of equations $AX = B$.

EXAMPLE 2

If we consider the system $AX = B$, as given in Example 1, we note that $B = A_1 + A_2$, in the notation of Theorem 1.3. It could then be deduced, without actual solving, that the system in that example is consistent and so solvable.

Problems 3.1

1. Verify that
$$\begin{bmatrix} 1 \\ 0 \\ -1 \end{bmatrix}$$
is a solution of
$$\begin{bmatrix} 1 & 2 & -1 \\ 3 & 1 & 2 \\ 2 & -1 & 1 \end{bmatrix} \begin{bmatrix} x \\ y \\ z \end{bmatrix} = \begin{bmatrix} 2 \\ 1 \\ 1 \end{bmatrix}$$

2. Verify that
$$\begin{bmatrix} 2 \\ 1 \\ 2 \end{bmatrix}$$
is a solution of $AX = 0$, where
$$A = \begin{bmatrix} 2 & 2 & -3 \\ -1 & 4 & -1 \\ 3 & 2 & -4 \end{bmatrix}$$

3. With A defined as in Prob. 2, and $B = [7\ 4\ 9]'$, verify that $[1\ 1\ -1]'$ is a solution of $AX = B$. Verify that $c[2\ 1\ 2]' + [1\ 1\ -1]'$ is also a solution, for any $c \in \mathbf{R}$.

4. (a) Express the following system of equations in the form $AX = B$:
$$\begin{aligned} 2x - 3y + z &= -1 \\ x - 2y + 4z &= 2 \\ 3x + y + z &= 9 \end{aligned}$$

   (b) Use an argument based on rank to decide the solvability of the system.

   (c) Verify that $[2\ 2\ 1]'$ is a solution of the system in (a) by direct substitution in the equations for x, y, z.

   (d) Use the matrix form $AX = B$ to check the solution given in (c).

5. Express the "system" $2x - 3y + z = 4$ in matrix form, and determine three solutions by inspection. What can you say about the multiplicity of the solutions of this equation?

6. In the proof of Theorem 1.3, verify that $B = c_1 A_1 + c_2 A_2 + \cdots + c_n A_n$ is equivalent to $AC = B$, where $C = [c_1 c_2 \cdots c_n]'$.

7. Explain why $A(cX) = c(AX)$, for any $c \in \mathbf{R}$, where A is a matrix and X a column matrix such that AX is defined.
8. Explain why the solutions of $AX = B$ do not constitute a vector space unless $B = 0$.

9. Use Theorem 1.3 (or its corollary) to determine c so that the following system is (a) consistent; (b) not consistent:

$$\begin{aligned} 2x \phantom{{}+3y} + 3z &= -1 \\ x + 3y - 6z &= c \\ x + y - z &= 1 \end{aligned}$$

10. If

$$A = \begin{bmatrix} 2 & 1 & 1 \\ 1 & 3 & -1 \\ 3 & 4 & 0 \end{bmatrix} \quad \text{and} \quad X = \begin{bmatrix} x \\ y \\ z \end{bmatrix}$$

find all solutions of $AX = 2X$.

11. With A and X defined as in Prob. 10, find all solutions of $AX = 3X$.

12. Solve the following system of equations by reducing the associated augmented coefficient matrix to Hermite normal form:

$$\begin{aligned} 2x - y - 2z &= 2 \\ x + y + 3z &= 0 \\ 3x - y - 2z &= 3 \end{aligned}$$

13. Use Cramer's rule to solve the system of equations given in Prob. 12.

14. Use a "rank" argument to prove that the following system of equations is inconsistent:

$$\begin{aligned} x \phantom{{}+y} + 2z &= 3 \\ -x + y + 2z &= 1 \\ 2x \phantom{{}+y} + 4z &= 5 \end{aligned}$$

15. Use trial and error to find a single solution of the system of equations

$$\begin{aligned} 2x - y - z &= 1 \\ x + y + z &= 2 \end{aligned}$$

Then solve the associated homogeneous system and use Theorem 1.2 to find an expression for the general solution of the given system.

16. Use the geometric interpretation of V_3 to describe the solution set, found in Prob. 15, for (a) the associated homogeneous system; (b) the given system.
17. The solution set of a system of linear equations consists of all vectors of the form $[c_1 \quad c_2 + 1 \quad 2c_2 - 1]'$, for arbitrary $c_1, c_2 \in \mathbf{R}$. Express these solutions in the form described in Theorem 1.2.
18. Use the directions given in Prob. 17 for solution vectors of the form $[a + b \quad a + 2 \quad 2a + b \quad b + 1]'$, for arbitrary $a, b \in \mathbf{R}$.
19. Use the directions given in Prob. 17 for solution vectors of the form $[a + b \quad 2a - b \quad a]'$, for arbitrary $a, b \in \mathbf{R}$.
20. If P_j is a solution of $AX = E_j$, where $\{E_i\}$ is the familiar basis of V_n with E_i expressed as a column matrix, prove that $\sum_{j=1}^{n} b_j P_j$ is a solution of $AX = B$, where $B = [b_1 \quad b_2 \quad \cdots \quad b_n]'$.
21. A polynomial of the form $a_1 x_1 + a_2 x_2 + \cdots + a_n x_n$ can be used to define a linear functional on V_n. Interpret the solution space of a system of linear homogeneous equations in terms of linear functionals.
22. If $AX = Y$, where A is the matrix of a linear transformation T, and X and Y are column matrices of appropriate lengths, interpret the given equation (a) for a fixed X; (b) for a fixed Y.

3.2 Elementary Matrices and Inverses

We have seen in Sec. 2.8 that the transformation equation Tα = β, indicating that the transformation T maps the vector α onto the vector β, can be expressed in equivalent matrix form as $AX = Y$, where A is the matrix of T and X and Y are the coordinate column vectors of α and β, respectively, relative to some basis (or bases). It was pointed out earlier that the close association between T and A leads some people even to think of A as a "matrix operator" that performs an operation on X that is equivalent to the one performed on α by T. In Sec. 1.5 we showed how to transform the augmented coefficient matrix of a system of linear equations into a unique row-echelon or Hermite normal form, from which we were able to obtain immediately all solutions of the system, provided there were any. We accomplished this transition into normal form by means of a sequence of elementary *row operations* on the matrix, and it is part of the purpose of this section to show that each of these elementary operations has a corresponding elementary matrix "operator" and that the composite of a sequence of elementary operations can be represented by a nonsingular matrix. To be precise, if A is the original matrix and P is the composite "operator" matrix, the product PA is a matrix in the desired Hermite normal form. For the most part, our discussions in this chapter will be centered on matrices rather than on linear transformations which they may represent. Each elementary row operation on a matrix can be interpreted as a transformation on a vector space (Prob. 7), but our present point of view will be matrix-theoretic. However, let us return to the beginning of the exposition!

We shall assume that the initial matrix A, on which elementary row operations are to be performed, is $m \times n$. There are three elementary row operations and three *elementary matrices* associated with them, each matrix being the result of applying an elementary row operation to the identity matrix I_m. To be specific, we recall that an elementary row operation of Type 1 interchanges the ith and jth rows of a matrix to which it is applied. If P_{ij} is the matrix that results after the ith and jth rows of I_m have been interchanged, it is clear that

$$P_{ij} = \begin{array}{c} \\ \\ i \\ \\ j \\ \\ \\ \end{array} \begin{bmatrix} 1 & \cdots & 0 & \cdots & 0 & \cdots & 0 \\ \cdot & & \cdot & & \cdot & & \cdot \\ 0 & \cdots & 0 & \cdots & 1 & \cdots & 0 \\ \cdot & & \cdot & & \cdot & & \cdot \\ 0 & \cdots & 1 & \cdots & 0 & \cdots & 0 \\ \cdot & & \cdot & & \cdot & & \cdot \\ 0 & \cdots & 0 & \cdots & 0 & \cdots & 1 \end{bmatrix} \begin{array}{c} i \quad\quad\quad j \\ \\ \\ \\ \\ \\ \\ \end{array}$$

It is now a simple matter to verify that $P_{ij}A$ is the matrix that results from interchanging the ith and jth rows of A. A simple calculation shows that $P_{ij}^2 = I_m$, so that P_{ij} is nonsingular with $P_{ij}^{-1} = P_{ij}$. The matrix P_{ij} is called an *elementary matrix of the first kind*.

An elementary row operation of Type 2 multiplies the elements of the ith row of a matrix to which it is applied by a real number $r \neq 0$. If we apply this operation to I_m, the result is to replace the 1 in the ith position in the diagonal of I_m by r. The resulting matrix $D_i(r)$ is an *elementary matrix of the second kind* and it is easy to see that

$$D_i(r) = \begin{array}{c} \\ \\ \\ i \\ \\ \\ \end{array} \begin{bmatrix} 1 & 0 & \cdots & 0 & \cdots & 0 & 0 \\ 0 & 1 & \cdots & 0 & \cdots & 0 & 0 \\ \cdot & \cdot & & \cdot & & \cdot & \cdot \\ 0 & 0 & \cdots & r & \cdots & 0 & 0 \\ \cdot & \cdot & & \cdot & & \cdot & \cdot \\ 0 & 0 & \cdots & 0 & \cdots & 1 & 0 \\ 0 & 0 & \cdots & 0 & \cdots & 0 & 1 \end{bmatrix}$$

It may be readily checked that $D_i(r)A$ is the matrix that results when the same elementary operation is applied to A. Moreover, since

$$[D_i(r)][D_i(r^{-1})] = I_m$$

we see that $D_i(r)$ is nonsingular with $[D_i(r)]^{-1} = D_i(r^{-1})$.

Finally, if we perform an elementary operation of Type 3 on the matrix I_m by adding an r-multiple of the jth row to the ith row, we obtain the matrix $S_{ij}(r)$, which is called an *elementary matrix of the third kind*. It is clear that

$$S_{ij}(r) = \begin{array}{c} \\ \\ j \\ \\ \\ i \\ \\ \\ \end{array} \begin{bmatrix} 1 & \cdots & 0 & \cdots & 0 & \cdots & 0 \\ \cdot & & \cdot & & \cdot & & \cdot \\ 0 & \cdots & 1 & \cdots & 0 & \cdots & 0 \\ \cdot & & \cdot & & \cdot & & \cdot \\ \cdot & & \cdot & & \cdot & & \cdot \\ 0 & \cdots & r & \cdots & 1 & \cdots & 0 \\ \cdot & & \cdot & & \cdot & & \cdot \\ 0 & \cdots & 0 & \cdots & 0 & \cdots & 1 \end{bmatrix} \begin{array}{c} j \quad\quad i \\ \end{array}$$

Since $[S_{ij}(r)][S_{ij}(-r)] = I_m$, we conclude that elementary matrices of the third kind are nonsingular with $S_{ij}^{-1}(r) = S_{ij}(-r)$. Moreover, it is a matter of a simple check to verify that $[S_{ij}(r)]A$ is the matrix that results when an r-multiple of the jth row is added to the ith row of A.

If $A_1, A_2, \ldots, A_k$ are elementary matrices that, when used as left multipliers in the order written, reduce a matrix A to a Hermite normal matrix N, it follows from what we have said that $A_k A_{k-1} \cdots A_2 A_1 A = N$. The matrix N, or any of the intermediate matrices, is said to be *row-equivalent* to A. If A is a nonsingular $n \times n$ matrix, so that its Hermite normal form is I_n, the result is that $A_k A_{k-1} \cdots A_2 A_1 A = I_n$, for elementary matrices $A_1, A_2, \ldots, A_k$. Hence, in view of the unique nature of an inverse matrix, $A^{-1} = A_k A_{k-1} \cdots A_2 A_1 = A_k A_{k-1} \cdots A_2 A_1 I_n$, and from this we make the following important observation:

> If a sequence of multiplications on the left by elementary matrices reduces a matrix A to the identity I, the same left matrix multiplications in the same order will reduce I to A^{-1}.

In practice, it is not necessary to identify all the elementary matrices involved in the composite operation, because it is what they *do* rather than what they *are* that is of importance. We clarify this procedure for finding the inverse of a nonsingular matrix by a very simple but effective device: *The nonsingular matrix A is bordered on its right (or left) by an identity matrix of the same size; and, if we include the rows of the identity matrix as extensions of the rows of A in the reduction of A to I, the final matrix that replaces the original identity matrix is A^{-1}*. We illustrate the process with an example.

EXAMPLE 1

Find the inverse of the nonsingular matrix

$$\begin{bmatrix} 2 & 1 & 1 \\ 1 & -1 & 1 \\ 2 & -2 & 3 \end{bmatrix}$$

Solution

We begin our operations on the matrix

$$\begin{bmatrix} 2 & 1 & 1 & 1 & 0 & 0 \\ 1 & -1 & 1 & 0 & 1 & 0 \\ 2 & -2 & 3 & 0 & 0 & 1 \end{bmatrix}$$

and perform a sequence of elementary row operations to produce the desired result. It will be clear which operations are performed at each step if we designate these operations, as in Sec. 1.5, according to type by (1), (2), or (3); a "product," such as (3)(3), will indicate the composite of two operations of Type 3. The sequence of matrices then follows.

$$\begin{bmatrix} 2 & 1 & 1 & 1 & 0 & 0 \\ 1 & -1 & 1 & 0 & 1 & 0 \\ 2 & -2 & 3 & 0 & 0 & 1 \end{bmatrix} \xrightarrow{(1)} \begin{bmatrix} 1 & -1 & 1 & 0 & 1 & 0 \\ 2 & 1 & 1 & 1 & 0 & 0 \\ 2 & -2 & 3 & 0 & 0 & 1 \end{bmatrix} \xrightarrow{(3)(3)}$$

$$\begin{bmatrix} 1 & -1 & 1 & 0 & 1 & 0 \\ 0 & 3 & -1 & 1 & -2 & 0 \\ 0 & 0 & 1 & 0 & -2 & 1 \end{bmatrix} \xrightarrow{(2)} \begin{bmatrix} 1 & -1 & 1 & 0 & 1 & 0 \\ 0 & 1 & -\frac{1}{3} & \frac{1}{3} & -\frac{2}{3} & 0 \\ 0 & 0 & 1 & 0 & -2 & 1 \end{bmatrix} \xrightarrow{(3)}$$

$$\begin{bmatrix} 1 & 0 & \frac{2}{3} & \frac{1}{3} & \frac{1}{3} & 0 \\ 0 & 1 & -\frac{1}{3} & \frac{1}{3} & -\frac{2}{3} & 0 \\ 0 & 0 & 1 & 0 & -2 & 1 \end{bmatrix} \xrightarrow{(3)(3)} \begin{bmatrix} 1 & 0 & 0 & \frac{1}{3} & \frac{5}{3} & -\frac{2}{3} \\ 0 & 1 & 0 & \frac{1}{3} & -\frac{4}{3} & \frac{1}{3} \\ 0 & 0 & 1 & 0 & -2 & 1 \end{bmatrix}$$

Thus, we conclude from our preceding discussion, that

$$A^{-1} = \begin{bmatrix} \frac{1}{3} & \frac{5}{3} & -\frac{2}{3} \\ \frac{1}{3} & -\frac{4}{3} & \frac{1}{3} \\ 0 & -2 & 1 \end{bmatrix}$$

We have seen above that the inverse of any nonsingular matrix A can be expressed in the form $A^{-1} = A_k A_{k-1} \cdots A_2 A_1$, for elementary matrices $A_1, A_2, \ldots, A_k$, so that $A = (A_k A_{k-1} \cdots A_2 A_1)^{-1} = A_1^{-1} A_2^{-1} \cdots A_k^{-1}$. (See Prob. 16 of Sec. 2.8.) Since it has been noted that the inverse of any elementary matrix is an elementary matrix of the same kind, the following theorem has been established.

Theorem 2.1 *Any nonsingular matrix can be expressed as a product of a finite number of elementary matrices.*

In other words, it may be of intuitive interest to regard elementary matrices as the "building blocks" out of which nonsingular matrices can be constructed. This result, however, has little practical value for us.

It may be asked what we have actually accomplished by the matrix description of the transition from the original matrix A to its normal form. We have already commented that we are not interested in the actual matrix representations of the elementary operations involved in the process. However, the fact that elementary matrices are nonsingular, and so also any product of such matrices, gives the following result which *is* of interest: *If N is the Hermite normal form of a matrix A, there exists a nonsingular matrix P such that $PA = N$.* Although we shall not go into the details here, it is not difficult to see that an elementary matrix as a *right* multiplier has the same effect on the *columns* of a matrix as we have just observed it to have on the *rows* of the matrix as a left multiplier. There then also exists a nonsingular matrix Q such that $AQ = N$, a result that parallels the one just obtained.

There is no really *easy* way to find the inverse of a nonsingular matrix of large order, but the method given in this section is usually to be preferred over the method by determinants (discussed in Sec. 1.6) if the order exceeds 3. *If the coefficient matrix of a system of linear equations is nonsingular*, it is of prime importance in the context of this chapter to observe that the solution is available as soon as the inverse of the matrix has been obtained. For, if $AX = B$ represents such a system, with A nonsingular, it follows that $A^{-1}(AX) = A^{-1}B$, whence $X = A^{-1}B$. The serious requirement imposed on a system of equations, subject to this method of solution, should be noted; we shall see later that the method by determinants has precisely the same requirement. In other words, the only method that is "guaranteed" to determine the solutions of a linear system of equations is that of Gauss reduction, as discussed in Sec. 1.5.

EXAMPLE 2

Solve the system of equations

$$\begin{aligned} 2x + y + z &= 1 \\ x - y + z &= 2 \\ 2x - 2y + 3z &= 3 \end{aligned}$$

Solution

The system of equations can be expressed in matrix form as $AX = B$, where

$$A = \begin{bmatrix} 2 & 1 & 1 \\ 1 & -1 & 1 \\ 2 & -2 & 3 \end{bmatrix} \quad X = \begin{bmatrix} x \\ y \\ z \end{bmatrix} \quad B = \begin{bmatrix} 1 \\ 2 \\ 3 \end{bmatrix}$$

Since A may be seen to be the matrix in Example 1, we already know that

$$A^{-1} = \begin{bmatrix} \tfrac{1}{3} & \tfrac{5}{3} & -\tfrac{2}{3} \\ \tfrac{1}{3} & -\tfrac{4}{3} & \tfrac{1}{3} \\ 0 & -2 & 1 \end{bmatrix}$$

Hence, it follows directly that

$$X = A^{-1}B = \begin{bmatrix} \tfrac{1}{3} & \tfrac{5}{3} & -\tfrac{2}{3} \\ \tfrac{1}{3} & -\tfrac{4}{3} & \tfrac{1}{3} \\ 0 & -2 & 1 \end{bmatrix} \begin{bmatrix} 1 \\ 2 \\ 3 \end{bmatrix} = \begin{bmatrix} \tfrac{5}{3} \\ -\tfrac{4}{3} \\ -1 \end{bmatrix}$$

Thus $x = \tfrac{5}{3}$, $y = -\tfrac{4}{3}$, $z = -1$ or, in the matrix notation of this chapter, $[x \ y \ z] = [\tfrac{5}{3} \ -\tfrac{4}{3} \ -1]$.

In a subsequent section, we shall return to a consideration of conditions that must be satisfied by a matrix if it is to be nonsingular. It will be discovered at that time that the nonvanishing of its determinant is both necessary and sufficient for a matrix to have this important property.

Problems 3.2

1. Determine A^{-1}, if

$$A = \begin{bmatrix} 1 & 1 & 1 \\ 1 & -1 & 2 \\ 2 & 3 & 1 \end{bmatrix}$$

2. Determine A^{-1}, if

$$A = \begin{bmatrix} 2 & 1 & 2 \\ 4 & 3 & 2 \\ 1 & 1 & 1 \end{bmatrix}$$

3. Determine A^{-1}, if

$$A = \begin{bmatrix} 1 & 0 & 0 & 0 \\ 2 & 1 & 1 & 1 \\ 1 & 1 & 2 & 1 \\ 1 & 0 & -1 & 1 \end{bmatrix}$$

4. Express A in Prob. 1 in two ways as a product of elementary matrices.
5. Express A in Prob. 2 in two ways as a product of elementary matrices.
6. Use an example to illustrate the fact that a product of elementary matrices is not necessarily an elementary matrix.

7. Interpret each elementary row operation on a matrix as a linear transformation of an appropriate vector space.
8. Verify that $P_{ij}^{-1} = P_{ij}$.
9. Verify that $[D_i(r)]^{-1} = D_i(r^{-1})$.
10. Verify that $[S_{ij}(r)]^{-1} = S_{ij}(-r)$.
11. Use a general 3×3 matrix A and the elementary 3×3 matrix P_{ij} to verify that $P_{ij}A$ is the matrix obtainable from A by interchanging its ith and jth rows. Generalize the proof for a general matrix A.
12. With A as in Prob. 11 and $D_i(r)$ the 3×3 elementary matrix of the second kind, verify that $D_i(r)A$ is the matrix obtainable from A by a multiplication of the ith row of A by r. Generalize the proof to apply to a general matrix A.
13. With A as in Prob. 11 and $S_{ij}(r)$ the 3×3 elementary matrix of the third kind, verify that $S_{ij}(r)A$ is the matrix obtainable from A by the addition of r times the elements of the jth row of A to the corresponding elements of its ith row.

14. Use the method of matrix inversion to solve the following system of equations:

$$\begin{aligned} 2x - y + z &= 3 \\ x + 2y - z &= -1 \\ 3x + y + 2z &= 2 \end{aligned}$$

15. Use the method of matrix inversion to solve the following system of equations:

$$x - 3y + 2z = 10$$
$$4x - 2y + z = 13$$
$$x + 4y + z = -5$$

16. If A and B are 2×2 matrices such that $AB = I$, prove directly that $BA = I$.

17. Decide whether the following vectors are linearly independent:
    (a) $(1,-1)$, $(2,3)$; (b) $(1,1,2)$, $(2,-1,1)$, $(3,2,1)$; (c) $(1,0,1,-1)$, $(0,1,-1,2)$, $(1,-1,2,1)$, $(3,-1,1,0)$.

18. Decide whether the following vectors are linearly independent:
    (a) $(2,1,2)$, $(-1,3,2)$, $(2,-2,1)$; (b) $(1,-1,2,-1)$, $(2,2,-1,3)$, $(1,1,-1,2)$, $(2,1,1,2)$.

19. Decide whether the following vectors are linearly independent, each vector being expressed in terms of the basis $\{\alpha_1, \alpha_2, \alpha_3, \alpha_4\}$ of the space:
    (a) $2\alpha_1 + \alpha_2 - \alpha_3 + \alpha_4$, $2\alpha_1 + \alpha_2 + 2\alpha_3 + 3\alpha_4$, $3\alpha_1 - \alpha_2 + \alpha_3 + 2\alpha_4$.
    (b) $2\alpha_1 + \alpha_2 + \alpha_4$, $3\alpha_1 - \alpha_2 + \alpha_3 + 2\alpha_4$, $-\alpha_1 - 3\alpha_2 + \alpha_3$.

20. If P_n designates the space of real polynomials of degree less than n in an indeterminate t, decide on the linear independence of the following vectors in the spaces indicated:
    (a) $2 + t - t^2$, $1 + 2t + t^2$, $3t - t^2$ in P_3.
    (b) $1 + 2t$, $2t + t^2$, $1 - 2t - 2t^2$ in P_3.
    (c) 1, $t - 1$, $t^2 + 1$, $t^3 - 2t$ in P_4.

21. If
$$A = \begin{bmatrix} 1 & 2 & -1 & 0 \\ 2 & 1 & 0 & 1 \\ 1 & -1 & 2 & 1 \end{bmatrix}$$

find the Hermite normal matrix N that is row-equivalent to A and a nonsingular matrix P such that $PA = N$.

22. If $AB = 0$, for a nonsingular matrix A, prove that $B = 0$.

23. Use elementary row operations to show that the matrix $\begin{bmatrix} a & b \\ c & d \end{bmatrix}$ is nonsingular if and only if $ad - bc \neq 0$.

24. Use the result in Prob. 23 to show that, if A is any 2×1 matrix and B is any 1×2 matrix, the product AB has no inverse.

25. Find two nonzero matrices of order 2 such that the square of each is the zero matrix.

26. A matrix is (lower) *triangular* if all its entries above the main diagonal are 0. Prove that the inverse of a nonsingular lower triangular matrix is also lower triangular.

27. Develop a procedure for matrix inversion, similar to the one described here, but which uses elementary operations on the columns instead of the rows. Then combine the two, transforming an "L array" $\begin{smallmatrix} I & \\ A & I \end{smallmatrix}$ into $\begin{smallmatrix} B & \\ I & C \end{smallmatrix}$ such that $BC = A^{-1}$. Check both procedures on any nonsingular matrix.

3.3 Determinants as Multilinear Functionals

In this section, we reintroduce the subject of determinants from a point of view that allows us to obtain some of the useful results more easily than with the classical approach. Before we do this, however, we include an important elementary theorem which can be readily obtained from the classical concept of a determinant. We first recall, from Sec. 1.6, that if $A = [a_{ij}]$ is an $n \times n$ matrix, we defined the *determinant* of A (denoted det A or $|A|$) as follows:

$$\det A = |A| = \sum_\pi (\operatorname{sgn} \pi)\, a_{1\pi(1)} a_{2\pi(2)} \cdots a_{n\pi(n)}$$

where the summation is over all permutations π of $\{1, 2, \ldots, n\}$.

Theorem 3.1 If A' is the transpose of the $n \times n$ matrix $A = [a_{ij}]$, then $\det A' = \det A$.

Proof

The proof of this theorem is not difficult but is nonetheless a bit "tricky." Hence a few preliminary comments are in order. We are trying to prove that

$$\begin{vmatrix} a_{11} & a_{12} & \cdots & a_{1n} \\ a_{21} & a_{22} & \cdots & a_{2n} \\ \vdots & & & \vdots \\ a_{n1} & a_{n2} & \cdots & a_{nn} \end{vmatrix} = \begin{vmatrix} a_{11} & a_{21} & \cdots & a_{n1} \\ a_{12} & a_{22} & \cdots & a_{n2} \\ \vdots & & & \vdots \\ a_{1n} & a_{2n} & \cdots & a_{nn} \end{vmatrix}$$

First, it is an immediate consequence of the definition recalled above that, *apart from the matter of algebraic sign to be attached to each term of the summations*, the expansions of the two determinants indicated above contain the same terms. In both cases, each of the $n!$ terms contains a factor a_{ij} with i an arbitrary *row* index and j an arbitrary *column* index, and with no repetitions. Hence the proof is simply a matter of settling the uncertainty of algebraic sign to be attached to each term. (An alternative proof of the theorem, based directly on this remark, is suggested in Prob. 14.) Our second comment is that, as π runs through all permutations of the set

SEC. 3.3 DETERMINANTS AS MULTILINEAR FUNCTIONALS

$\{1, 2, \ldots, n\}$, so does π^{-1}, and these two collections of $n!$ permutations are identical. This follows directly from the basic concept of a permutation. It will be convenient, in the proof below, to adopt the symbol Π for products just as the familiar Σ is used for sums. Now to the proof of the theorem!

A typical term of the summation for $\det A$ is $(\operatorname{sgn} \pi) \prod_{i=1}^{n} a_{i\pi(i)}$. If π^{-1} is the permutation inverse to π and $\pi(i) = j$, then $i = \pi^{-1}(j)$; and, for any fixed i, the index j takes on all integral values from 1 to n, as π runs through all permutations of $\{1, 2, \ldots, n\}$. As we turn to A', we first note that $A' = [a'_{ij}]$, where $a'_{ij} = a_{ji}$. In view of our second comment above, we may now express the determinant of A' so that $\det A' = \sum_{\pi^{-1}} (\operatorname{sgn} \pi^{-1}) a'_{1\pi^{-1}(1)} a'_{2\pi^{-1}(2)} \cdots a'_{n\pi^{-1}(n)}$. A typical term of this sum is $(\operatorname{sgn} \pi^{-1}) \prod_{j=1}^{n} a'_{j\pi^{-1}(j)} = (\operatorname{sgn} \pi^{-1}) \prod_{j=1}^{n} a_{\pi^{-1}(j)j} = (\operatorname{sgn} \pi) \prod_{i=1}^{n} a_{i\pi(i)}$, since $\operatorname{sgn} \pi^{-1} = \operatorname{sgn} \pi$ (Prob. 12). But this shows that a typical term of $\det A'$ is identical with the typical term of $\det A$; and so, since the summations are over the same set of permutations, the two sums are the same. The proof of the theorem is then complete. ∎

We now proceed to the new view of a determinant, announced at the beginning of this section, as a *multilinear* functional. A *linear* functional is a scalar-valued linear function on a vector space, whereas the classical definition describes a determinant as a scalar-valued function on square matrices —which may sometimes be considered vectors. However, it is easy to see from simple examples that, except for special cases, $|A + B| \neq |A| + |B|$, for $n \times n$ matrices A and B, from which we conclude that the determinant function is not linear, *as a function on matrices*. However, an $n \times n$ matrix may be regarded as a composite of n (horizontal) rows or n (vertical) columns of n-tuples, each of which can be considered an element of the coordinate vector space V_n. In Chap. 2, we adopted this point of view of a matrix on several occasions, but we give a further illustration. The matrix

$$\begin{bmatrix} 1 & 2 & -1 \\ 2 & 3 & -4 \\ 2 & 0 & 1 \end{bmatrix}$$

can be considered to be made up of the three row vectors (1 2 −1), (2 3 −4), (2 0 1) and the three column vectors (1 2 2), (2 3 0), (−1 −4 1). Although we are indifferent whether row vectors or column vectors are to be considered the components of a matrix, for definiteness in the present discussion we shall adopt the "row" point of view and regard an $n \times n$ matrix A as a composite of its n rows $X_1, X_2, \ldots, X_n$. It is clear that the function D, defined so that $D(X_1, X_2, \ldots, X_n) = \det A$, is a scalar-valued function on the product set $V_n \times V_n \times \cdots \times V_n$, with n copies of V_n, and the following elementary theorem shows that D is linear in each argument.

Theorem 3.2 *If $D(X_1, X_2, \ldots, X_n) = \det A$, where A is the matrix whose ordered rows are $X_1, X_2, \ldots, X_n \in V_n$, the function D is linear in each argument (or multilinear). That is, $D(X_1, \ldots, X_{i-1}, cX_i + d\tilde{X}_i, X_{i+1}, \ldots, X_n) = cD(X_1, \ldots, X_{i-1}, X_i, X_{i+1}, \ldots, X_n) + dD(X_1, \ldots, X_{i-1}, \tilde{X}_i, X_{i+1}, \ldots, X_n)$, $i = 1, 2, \ldots, n$, for arbitrary $c, d \in \mathbf{R}$, and $\tilde{X}_i \in V_n$.*

Proof
If $X_i = [a_{i1} a_{i2} \cdots a_{in}]$ and $\tilde{X}_i = [\tilde{a}_{i1} \tilde{a}_{i2} \cdots \tilde{a}_{in}]$, then $cX_i + d\tilde{X}_i = [ca_{i1} + d\tilde{a}_{i1} \quad ca_{i2} + d\tilde{a}_{i2} \cdots ca_{in} + d\tilde{a}_{in}]$. But now $D(X_1, \ldots, X_{i-1}, cX_i + d\tilde{X}_i, X_{i+1}, \ldots, X_n) = \sum_\pi (\text{sgn } \pi) a_{1\pi(1)} a_{2\pi(2)} \cdots (ca_{i\pi(i)} + d\tilde{a}_{i\pi(i)}) \cdots a_{n\pi(n)}$
$= \sum_\pi (\text{sgn } \pi) a_{1\pi(1)} a_{2\pi(2)} \cdots ca_{i\pi(i)} \cdots a_{n\pi(n)} + \sum_\pi (\text{sgn } \pi) a_{1\pi(1)} a_{2\pi(2)} \cdots d\tilde{a}_{i\pi(i)} \cdots a_{n\pi(n)}$ and, on factoring out c from the first and d from the second of these summations (Prob. 13), we obtain the desired result. ∎

Corollary *If $X_j \in \{X_1, X_2, \ldots, X_n\}$, then $D(X_1, \ldots, X_{i-1}, cX_i + dX_j, X_{i+1}, \ldots, X_n) = cD(X_1, \ldots, X_{i-1}, X_i, X_{i+1}, \ldots, X_n)$, provided $j \neq i$.*

Proof
This is a consequence of Theorem 3.2 and Property 3 of a determinant, as described in Sec. 1.6. ∎

We now state that the function D, as defined in Theorem 3.2, has the following three properties (which we do not claim to be independent):

(1) D is multilinear, i.e. linear in each of its n arguments.
(2) $D(X_1, X_2, \ldots, X_n) = 0$, if $X_i = X_j$, for $i \neq j$, and $i, j = 1, 2, \ldots, n$.
(3) $D(E_1, E_2, \ldots, E_n) = 1$, where $\{E_1, E_2, \ldots, E_n\}$ is the familiar basis of V_n.

The verification of Property 1 was done in Theorem 3.2, Property 2 is a consequence of the third property of a determinant in Sec. 1.6, and Property 3 follows very easily (Prob. 11) from the definition of a determinant. A fourth property of D is a simple deduction from (1) and (2) and is now stated.

(4) $D(X_1, X_2, \ldots, X_n)$ is altered only in algebraic sign if two of the vectors $X_1, X_2, \ldots, X_n$ are interchanged.

Proof
For $D(X_1, \ldots, X_i, \ldots, X_j, \ldots, X_n) + D(X_1, \ldots, X_j, \ldots, X_i, \ldots, X_n) = D(X_1, \ldots, X_i + X_j, \ldots, X_j + X_i, \ldots, X_n)$, by (1) and (2). We now use (2) to conclude that $D(X_1, \ldots, X_i, \ldots, X_j, \ldots, X_n) = -D(X_1, \ldots, X_j, \ldots, X_i, \ldots, X_n)$. ∎

It is an interesting fact, and one which we shall find useful in a later discussion, that not only are Properties 1, 2, and 3 possessed by the function D, but D is uniquely characterized by these three properties, as asserted in the following theorem.

Theorem 3.3 *If D is a function, defined on the set of all n-tuples of vectors from V_n so that Properties 1, 2, and 3 are possessed, then $D(X_1, X_2, \ldots, X_n) = \det A$, where A is the matrix whose ordered rows (columns) are $X_1, X_2, \ldots, X_n$.*

Proof
If we express each X_i in terms of the $\{E_i\}$ basis of V_n, the result is $X_i = a_{i1}E_1 + a_{i2}E_2 + \cdots + a_{in}E_n$, $i = 1, 2, \ldots, n$. Then, in view of the multilinear nature of D in Property 1, we may express $D(X_1, X_2, \ldots, X_n)$ as a linear combination of n^n terms, a typical one of which has the form $a_{1i_1}a_{2i_2} \cdots a_{ni_n} D(E_{i_1}, E_{i_2}, \ldots, E_{i_n})$, where $i_1, i_2, \ldots, i_n$ are subscripts, *in general not distinct*, from the set $\{1, 2, \ldots, n\}$. However, in view of Property 2, those terms in which these subscripts are *not* all distinct may be eliminated, and there will remain at most $n!$ nonzero terms of the above form in which $i_1, i_2, \ldots, i_n$ are *all distinct* and hence arrangements of $1, 2, \ldots, n$. We may now use Property 4, which was shown to be a consequence of the other three properties, to replace $D(E_{i_1}, E_{i_2}, \ldots, E_{i_n})$ by $\pm D(E_1, E_2, \ldots, E_n)$, the algebraic sign being $+$ or $-$ according as the arrangement $i_1 i_2 \cdots i_n$ contains an even or odd number of inversions. But then $D(E_{i_1}, E_{i_2}, \ldots, E_{i_n}) = (\text{sgn } \pi) D(E_1, E_2, \ldots, E_n)$, where π is the permutation $i_1 i_2 \cdots i_n$. We now use Property 3 to see that $D(E_1, E_2, \ldots, E_n) = 1$ and conclude from this that $D(X_1, X_2, \ldots, X_n) = \sum_\pi (\text{sgn } \pi) a_{1\pi(1)} a_{2\pi(2)} \cdots a_{n\pi(n)} = \det A$, where A is the matrix whose ordered rows are $X_1, X_2, \ldots, X_n$, and the summation is taken over all permutations π of $1, 2, \ldots, n$. In view of Theorem 3.1, the vectors $X_1, X_2, \ldots, X_n$ can be considered the ordered columns of A, if desired, and the proof of the theorem is complete. ∎

Although the proof of Theorem 3.3 is complete, it may help to clarify some of the reduction involved if we repeat the arguments for the case $n = 2$. In this case, we are interested in $D(X_1, X_2)$, where $X_1 = [a_{11} \quad a_{12}]$ and $X_2 = [a_{21} \quad a_{22}]$. In terms of the basis $\{E_1, E_2\}$ of V_2, we have $X_1 = a_{11}E_1 + a_{12}E_2$

and $X_2 = a_{21}E_1 + a_{22}E_2$. Then, the multilinear (bilinear) Property 1 makes it possible to write $D(X_1,X_2) = D(a_{11}E_1 + a_{12}E_2, a_{21}E_1 + a_{22}E_2) = a_{11}D(E_1, a_{21}E_1 + a_{22}E_2) + a_{12}D(E_2, a_{21}E_1 + a_{22}E_2) = a_{11}[a_{21}D(E_1,E_1) + a_{22}D(E_1,E_2)] + a_{12}[a_{21}D(E_2,E_1) + a_{22}D(E_2,E_2)] = a_{11}a_{21}D(E_1,E_1) + a_{11}a_{22}D(E_1,E_2) + a_{12}a_{21}D(E_2,E_1) + a_{12}a_{22}D(E_2,E_2)$. But, by Property 2, $D(E_1,E_1) = D(E_2,E_2) = 0$, and, by Property 3, $D(E_1,E_2) = 1$, whereas Property 4 leads to $D(E_2,E_1) = -D(E_1,E_2) = -1$. Hence, $D(X_1,X_2) = a_{11}a_{22}D(E_1,E_2) - a_{12}a_{21}D(E_1,E_2) = a_{11}a_{22} - a_{12}a_{21} = \det A$, where $A = \begin{bmatrix} a_{11} & a_{12} \\ a_{21} & a_{22} \end{bmatrix}$.

Inasmuch as we have been led to the result in Theorem 3.3 *entirely on the assumption that Properties 1, 2, and 3 are held by D*, it can be asserted that the function D, defined so that $D(X_1,X_2, \ldots, X_n) = \det A$, is the *unique* function on the set of n-tuples or vectors from V_n that possesses these properties. This allows us to make the following equivalent definition of the *determinant* function:

> The function D, defined on $V_n \times V_n \times \cdots \times V_n$ (n copies) so as to possess Properties 1, 2, and 3, is said to be the determinant function. In particular, if $X_1, X_2, \ldots, X_n$ are the constituent row (column) vectors of an $n \times n$ matrix A, then $D(X_1,X_2, \ldots, X_n) = \det A = |A|$.

Problems 3.3

1. Use Property 1 to express $D(X_1 + X_2, Y_1 + Y_2)$ as a sum of four terms.
2. Use Property 1 to express $D(X_1 + X_2, Y_1 + Y_2, Z_1 + Z_2)$ as a sum of eight terms.
3. If $X = [x_1 \ x_2]$ and $Y = [y_1 \ y_2]$, verify directly that $D(X,Y) = x_1y_2 - x_2y_1$ has Properties 1, 2, and 3 of a determinant function.
4. Use Properties 1, 2, and 3 to develop the expansion for $D(X,Y,Z)$, where $X = [x_1 \ x_2 \ x_3]$, $Y = [y_1 \ y_2 \ y_3]$, $Z = [z_1 \ z_2 \ z_3]$.
5. Verify that the expansion for $D(X,Y,Z)$, as obtained in Prob. 4, has Properties 1, 2, and 3 of a determinant function.
6. Use Properties 1, 2, and 3 to see that $D(X_1,X_2, \ldots, X_n) = 0$, if $X_j = 0$ for $j = 1, 2, \ldots, n$. [Cf. Theorem 7.3 (Chap. 1).]

7. If the vectors $X_1, X_2, \ldots, X_n \in V_n$ are linearly dependent, use Properties 1, 2, and 3 to prove that $D(X_1, X_2, \ldots, X_n) = 0$.
8. Verify that (2) is a consequence of (4) as properties of D.
9. Explain why there are at most $n!$ nonzero terms of the form $a_{1i_1} a_{2i_2} \cdots a_{ni_n} D(E_{i_1}, E_{i_2}, \ldots, E_{i_n})$ in the expansion of $D(X_1, X_2, \ldots, X_n)$.
10. Explain why Property 1 of D requires that $D(X_1, \ldots, aX_i, \ldots, X_n) = aD(X_1, \ldots, X_i, \ldots, X_n)$, for any $a \in \mathbf{R}$.
11. Use the definition in Sec. 1.6 to verify that $\det I_n = 1$, for any n.
12. If π is any permutation of $1, 2, \ldots, n$, explain why $\operatorname{sgn} \pi^{-1} = \operatorname{sgn} \pi$.
13. Complete the computation as suggested in the proof of Theorem 3.2.
14. The definition of $\det A$, as given in Sec. 1.6, requires that $\det A' = \sum_{\pi} (\operatorname{sgn} \pi) a_{i_1 1} a_{i_2 2} \cdots a_{i_n n}$, where π is a typical arrangement $i_1 i_2 \cdots i_n$ of $1, 2, \ldots, n$. If the factors in each term are rearranged so that the row indices are in natural order, let the resulting product be designated $a_{1j_1} a_{2j_2} \cdots a_{nj_n}$, with π' indicating the arrangement $j_1 j_2 \cdots j_n$. Then prove that π and π' have the same parity and see that this gives a proof of Theorem 3.1.

15. If $X = [x_1 \ x_2]$ and $Y = [y_1 \ y_2]$ are considered geometric plane vectors emanating from the origin of a cartesian coordinate system, verify that $|D(X_1, X_2)|$ is the area of the parallelogram determined by X and Y. What is the corresponding result in 3-space?
16. Use the procedure in this section, *rather than the formula*, to determine $D(X, Y)$, where $X = [1 \ 2]$ and $Y = [-2 \ 1]$.
17. Use the procedure in this section, *rather than the formula*, to determine $D(X, Y, Z)$, where $X = [1 \ 1 \ 1]$, $Y = [2 \ -1 \ 1]$, $Z = [2 \ 0 \ 1]$.
18. Prove that $D(X_1 + aX_2, X_2 + bX_3, X_3 + cX_1) = (abc + 1) D(X_1, X_2, X_3)$, for $a, b, c \in \mathbf{R}$.
19. If $X_i = [a_{i1} \ a_{i2} \ a_{i3} \ a_{i4} \ a_{i5} \ a_{i6}]$, $i = 1, 2, 3, 4, 5, 6$, are vectors in V_6, determine the sign to be attached to the following terms of $D(X_1, X_2, X_3, X_4, X_5, X_6)$: (a) $a_{11} a_{23} a_{35} a_{42} a_{56} a_{64}$; (b) $a_{16} a_{23} a_{32} a_{41} a_{55} a_{64}$.
20. A *permanent* P is similar to a determinant except that the sign factor $(\operatorname{sgn} \pi)$ does not occur. Decide which of the four basic properties of a determinant function hold for P.

21. Refer to Prob. 20 and find $P(X,Y,Z)$, where $X = [1 \quad 2 \quad 1]$, $Y = [2 \quad -1 \quad 2]$, $Z = [0 \quad 1 \quad 2]$.
22. If $X_i = c_{i1}Y_1 + c_{i2}Y_2 + \cdots + c_{in}Y_n$, for real numbers $c_{i1}, c_{i2}, \ldots, c_{in}$ and vectors $Y_1, Y_2, \ldots, Y_n$, prove that $D(X_1, X_2, \ldots, X_n) = \sum_\pi (\text{sgn } \pi) c_{1i_1} c_{2i_2} \cdots c_{ni_n} D(Y_1, Y_2, \ldots, Y_n)$, the summation being taken over all permutations $\pi (= i_1 i_2 \cdots i_n)$ of $1, 2, \ldots, n$.
23. Refer to Prob. 22, let $C_i = [c_{i1} \quad c_{i2} \quad \cdots \quad c_{in}]$, $i = 1, 2, \ldots, n$, and prove that $D(X_1, X_2, \ldots, X_n) = D(C_1, C_2, \ldots, C_n) D(Y_1, Y_2, \ldots, Y_n)$.

3.4 An Alternative Method for Evaluating a Determinant

At this point it may be well to recall that our principal interest in determinants arises from their connection with the solutions of systems of linear equations. These linear systems arise, in the present context, mostly from attempts to check the linear independence of n vectors in an n-dimensional vector space. In other words, determinants are involved in a search for a basis of a vector space. It will be remembered, from the earlier discussion in Sec. 1.6, that the method of expansion of a determinant by minors and cofactors was quite unwieldy if the order of the associated matrix is above 3, except for special cases. In this section we shall present an alternative method for evaluating a determinant, making use of the properties of elementary matrices to reduce the matrix to a form from which the determinant can be easily calculated. But first we prove a theorem which is of importance in its own right. If $AB = C$, for square matrices A, B, C, it is only natural to inquire how det C is related to det A and det B. This query is answered in Theorem 4.1, after the proof of a preliminary lemma which may be seen to be a generalization of Prob. 23 of Sec. 3.3.

Lemma If f is a multilinear functional on $V_n \times V_n \times \cdots \times V_n$ (n copies), that also possesses Property 2 of the determinant function D, then $f(X_1, X_2, \ldots, X_n) = D(X_1, X_2, \ldots, X_n) f(E_1, E_2, \ldots, E_n)$, where $\{E_i\}$ is the familiar basis of V_n.

Proof
If $f(E_1, E_2, \ldots, E_n) = 1$, then $f = D$ and the lemma is proved. If, on the other hand, $f(E_1, E_2, \ldots, E_n) = k \neq 1$, let us consider the function F, defined on the same domain as f, such that $F(X_1, X_2, \ldots, X_n) = [D(X_1, X_2, \ldots, X_n) - f(X_1, X_2, \ldots, X_n)]/(1 - k)$. Since F clearly possesses Property 3 of a determinant function, as well as Properties 1 and 2, the

SEC. 3.4 AN ALTERNATIVE METHOD FOR EVALUATING A DETERMINANT 161

unique characterization of D by these properties (Theorem 3.3) leads again to $F = D$. But now, on solving this equality for $f(X_1, X_2, \ldots, X_n)$, we find that $f(X_1, X_2, \ldots, X_n) = D(X_1, X_2, \ldots, X_n)k = D(X_1, X_2, \ldots, X_n)f(E_1, E_2, \ldots, E_n)$, as asserted. ∎

Theorem 4.1 *If A and B are $n \times n$ matrices, det AB = (det A)(det B).*

Proof
Let $A = [a_{ij}]$ and $B = [b_{ij}]$ be any two $n \times n$ matrices, the ordered *columns* of B being the vectors $X_1, X_2, \ldots, X_n$. There is no loss in generality if we regard A as the matrix of a linear operator T on V_n, with respect to the $\{E_i\}$ basis of this space. We now define the function f so that $f(X_1, X_2, \ldots, X_n) = D(\mathsf{T}X_1, \mathsf{T}X_2, \ldots, \mathsf{T}X_n)$, noting that f inherits the characteristic Properties 1 and 2 of D. But now the lemma is applicable and so $f(X_1, X_2, \ldots, X_n) = D(X_1, X_2, \ldots, X_n)f(E_1, E_2, \ldots, E_n)$, whence $D(\mathsf{T}X_1, \mathsf{T}X_2, \ldots, \mathsf{T}X_n) = D(X_1, X_2, \ldots, X_n)D(\mathsf{T}E_1, \mathsf{T}E_2, \ldots, \mathsf{T}E_n)$. Since A represents the operator T, the matrix whose ordered columns are $\mathsf{T}X_1, \mathsf{T}X_2, \ldots, \mathsf{T}X_n$ is AB; the matrix whose ordered columns are $\mathsf{T}E_1, \mathsf{T}E_2, \ldots, \mathsf{T}E_n$ is A (Prob. 13). We know from Theorem 3.1 and the discussion in Sec. 3.3 that $D(X_1, X_2, \ldots, X_n) = $ det B, and the above equation now yields det $(AB) = $ (det B)(det A) = (det A)(det B), as desired. ∎

There are many proofs of Theorem 4.1, one of which can be obtained (rather laboriously) from the classical definition of a determinant as given in Sec. 1.6. Another proof is essentially contained in Prob. 23 of Sec. 3.3. In the notation of that problem, we note that $A = BC$, where A, B, C are, respectively, the matrices whose row vectors are X_i, Y_i, C_i. The details of this proof are left to the reader (Prob. 12). We now show how it is possible to use Theorem 4.1 to evaluate determinants by a method that is often more palatable than the one based on the classical definition.

In order to understand better the procedure we are about to outline, the reader should recall (see Sec. 3.2) the relationship that exists between the elementary matrices P_{ij}, $D_i(r)$, and $S_{ij}(r)$ and the identity matrix I of the same order. It is easy to see (Prob. 11 of Sec. 3.3) and assumed to be well known that det $I = 1$, for any identity matrix I. It follows immediately, from Property 4 of a determinant function, that det $P_{ij} = -1$; from Property 1 that det $D_i(r) = r$; and, from Properties 1 and 2, that det $S_{ij}(r) = 1$. The next theorem, on which our process for the evaluation of a determinant is to be based, is then a consequence of these observations and Theorem 4.1.

Theorem 4.2 *If an elementary (row or column) operation is performed on a square matrix, the value of its determinant is multiplied by -1, r, or 1 according as the operation is of Type 1,*

Type 2, *or Type* 3. In terms of elementary matrices, this result may be stated as follows: $\det P_{ij}A = \det AP_{ij} = -\det A$; $\det D_i(r)A = \det AD_i(r) = r \det A$; $\det S_{ij}(r)A = \det AS_{ij}(r) = \det A$.

Since the effect of an elementary operation on the determinant of a matrix is established by this theorem, any matrix can be transformed by elementary operations to some simple form whose determinant is easily determined, and from this the determinant of the original matrix can be obtained. For example, we could reduce the matrix to Hermite normal form (which is diagonal if the original matrix is square, as it is for the cases under discussion), but it is usually not necessary to carry the reduction so far. For example, if we reduce a matrix to either *upper* or *lower triangular* form (i.e., all entries respectively *below* or *above* the main diagonal are 0), the determinant of such a simplified matrix is easily seen to be the product of the diagonal entries. We shall illustrate this method for evaluating a determinant with some examples, the symbols (1), (2), (3) above the arrows indicating, as in Sec. 3.2, which types of operations were used at the various stages. Although column operations are quite permissible (see Sec. 3.2), we shall always understand row operations in the examples.

EXAMPLE 1

Find $\begin{vmatrix} 3 & 2 & 1 \\ 1 & 4 & 1 \\ 2 & 1 & 4 \end{vmatrix}$.

Solution

We first indicate the sequence of elementary row operations used in the reduction of the matrix.

$$\begin{bmatrix} 3 & 2 & 1 \\ 1 & 4 & 1 \\ 2 & 1 & 4 \end{bmatrix} \xrightarrow{(1)} \begin{bmatrix} 1 & 4 & 1 \\ 3 & 2 & 1 \\ 2 & 1 & 4 \end{bmatrix} \xrightarrow{(3)(3)} \begin{bmatrix} 1 & 4 & 1 \\ 0 & -10 & -2 \\ 0 & -7 & 2 \end{bmatrix}$$

and it is clear, if the expansion is made in terms of the cofactors of the first column, that the determinant of the final matrix is $1(-20 - 14) = -34$. Since we used one elementary operation of Type 1 (which changes the sign of a determinant) and two elementary operations of Type 3 (which leaves a determinant unchanged) it follows that the determinant of the original matrix must have been $-(-34) = 34$.

SEC. 3.4 AN ALTERNATIVE METHOD FOR EVALUATING A DETERMINANT 163

EXAMPLE 2

Find $\begin{vmatrix} 1 & 3 & 4 & 0 \\ 1 & 2 & 0 & 1 \\ 0 & 1 & 5 & 6 \\ 1 & 2 & 3 & 4 \end{vmatrix}.$

Solution
As in Example 1, we reduce the original matrix to a simple form (in this case upper triangular), as indicated below:

$$\begin{bmatrix} 1 & 3 & 4 & 0 \\ 1 & 2 & 0 & 1 \\ 0 & 1 & 5 & 6 \\ 1 & 2 & 3 & 4 \end{bmatrix} \xrightarrow{(3)(3)} \begin{bmatrix} 1 & 3 & 4 & 0 \\ 0 & -1 & -4 & 1 \\ 0 & 1 & 5 & 6 \\ 0 & -1 & -1 & 4 \end{bmatrix}$$

$$\xrightarrow{(3)(3)} \begin{bmatrix} 1 & 3 & 4 & 0 \\ 0 & -1 & -4 & 1 \\ 0 & 0 & 1 & 7 \\ 0 & 0 & 3 & 3 \end{bmatrix} \xrightarrow{(3)} \begin{bmatrix} 1 & 3 & 4 & 0 \\ 0 & -1 & -4 & 1 \\ 0 & 0 & 1 & 7 \\ 0 & 0 & 0 & -18 \end{bmatrix}$$

The determinant of this final upper triangular matrix is the product of its diagonal entries: $(1)(-1)(1)(-18) = 18$. Since all elementary operations were of Type 3, the desired determinant (of the original matrix) is also 18.

Although we have used somewhat simple matrices of orders 3 and 4 for illustrative purposes, it should be clear that the superiority of this method over that given in Sec. 1.6 increases with the order of the matrix. Elementary operations of Type 2 were not used in either example, but it should be noted that *if such operations are used* the determinant of the original matrix may differ from that of the final matrix by a factor other than -1. Before our next very important theorem, we need a definition.

Definition   If $A = [a_{ij}]$, with A_{ij} the cofactor of a_{ij}, the classical adjoint of A, designated adj A, is the matrix $[A'_{ij}]$, where $A'_{ij} = A_{ji}$, $i, j = 1, 2, \ldots, n$.

In other words, to obtain adj A from A, replace each entry of A by its cofactor and then transpose the resulting matrix.

Theorem 4.3   An $n \times n$ matrix A is nonsingular if and only if $|A| \neq 0$. If $|A| \neq 0$, $A^{-1} = (1/|A|)$ adj A.

Proof
If A is nonsingular, A^{-1} exists such that $AA^{-1} = I$. But now, by Theorem

4.1, we see that $|A| \, |A^{-1}| = |I| = 1$, whence $|A| \neq 0$. Conversely, if $|A| \neq 0$, we shall show that $(1/|A|)$ adj A has the property required of A^{-1}. Thus, $A(\text{adj } A) = [c_{ki}]$, where $c_{ki} = \sum_{j=1}^{n} a_{kj} A'_{ji} = \sum_{j=1}^{n} a_{kj} A_{ij}$. By Property 4 in Sec. 1.6, $c_{ki} = 0$, if $k \neq i$; and the classical definition of a determinant implies that $c_{ii} = |A|$, for $i = 1, 2, \ldots, n$. Hence,

$$A(\text{adj } A) = \begin{vmatrix} |A| & 0 & 0 & \cdots & 0 \\ 0 & |A| & 0 & \cdots & 0 \\ \cdots & \cdots & \cdots & \cdots & \cdots \\ 0 & 0 & 0 & \cdots & |A| \end{vmatrix} = |A|I$$

so that $A[(1/|A|) \text{ adj } A] = I$, as asserted. A similar proof (which is not really needed) shows that $[(1/|A|) \text{ adj } A]A = I$, and the uniqueness of an inverse leads to the conclusion that $A^{-1} = (1/|A|)$ adj A. Hence A is nonsingular. ∎

Although Theorem 4.3 supplies a formula for the inverse of any nonsingular matrix, this method for finding an inverse has the same undesirable features as those inherent in the evaluation of a determinant by the classical definition. In other words, except for matrices of small order, the more practical method of matrix inversion is that given in Sec. 3.2. There are many methods of inverting a matrix, and in any given instance the technique most efficient will depend on the nature of the matrix. Books have been written on this subject, with many paying special attention to aspects of a computational nature. (See Selected Readings at the end of the chapter.) However, we shall not pursue the topic further: for our purposes here, it is of more importance to know *if* the inverse of a matrix exists rather than how to determine the inverse of a nonsingular matrix. The following theorem is a very important one, because it focuses attention on the point of most interest in our entire discussion of determinants.

Theorem 4.4 *A set of n vectors from V_n are linearly independent if and only if $|A| \neq 0$, where A is the matrix whose rows (columns) may be identified with these vectors.*

Proof
That we are indifferent whether the given vectors constitute the rows or columns of A is a consequence of Theorem 3.1. If $|A| \neq 0$, it is stated in Theorem 7.3 (Chap. 1) that the columns (and hence also the rows) of A are linearly independent vectors of V_n. An alternative argument was requested in Prob. 7 of Sec. 3.3, but we outline the details. Thus, let us assume that the constituent row (or column) vectors $X_1, X_2, \ldots, X_n$ of A are linearly dependent. Then, for some j, $1 \leq j \leq n$, there exist real numbers $c_1, c_2, \ldots, c_n$, not all 0, such that $X_j = c_1 X_1 + c_2 X_2 + \cdots + c_{j-1} X_{j-1}$ [see

SEC. 3.4 AN ALTERNATIVE METHOD FOR EVALUATING A DETERMINANT

Theorem 7.1 (Chap. 1)]. The multilinear property of the determinant function D allows us to write $D(X_1, \ldots, X_{j-1}, X_j, X_{j+1}, \ldots, X_n) = D(X_1, \ldots, X_{j-1}, c_1X_1 + c_2X_2 + \cdots + c_{j-1}X_{j-1}, X_{j+1}, \ldots, X_n) = c_1 D(X_1, \ldots, X_{j-1}, X_1, X_{j+1}, \ldots, X_n) + c_2 D(X_1, \ldots, X_{j-1}, X_2, X_{j+1}, \ldots, X_n) + \cdots + c_{j-1} D(X_1, \ldots, X_{j-1}, X_{j-1}, X_{j+1}, \ldots, X_n)$. But each of these terms is 0, by Property 2 of D, and so $D(X_1, X_2, \ldots, X_n) = |A| = 0$. It follows that, if $|A| \neq 0$, the row (or column) vectors comprising A must be linearly independent. Conversely, let $|A| \neq 0$, where $X_1, X_2, \ldots, X_n$ are the constituent rows of A. If some X_j is a linear combination of the other row vectors, it is an elementary consequence of Properties 1 and 2 of the determinant function that $D(X_1, X_2, \ldots, X_n) = 0$. Thus, if the (row) rank of A is less than n, $\det A = |A| = 0$. We must then show that, if the rank of A is n, $\det A \neq 0$. However, if this rank is n, the vectors $X_1, X_2, \ldots, X_n$ form a basis for the row space; and, for each k, $1 \leq k \leq n$, $E_k = b_{k1}X_1 + b_{k2}X_2 + \cdots + b_{kn}X_n$ with $b_{k1}, b_{k2}, \ldots, b_{kn} \in \mathbf{R}$. If we define B to be the matrix $[b_{ij}]$, these expressions for E_k imply that BA is the matrix whose rows are $E_1, E_2, \ldots, E_n$. But $D(E_1, E_2, \ldots, E_n) = 1$ and, by Theorem 4.1, $|BA| = \det BA = (\det B)(\det A)$. Hence $(\det A)(\det B) = 1$, and so $\det A = |A| \neq 0$, as required. In view of the remark made in the introductory sentence of the proof, the argument is now complete. ∎

With this result, we have completed our principal objective in this chapter. We are now in a position *to find the inverse of any nonsingular matrix* and, if desired, to use the matrix-inversion method to solve a system of linear equations having a nonsingular coefficient matrix and also *to test the linear independence of any collection of vectors drawn from a vector space.*

Problems 3.4

1. Determine $|A|$, $|B|$, and $|AB|$, checking Theorem 4.1, where

    (a) $A = \begin{bmatrix} 1 & -1 & 1 \\ 2 & 2 & 1 \\ 1 & 1 & 1 \end{bmatrix} \quad B = \begin{bmatrix} 2 & 1 & 2 \\ 1 & -1 & 1 \\ 2 & 3 & 4 \end{bmatrix}$

    (b) $A = \begin{bmatrix} 2 & 0 & 1 \\ 1 & 2 & -1 \\ 2 & 0 & 4 \end{bmatrix} \quad B = \begin{bmatrix} 2 & 4 & -1 \\ 2 & 5 & 0 \\ 2 & 1 & 3 \end{bmatrix}$

2. Use a determinant to check the linear independence of the following vectors of V_4: (a) $\{(1,0,1,1), (1,1,-1,2), (0,1,1,1), (2,-1,2,-2)\}$; (b) $\{(1,2,-2,3), (-1,2,4,5), (1,2,1,1), (2,1,1,4)\}$.

3. Expand $\{(1,2,-1,2), (3,2,-1,2)\}$ to a basis of V_4.
4. *Do not solve*, but decide whether the following system of equations has a nontrivial solution:

$$2x - 3y + 3z = 0$$
$$x + 2y - 4z = 0$$
$$3x - 4y + z = 0$$

5. *Do not solve*, but decide whether the following system of equations has a nontrivial solution:

$$x - 4y + 2z + 2w = 0$$
$$3x - y + z - w = 0$$
$$x + 2y - z - w = 0$$
$$2x - y + z = 0$$

6. Decide whether the vector $(1,3,-1)$ is in the space $\mathcal{L}\{(2,1,-1), (1,3,2)\}$. Find two more vector triples, one of which is in the space and one of which is not.
7. Use matrix inversion to solve the following system of equations:

$$x + 2y + 3z = 7$$
$$2x + y + 2z = 6$$
$$-2x + y - z = -4$$

8. Find adj A, if

(a) $A = \begin{bmatrix} 1 & 2 & -1 \\ 0 & 1 & 3 \\ 2 & 3 & -3 \end{bmatrix}$ (b) $A = \begin{bmatrix} 2 & 4 & 1 \\ 1 & -1 & 0 \\ 3 & 4 & 1 \end{bmatrix}$

(c) $A = \begin{bmatrix} 2 & -1 & 2 \\ 6 & 1 & 3 \\ 1 & 2 & -1 \end{bmatrix}$

In each case, find A^{-1} and check, provided $|A| \neq 0$.

9. Do the necessary computations to find the determinant of each of the elementary matrices.
10. Explain why the determinant of a matrix in upper or lower triangular form is the product of its diagonal entries.
11. If $AA' = I$, for any square matrix A, explain why $|A| = \pm 1$.
12. If you have not already done so, work Prob. 23 of Sec. 3.3 to obtain another proof of Theorem 4.1, as suggested in the text.

13. Substantiate the assertions, made in the proof of Theorem 4.1, concerning the matrices whose ith columns ($i = 1, 2, \ldots, n$) are TX_i and TE_i, respectively.

14. Explain why (a) $|rA| = r^n|A|$, for any $r \in \mathbf{R}$ and any $n \times n$ matrix A; (b) det $A^{-1} = (\det A)^{-1}$, for any nonsingular matrix A.

15. If A, B, C are nonzero matrices such that $AB = O$, and $C^n = O$ for some positive integer $n > 1$, A and B are called matrix *divisors of zero* and C is said to be *nilpotent*. Prove that no matrix divisor of zero nor any nilpotent matrix is nonsingular.

16. Without evaluating, show that

    (a) $\begin{vmatrix} 1 & a & bc \\ 1 & b & ca \\ 1 & c & ab \end{vmatrix} = \begin{vmatrix} 1 & a & a^2 \\ 1 & b & b^2 \\ 1 & c & c^2 \end{vmatrix}$

    (b) $\begin{vmatrix} a-b & b-c & c-a \\ 1 & 1 & 1 \\ a & b & c \end{vmatrix} = \begin{vmatrix} a & b & c \\ 1 & 1 & 1 \\ b & c & a \end{vmatrix}$

17. Determine a necessary and sufficient condition in order that $|A + B| = |A| + |B|$, for 2×2 matrices $A = [a_{ij}]$ and $B = [b_{ij}]$.

18. Prove that $\begin{bmatrix} x & y \\ 0 & z \end{bmatrix}$ and $\begin{bmatrix} x' & y' \\ 0 & z' \end{bmatrix}$ are commutative matrices (under multiplication) if and only if $\begin{vmatrix} y & x-z \\ y' & x'-z' \end{vmatrix} = 0$.

19. Use the procedure based on Theorem 4.2 to prove that the determinant of each of the following matrices is 0:

    (a) $\begin{bmatrix} 2 & 3 & 1 \\ 2 & 1 & 1 \\ 1 & -1 & 2 \end{bmatrix}$
    (b) $\begin{bmatrix} a+1 & 3a & b+2a & b+1 \\ 2b & b+1 & 2-b & 1 \\ a+2 & 0 & 1 & a+3 \\ b-1 & 1 & a+2 & a+b \end{bmatrix}$

20. If A, B, C, O are square submatrices of compatible orders, prove that $\begin{vmatrix} A & O \\ C & B \end{vmatrix} = |A| |B|$, in which O denotes a zero submatrix.

21. Let X_1, X_2, X_3 be linearly independent vectors, and define $X_1' = a_1 X_1 + b_1 X_2 + c_1 X_3$, $X_2' = a_2 X_1 + b_2 X_2 + c_2 X_3$, $X_3' = a_3 X_1 + b_3 X_2 + c_3 X_3$. Then prove that X_1', X_2', X_3' are linearly independent if and only if

$$\begin{vmatrix} a_1 & b_1 & c_1 \\ a_2 & b_2 & c_2 \\ a_3 & b_3 & c_3 \end{vmatrix} \neq 0$$

22. Evaluate the following determinant:

$$\begin{vmatrix} 1 & 0 & 0 & 1 & -1 & 1 \\ 0 & 0 & 1 & 0 & 2 & 0 \\ 1 & 0 & 0 & 0 & 0 & 1 \\ 0 & 0 & 0 & 1 & 1 & 0 \\ 1 & 0 & 2 & 0 & 0 & 1 \\ 0 & 0 & 1 & 1 & 0 & 0 \end{vmatrix}$$

23. Show that the matrix-inversion method, involving the classical adjoint of the matrix, does, in fact, give Cramer's rule.

24. If A is an $n \times n$ matrix, an $r \times r$ submatrix of A is obtained by the deletion of any $n - r$ rows and $n - r$ columns of A, $1 \leq r < n$. The *determinant rank* of A is the order of the largest submatrix with nonzero determinant, obtained in this way. Prove that the determinant rank is equal to the rank of A.

25. If A is an $n \times n$ matrix, prove that there are at most n distinct numbers c (real or complex) such that $|A - cI| = 0$.

26. Use any method to prove that

$$\begin{vmatrix} 1 & 1 & 1 & \cdots & 1 \\ 1 & 2 & 2^2 & \cdots & 2^{n-1} \\ 1 & 3 & 3^2 & \cdots & 3^{n-1} \\ \cdots & \cdots & \cdots & \cdots & \cdots \\ 1 & n & n^2 & \cdots & n^{n-1} \end{vmatrix} = 1!2!3! \cdots (n-1)!$$

3.5 An Introduction to Alternating Multilinear Forms

The material in these two final sections of the chapter is not essential to the sequence of the book. However, it contains the germ of a theory that is becoming more and more prominent in the context of elementary studies. It can be considered either a generalization of the theory of determinants, as we have developed this theory in the chapter, or as an abstract theory

SEC. 3.5 AN INTRODUCTION TO ALTERNATING MULTILINEAR FORMS 169

which specializes to that of determinants in an important special case. In brief, we shall study functions much like the determinant function D but which are required to satisfy only conditions 1 and 2 of that function. The lemma prior to Theorem 4.1 will then connect such functions with D.

The domain of the functions under discussion is the set $V \times V \times \cdots \times V$, in which there are n identical vector spaces V in this product set. It is clear how to translate Properties 1 and 2 of D, as defined in Sec. 3.3, into the present more general situation. A function that satisfies Property 1 is, as before, called *multilinear*, and it is said to be *alternating* if condition 2 is satisfied. Our general study is then to be in the context of multilinear functions on $\prod_{i=1}^{n} V_i$, where $V_i = V$, and V is an arbitrary vector space of dimension m. If f_1, f_2 are two such functions, we define $a_1 f_1 + a_2 f_2$ and af, for $a_1, a_2, a \in \mathbf{R}$, so that $(a_1 f_1 + a_2 f_2)(\alpha_1, \alpha_2, \ldots, \alpha_n) = a_1 f_1(\alpha_1, \alpha_2, \ldots, \alpha_n) + a_2 f_2(\alpha_1, \alpha_2, \ldots, \alpha_n)$ and $(af)(\alpha_1, \alpha_2, \ldots, \alpha_n) = af(\alpha_1, \alpha_2, \ldots, \alpha_n)$, for $\alpha_1, \alpha_2, \ldots, \alpha_n \in V$. It is easy to see (Prob. 6) that, with this definition of addition, the set of all such multilinear functions forms a vector space. It can be shown (Prob. 18) that the dimension of this space is m^n, but, since we shall not need this fact, we omit the proof. The *value* of one of these multilinear functions f [that is, $f(\alpha_1, \alpha_2, \ldots, \alpha_n)$] is called an *n-linear form* on V, but we shall often not distinguish, notationally or otherwise, between a multilinear function f and the n-linear form $f(\alpha_1, \alpha_2, \ldots, \alpha_n)$. Our remarks will usually be phrased in the language of forms.

Since the domain of a multilinear function f is a set of n-tuples of vectors, with subscripts ordered from 1 to n, it is meaningful to speak of permutations of these subscripts $\{1, 2, \ldots, n\}$. If f is an n-linear form and π is a permutation of $\{1, 2, \ldots, n\}$, we define πf to be the form such that $\pi f(\alpha_1, \alpha_2, \ldots, \alpha_n) = f(\alpha_{\pi(1)}, \alpha_{\pi(2)}, \ldots, \alpha_{\pi(n)})$, and it is easy to see that πf is also n-linear (Prob. 7). The two following definitions will doubtless seem quite appropriate.

Definitions   (1) An n-linear form f is said to be *symmetric* if $\pi f = f$, for every permutation π of $\{1, 2, \ldots, n\}$.

(2) An n-linear form f is said to be *skew-symmetric* if $\pi f = -f$, for every odd permutation π of $\{1, 2, \ldots, n\}$; that is, $\pi f = (\operatorname{sgn} \pi) f$, for every permutation π.

It is quite an elementary exercise (Prob. 8) to verify that the set of all symmetric n-linear forms and the set of all skew-symmetric n-linear forms constitute subspaces of the vector space of all n-linear forms.

It is not difficult to obtain nontrivial examples of these two kinds of forms. If $n = 2$, and f_1 and f_2 are linear functionals (or 1-forms) on V, we can define $f(\alpha_1, \alpha_2) = f_1(\alpha_1) f_2(\alpha_2) + f_1(\alpha_2) f_2(\alpha_1)$, and it is clear (Prob. 10) that f is symmetric. Likewise, if we define $F(\alpha_1,\alpha_2) = f_1(\alpha_1)f_2(\alpha_2) - f_1(\alpha_2)f_2(\alpha_1)$, it is evident that F is skew-symmetric. We note, in passing, that if $\alpha_1 = \alpha_2 = \alpha$, then $F(\alpha,\alpha) = 0$. That this is no exceptional phenomenon is the message given in the following theorem.

Theorem 5.1   An n-linear form is alternating if and only if it is skew-symmetric.

Proof

Let $f(\alpha_1, \ldots, \alpha_i, \ldots, \alpha_j, \ldots, \alpha_n)$ be an alternating n-linear form on V. The multilinear property of f makes it possible to write $f(\alpha_1, \ldots, \alpha_i + \alpha_j, \ldots, \alpha_j + \alpha_i, \ldots, \alpha_n) = f(\alpha_1, \ldots, \alpha_i, \ldots, \alpha_i, \ldots, \alpha_n) + f(\alpha_1, \ldots, \alpha_i, \ldots, \alpha_j, \ldots, \alpha_n) + f(\alpha_1, \ldots, \alpha_j, \ldots, \alpha_i, \ldots, \alpha_n) + f(\alpha_1, \ldots, \alpha_j, \ldots, \alpha_j, \ldots, \alpha_n)$, where $1 \leq i < j \leq n$. Since f is assumed to be alternating, this equation reduces to $0 = f(\alpha_1, \ldots, \alpha_j, \ldots, \alpha_i, \ldots, \alpha_n) + f(\alpha_1, \ldots, \alpha_i, \ldots, \alpha_j, \ldots, \alpha_n)$, whence $f(\alpha_1, \ldots, \alpha_i, \ldots, \alpha_j, \ldots, \alpha_n) = -f(\alpha_1, \ldots, \alpha_j, \ldots, \alpha_i, \ldots, \alpha_n)$. This means that any transposition of indices i and j changes f into $-f$; and so, if π is any odd permutation of $\{1,2, \ldots, n\}$, we conclude that $\pi f = -f$. Hence f is skew-symmetric. Conversely, let us suppose that f is skew-symmetric, so that $f(\alpha_1, \ldots, \alpha_i, \ldots, \alpha_j, \ldots, \alpha_n) = -f(\alpha_1, \ldots, \alpha_j, \ldots, \alpha_i, \ldots, \alpha_n)$, for $1 \leq i < j \leq n$. Then, if $\alpha_i = \alpha_j$, it follows that $f(\alpha_1, \ldots, \alpha_i, \ldots, \alpha_j, \ldots, \alpha_n) = -f(\alpha_1, \ldots, \alpha_i, \ldots, \alpha_j, \ldots, \alpha_n) = 0$, so that f is alternating. ∎

In view of Theorem 5.1, it is appropriate to consider the property that $(ij)f = -f$ [in which (ij) is the transposition that interchanges i with j] as equivalent to the one given originally to define an alternating form. This definition is, in fact, the more intuitive one.

Theorem 5.2   Any two alternating n-linear forms on a finite-dimensional vector space V are linearly dependent.

Proof

Let us suppose that f_1 and f_2 are alternating n-linear forms on V, with $\{\alpha_1,\alpha_2, \ldots, \alpha_m\}$ a basis for V. We consider $f_1(\beta_1,\beta_2, \ldots, \beta_n)$ and $f_2(\beta_1,\beta_2, \ldots, \beta_n)$, for arbitrary vectors $\beta_1,\beta_2, \ldots, \beta_n \in V$, and show that f_1 and f_2 are linearly dependent. We may assume that $\beta_j = \sum_{i=1}^{m} a_{ji}\alpha_i$, for real numbers a_{ji}, $i = 1, 2, \ldots, m$ and $j = 1, 2, \ldots, n$. The proof of the theorem is now essentially the same as that of Theorem 3.3, with the exception that we do not assume that f has Property 3 of D. As a result of the multilinear nature of the functions, $f_1(\beta_1,\beta_2, \ldots, \beta_n)$ and $f_2(\beta_1,\beta_2, \ldots, \beta_n)$ are the *same* linear combination of terms of the form $f_1(\alpha_{i_1},\alpha_{i_2}, \ldots, \alpha_{i_n})$ and $f_2(\alpha_{i_1},\alpha_{i_2}, \ldots, \alpha_{i_n})$, respectively, for permutations $i_1 i_2 \cdots i_n$ of $\{1,2, \ldots, n\}$.

Moreover, if we let π designate such a permutation, the skew-symmetric property of f_1 and f_2 allows us, by Theorem 5.1, to express $f_1(\alpha_{i_1}, \alpha_{i_2}, \ldots, \alpha_{i_n}) = (\text{sgn } \pi) f_1(\alpha_1, \alpha_2, \ldots, \alpha_n)$ and $f_2(\alpha_{i_1}, \alpha_{i_2}, \ldots, \alpha_{i_n}) = (\text{sgn } \pi) f_2(\alpha_1, \alpha_2, \ldots, \alpha_n)$. But $f_1(\alpha_1, \alpha_2, \ldots, \alpha_n)$ and $f_2(\alpha_1, \alpha_2, \ldots, \alpha_n)$ are real numbers, and so "linearly dependent" on each other; i.e., there exist real numbers a_1 and a_2, not both 0, such that $a_1 f_1(\alpha_1, \alpha_2, \ldots, \alpha_n) + a_2 f_2(\alpha_1, \alpha_2, \ldots, \alpha_n) = 0$. Since the terms of $a_1 f_1(\beta_1, \beta_2, \ldots, \beta_n) + a_2 f_2(\beta_1, \beta_2, \ldots, \beta_n)$ can be so grouped that $a_1 f_1(\alpha_1, \alpha_2, \ldots, \alpha_n) + a_2 f_2(\alpha_1, \alpha_2, \ldots, \alpha_n)$ is a factor of each term in the new grouping, it is a consequence that the sum is 0. Hence $a_1 f_1 + a_2 f_2 = 0$, thereby completing the proof. An alternative proof, based on the lemma in Sec. 3.4, is suggested in Prob. 9. ∎

It follows from this theorem that the subspace of alternating n-linear forms has dimension not exceeding 1, but it must be emphasized that we should not conclude from this that the dimension of this subspace is in fact 1; for we have given no proof of the existence, in general, of any such nonzero forms. In fact, if $n > m$, it can be shown (Prob. 21) that any n-linear alternating form must be the zero form. In Sec. 3.6, however, we shall show that, in the important case in which $n = m$, the space of n-linear forms on an n-dimensional vector space is of dimension 1.

Problems 3.5

1. Determine which of the following forms f are 3-linear on V_3, with each triplet of vectors X_1, X_2, X_3 considered to form the ordered rows of a matrix $A = [a_{ij}]$:
   (a) $f(X_1, X_2, X_3) = a_{11} + a_{22} + a_{33}$
   (b) $f(X_1, X_2, X_3) = a_{11} a_{22} a_{33}$
   (c) $f(X_1, X_2, X_3) = a_{11}^2 + 2 a_{11} a_{12}$
   (d) $f(X_1, X_2, X_3) = 1$
   (e) $f(X_1, X_2, X_3) = 0$

2. Determine all 2-linear forms on **R**. Are any of these alternating?

3. Explain what is meant by the remark that two real numbers are "linearly dependent" on each other. Illustrate with examples.

4. Find an example of a symmetric 3-linear form on a vector space V, using for the construction a 2-linear and a 1-linear form. [*Hint:* Let g and h be the 2-linear and 1-linear forms, respectively. Then define f so that $f(\alpha_1, \alpha_2, \alpha_3) = g(\alpha_1, \alpha_2) h(\alpha_3)$ and finally $F(\alpha_1, \alpha_2, \alpha_3) = \sum_\pi \pi f(\alpha_1, \alpha_2, \alpha_3)$, with $\alpha_1, \alpha_2, \alpha_3 \in V$, and the summation is over all $\pi \in S_3$.]

5. Expand on the hint given in Prob. 4 to construct a symmetric r-linear form from a k-linear form and a $(r - k)$-linear form, with $1 < k < r$.

6. Prove that the set of all n-linear forms on a vector space V is a vector space.

7. If f is an n-linear form, prove that πf is also an n-linear form, for any $\pi \in S_n$.

8. Prove that the set of all symmetric (skew-symmetric) n-linear forms on a vector space V is a vector space.

9. Use the lemma in Sec. 3.4 to prove Theorem 5.2, by considering a coordinate space isomorphic to V.

10. Verify that the 2-forms f and F, constructed in the text from linear functionals f_1 and f_2, are, respectively, symmetric and skew-symmetric.

11. If f is any n-linear form, show that $\sum_{\pi} \pi f$ is a symmetric form, if the summation is over all permutations $\pi \in S_n$.

12. Imitate the proof of Theorem 3.3, and complete the expansion of $f_1(\beta_1, \beta_2, \ldots, \beta_n)$ and $f_2(\beta_1, \beta_2, \ldots, \beta_n)$, as indicated in the proof of Theorem 5.2.

13. If f is a nonzero alternating n-linear form on a vector space V, with $\alpha_1, \alpha_2, \ldots, \alpha_n$ linearly independent elements of V, prove that $f(\alpha_1, \alpha_2, \ldots, \alpha_n) \neq 0$. [*Hint:* Imitate the proof of Theorem 5.2 and obtain the desired result by "contradiction."]

14. There exist fields of "characteristic" 2, in which $1 + 1 = 0$. If the scalars of the basic vector space V are drawn from such a field rather than from **R**, explain why the notions of *alternating* and *skew-symmetric* are no longer equivalent.

15. Determine all 2-linear forms on the row space of a 2×2 matrix. [*Hint:* If $A = [a_{ij}]$, with X_1, X_2 the rows of A, then $X_1 = a_{11}E_1 + a_{12}E_2$ and $X_2 = a_{21}E_1 + a_{22}E_2$.]

16. If we require the 2-linear forms in Prob. 15 to be alternating, what is the result?

17. Assume the analogous operations for matrices with polynomial entries as for real entries, and determine an alternating 3-linear form on the row space of the matrix

$$A = \begin{bmatrix} x & 0 & x \\ 0 & 2 & 0 \\ x^2 & 0 & -x^3 \end{bmatrix}$$

18. Prove that the space of n-linear forms on a vector space V of dimension m has dimension m^n.
19. Find the dimension of the space of symmetric n-linear forms on a space V of dimension m.
20. Find the dimension of the space of skew-symmetric n-linear forms on a space V of dimension m.
21. Prove that an alternating n-linear form on a vector space of dimension m, with $m < n$, must be the zero form.

3.6 Determinants Discovered Anew

We first prove a theorem that substantiates the remark made at the end of Sec. 3.5. It must be understood that in this present study *we do not assume the prior existence of a determinant function D.*

Theorem 6.1 *The vector space of alternating n-linear forms on a vector space V of dimension n (> 0) is of dimension 1.*

Proof

The proof is one of the *existence* of an alternating n-linear form on V that is nonzero, because the result then follows directly from Theorem 5.2. We know that linear functionals (and so 1-linear forms) on V exist, because the dual space V^* of such functionals has dimension n. Moreover, these linear functionals are alternating in a trivial sort of way. We shall show that, if $1 \leq k \leq n$, there exists at least one nonzero alternating k-linear form on V, by using induction on k. In our preliminary discussion below, we assume $n \geq 3$.

We start the inductive construction by making use of a nonzero 1-form or linear functional f_1 to construct a nonzero 2-form. Since the dimension of V^* is n, there are plenty of nonzero 1-forms available. If f_1 is any nonzero 1-form, we then select a vector $\alpha_1^0 \in V$ such that $f_1(\alpha_1^0) \neq 0$, and another vector $\alpha_2^0 \in V$ linearly independent of α_1^0. The one-dimensional spaces generated by each of α_1^0 and α_2^0 have only the zero vector in common, and it follows from Theorem 5.2 (Chap. 2) that the annihilators of these two spaces are distinct. There then exists a linear functional f_1^0 such that $f_1^0(\alpha_1^0) = 0$ and $f_1^0(\alpha_2^0) \neq 0$. With (12) designating the transposition that interchanges 1 and 2, we now use a notation introduced in Sec. 3.5 to define $f_2(\alpha_1, \alpha_2) = (12)f_1(\alpha_1)f_1^0(\alpha_2) - f_1(\alpha_1)f_1^0(\alpha_2) = f_1(\alpha_2)f_1^0(\alpha_1) - f_1(\alpha_1)f_1^0(\alpha_2)$. (It may be observed that, except for notation, this is the negative of the alternating 2-form F that was used for illustrative purposes in Sec. 3.5.) Moreover, $f_2(\alpha_1^0, \alpha_2^0) = -f_1(\alpha_1^0)f_1^0(\alpha_2^0) \neq 0$, by the choice of f_1, α_1^0, and α_2^0, so that f_2 is nonzero.

With a nonzero 2-form f_2 at our disposal, the construction of a nonzero 3-form proceeds along similar lines. Since f_2 is nonzero, there exist vectors

$\alpha_1, \alpha_2 \in V$ such that $f_2(\alpha_1, \alpha_2) \neq 0$. The same "annihilator" argument used above now implies the existence of a linear functional f_{12}^0 and a vector $\alpha_3 \in V$ such that $\alpha_3 \notin \mathcal{L}\{\alpha_1, \alpha_2\}, f_{12}^0(\alpha_1) = f_{12}^0(\alpha_2) = 0$, but $f_{12}^0(\alpha_3) \neq 0$. The desired 3-linear form f_3 is defined so that $f_3(\alpha_1, \alpha_2, \alpha_3) = \sum_{i=1}^{2} (i3) f_2(\alpha_1, \alpha_2) f_{12}^0(\alpha_3) - f_2(\alpha_1, \alpha_2) f_{12}^0(\alpha_3) = f_2(\alpha_3, \alpha_2) f_{12}^0(\alpha_1) + f_2(\alpha_1, \alpha_3) f_{12}^0(\alpha_2) - f_2(\alpha_1, \alpha_2) f_{12}^0(\alpha_3)$. It is immediate that $f_3(\alpha_1^0, \alpha_2^0, \alpha_3^0) \neq 0$, and there is no difficulty in verifying that f_3 is alternating (Prob. 10).

It is now easy to generalize this method of construction, with the assumption of the existence of a nonzero alternating k-linear form f_k, $1 \leq k < n$. There exist vectors $\alpha_1, \alpha_2, \ldots, \alpha_k \in V$ such that $f_k(\alpha_1, \alpha_2, \ldots, \alpha_k) \neq 0$, since f_k is nonzero, and also a nonzero linear functional $h \in V^*$ and a nonzero vector $\alpha_{k+1} \in V$ such that $h(\alpha_1) = h(\alpha_2) = \cdots = h(\alpha_k) = 0$ but $h(\alpha_{k+1}) \neq 0$. The $(k + 1)$-form f_{k+1} is now defined so that $f_{k+1}(\alpha_1, \alpha_2, \ldots, \alpha_{k+1}) = \sum_{i=1}^{k} (i\ k+1) f_k(\alpha_1, \alpha_2, \ldots, \alpha_k) h(\alpha_{k+1}) - f_k(\alpha_1, \alpha_2, \ldots, \alpha_k) h(\alpha_{k+1})$. The proof that f_{k+1} is a nonzero alternating $(k + 1)$-form is left (Prob. 11) to the reader; the proof that f_{k+1} is nonzero is easy, but the reader is cautioned to consider all the various dispositions of the indices i and j when he assumes $\alpha_i = \alpha_j$ in the proof that f_{k+1} is alternating. With the filling in of these details, the proof of the theorem will have been completed. ∎

This concludes our general discussion of multilinear algebra, a topic with many very diverse applications. Our interest in it, however, stems from the fact that it can be specialized to obtain the theory of determinants. We now take a brief look at this application.

One result of Theorem 6.1 is that *any* nonzero alternating n-linear form constitutes a basis for the space of these forms on a vector space V of dimension $n > 0$. Hence if T is an arbitrary linear operator on V, it is possible to define an operator T' on the space of all alternating n-linear forms f on V such that

$$(T'f)(\alpha_1, \alpha_2, \ldots, \alpha_n) = f(T\alpha_1, T\alpha_2, \ldots, T\alpha_n)$$

for arbitrary $\alpha_1, \alpha_2, \ldots, \alpha_n \in V$. It is not difficult (Prob. 15) to verify that T'f is also an alternating n-linear form, so that T' is, in fact, an "operator" as we have been using this word. We do not name the operator T' in general, but, for the case $n = 1$, it is clearly the *transpose* of T, as defined in Sec. 2.10 In view of our comment at the beginning of this paragraph, we are now able to assert that T'$f = \delta f$, for some $\delta \in \mathbf{R}$. The number δ, which is independent of f, is called the *determinant* of T, and we write $\delta = \det$ T. The function det then associates with every linear operator T on V the real number det T. At the end of this section, we shall give a brief exposition of the fact that the two concepts of a determinant, the one a function on linear operators and the other a function on square matrices, are very closely related. In fact, if A is any transformation matrix of T, we shall see that det T = det A.

It is easy to obtain some of the elementary properties of the det function, as we have defined it here. For example, suppose T is the operator that multiplies each vector $\alpha \in V$ by the real number c. Then $(T'f)(\alpha_1, \alpha_2, \ldots, \alpha_n) = f(T\alpha_1, T\alpha_2, \ldots, T\alpha_n) = f(c\alpha_1, c\alpha_2, \ldots, c\alpha_n) = c^n f(\alpha_1, \alpha_2, \ldots, \alpha_n)$ for every alternating n-linear form f. It then follows that det T = c^n; and, as special cases, det I = 1 and det 0 = 0. The following "product" theorem is also quite easy to obtain from our present point of view.

Theorem 6.2   *If T_1 and T_2 are linear operators on an n-dimensional vector space V, with $n > 0$, then det $(T_1 T_2) = (det\ T_1)(det\ T_2)$.*

Proof

Let T = $T_1 T_2$, with f any nonzero alternating n-linear form on the n-dimensional vector space V. Then $(T'f)(\alpha_1, \alpha_2, \ldots, \alpha_n) = f(T\alpha_1, T\alpha_2, \ldots, T\alpha_n) = f[(T_1 T_2)\alpha_1, (T_1 T_2)\alpha_2, \ldots, (T_1 T_2)\alpha_n] = (T_1'f)(T_2\alpha_1, T_2\alpha_2, \ldots, T_2\alpha_n) = (T_2'T_1'f)(\alpha_1, \alpha_2, \ldots, \alpha_n)$, and from this we conclude that $T' = T_2'T_1'$. But $T'f = (det\ T)f$, and $(T_2'T_1')f = T_2'(T_1'f) = T_2'[(det\ T_1)f] = (det\ T_1)(T_2'f) = (det\ T_1)(det\ T_2)f = [(det\ T_2)(det\ T_1)]f$. We have already implied that det T is uniquely (Prob. 14) associated with T, and so we conclude that det T = $(det\ T_1)(det\ T_2)$. ∎

Finally, we show that the concept of a determinant, as developed in this section, is equivalent to the one developed earlier. The existence of a *basis* of V is the cohesive factor that unifies the various concepts. Suppose A is the matrix of a linear operator T on the n-dimensional space V, relative to the basis $\{\alpha_1, \alpha_2, \ldots, \alpha_n\}$, $n > 0$. Then, if f is any nonzero alternating n-linear form on V, we have seen that

$$(det\ T)f(\alpha_1, \alpha_2, \ldots, \alpha_n) = f(T\alpha_1, T\alpha_2, \ldots, T\alpha_n)$$

If $A = [a_{ij}]$, the relationship between A and T requires that $T\alpha_j = \sum_{i=1}^{n} a_{ij}\alpha_i$, with $j = 1, 2, \ldots, n$. Hence, on substitution of these expressions in the right member of the preceding equality, and using the fact that f is both n-linear and alternating (as in the proof of Theorem 3.3) we find (Prob. 16) that $(det\ T)f(\alpha_1, \alpha_2, \ldots, \alpha_n) = \left[\sum_\pi (sgn\ \pi) a_{1\pi(1)} a_{2\pi(2)} \cdots a_{n\pi(n)}\right] f(\alpha_1, \alpha_2, \ldots, \alpha_n)$, where the summation is over all $\pi \in S_n$. Since f is nonzero, we may conclude from Prob. 13 of Sec. 3.5 that $det\ T = \sum_\pi (sgn\ \pi) a_{1\pi(1)} a_{2\pi(2)} \cdots a_{n\pi(n)} = |A|$. It follows that the determinant of a linear operator T on a vector space V and the determinant of any matrix A that represents T are equal. Although it is necessary to use a basis of V to obtain A, it is an easy consequence of Theorem 9.3 (Chap. 2) and Theorem 4.1 that the determinants of all such (similar) matrices A are equal. The assertion that det T = det A is then quite in order.

Problems 3.6

Note For the problems in this collection, it is to be understood that the definition of det, as given in this section, is to be used.

1. Explain why the known existence of the determinant function D would obviate Theorem 6.1 in the case where $V = V_n$.
2. If T_1 and T_2 are nonzero linear operators on a finite-dimensional vector space V with $T_1T_2 = 0$, explain why det T_1 = det $T_2 = 0$.
3. Use direct arguments to verify that det T is 0 or 1 according as T is the zero or identity operator on a vector space.
4. Find det T, where T is the operator on the space V of plane vectors which rotates each $\alpha \in V$ through an angle of (a) 60°; (b) 90°.
5. Find det D, if D is the differentiation operator on the space P_3 of polynomials of degree less than 3 in a real variable t, such that $Df(t) = f'(t)$.
6. Find det T, if T is the operator on V_2 such that, for each $(x_1,x_2) \in V_2$, T maps (x_1,x_2) onto (a) $(x_1 + x_2, x_1 - x_2)$; (b) $(x_1 - x_2, 0)$.
7. Select an explicit nonzero functional f_1 and vectors $\alpha_1^0, \alpha_2^0 \in V_2$, and then use the method of Theorem 6.1 to construct a nonzero alternating 2-form on V_2.
8. Use the results in Probs. 4, 5, and 6 to check the equality $f(T\alpha_1, T\alpha_2, \ldots, T\alpha_n) = (\det T)f(\alpha_1, \alpha_2, \ldots, \alpha_n)$, for any alternating n-linear form f on a vector space V, with $\alpha_1, \alpha_2, \ldots, \alpha_n \in V$.

9. If V_1 and V_2 are nonzero subspaces of a space V, such that $V_1 \cap V_2 = 0$, review the argument that there exist vectors $\alpha_1 \in V_1$ and $\alpha_2 \in V_2$ and a linear functional f on V such that $f(\alpha_1) = 0$ and $f(\alpha_2) \neq 0$.
10. Prove that the form f_3, in the preliminary part of the proof of Theorem 6.1, is alternating.
11. Verify that the form f_{k+1}, in the proof of Theorem 6.1, is nonzero and alternating. [*Hint:* To prove f_{k+1} alternating, let $\alpha_i = \alpha_j$ and consider (1) $j = k + 1$ and (2) $j \leq k$.]
12. If you have not already done so, do Prob. 13 of Sec. 3.5. [*Hint:* Cf. Proof of Theorem 4.4.]

13. Complete the inductive proof as suggested in the proof outline of Theorem 6.1.
14. Explain why det T is uniquely determined by T.
15. Prove that $T'f$, as defined in this section, is an alternating n-linear form for any alternating n-linear form f and linear operator T on a vector space V.
16. Complete the details of the argument outlined in the final paragraph of this section, from the point at which the substitutions are made for $T\alpha_j, j = 1, 2, \ldots, n$.

17. Prove that det T $\neq$ 0, if and only if T is an invertible operator.
18. If T_1 and T_2 are linear operators such that $T_1T_2 = I$, prove that both T_1 and T_2 are invertible.
19. If T is a linear operator that maps one basis of an n-dimensional vector space V onto another, prove that det T $\neq$ 0 if $n > 0$.
20. If T is the operator in Prob. 19, with $T\alpha_i = \beta_i$, for bases $\{\alpha_i\}$ and $\{\beta_i\}$, prove that the form f, defined so that $f(\beta_1,\beta_2, \ldots, \beta_n) = \det T$, is n-linear and alternating. Does the result require that $\{\beta_i\}$ be a basis?
21. If A, B, C, D are $n \times n$ submatrices, prove that $\begin{vmatrix} A & B \\ C & D \end{vmatrix} = |AD - BC|$, where (a) A, B, C, D are permutable; (b) $CD = DC$, and D is nonsingular.

Selected Readings

Dwyer, P. S.: "Linear Computations," John Wiley & Sons, Inc., New York, 1951.

Fox, L.: "An Introduction to Numerical Linear Algebra," Oxford University Press, Fair Lawn, N.J., 1965.

Hohn, F. E.: "Elementary Matrix Algebra," The Macmillan Company, New York, 1964. (Note, especially, the bibliography, pp. 376–383.)

Householder, A. S.: "The Theory of Matrices in Numerical Analysis," Blaisdell-Ginn, New York, 1964.

Sheffield, R. D.: A General Theory for Linear Systems, *Amer. Math. Monthly*, **65** (2):109–111 (February, 1958).

Varga, R. S.: "Matrix Iterative Analysis," Prentice-Hall, Inc., Englewood Cliffs, N.J., 1962.

Von Neumann, J., and H. H. Goldstine: Numerical Inverting of Matrices of High Order, *Bull. Am. Math. Soc.*, **53**: 1021–1099 (1947).

4

INNER-PRODUCT SPACES

4.1 Inner Products in V_3

Up to this stage in our study of vector spaces, there has been no mention of the concepts of *length* and *direction*, as applied to an abstract vector. This may be quite surprising, in view of the all-important role played by these notions in any study of the line vectors of physics or geometry. However, it may be recalled that, in our review of these vectors in Sec. 1.2, we made no mention even there of a *measure* of length or direction but confined our remarks to an intuitive notion of when such measures would be *equal*. It is well known that measures exist for both length and direction of line vectors, and in this chapter we shall generalize these ideas and study abstract vector spaces in which a *metric* has been introduced. It will then make sense to speak of the "length" of a vector and the "angle" between two vectors in such a vector space. The basic function to be introduced is an *inner product*. We shall first see how this function arises in a natural way in the geometry of 3-space and then use its properties in this familiar environment as a guide to our abstract definition. For the discussions of this section, it would be well for the reader to review the distinction made in Sec. 1.2 between directed line *segments* and line *vectors* (Prob. 7) and also our dual use (see Sec. 1.3) of a

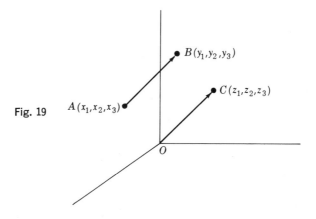

Fig. 19

triple of real numbers to designate both a *point* of geometric space and an *element*, *vector*, or *point* of the vector space V_3.

In line with this proposal, let us then take a look at the geometry of ordinary 3-space, as coordinatized by a cartesian coordinate system. In Fig. 19 we have shown a line vector AB in such a space, the coordinates of the endpoints A and B being indicated. The fact that we are using a cartesian coordinate system (see Sec. 1.3) leads immediately to certain results that are doubtless familiar to all readers.

First, we can use the Pythagorean theorem to obtain the *length* $|AB|$ of AB as $[(y_1 - x_1)^2 + (y_2 - x_2)^2 + (y_3 - x_3)^2]^{\frac{1}{2}}$. If we replace AB by the equivalent vector OC drawn from the origin O, we see that $|AB| = |OC| = (z_1^2 + z_2^2 + z_3^2)^{\frac{1}{2}}$, where (z_1, z_2, z_3) is the coordinate triple of C. Or, alternatively, if we regard $X = (x_1, x_2, x_3)$ as an arbitrary vector of the coordinate space V_3, we are asserting that the *length* $|X|$ of X is given by

$$|X| = (x_1^2 + x_2^2 + x_3^2)^{\frac{1}{2}}$$

The *direction* of a line vector, which we regarded as purely intuitive in Sec. 1.2, can also be described in terms of the *direction angles* or *direction cosines* of solid geometry. Figure 20 is essentially a reproduction of Fig. 19 but includes the direction angles a, b, c of the equivalent vectors AB and OC. The cosines of these angles are the direction cosines of AB and OC, regarded as either line *vectors* or directed line *segments*, and it is easy to obtain the following results: $\cos a = z_1/(z_1^2 + z_2^2 + z_3^2)^{\frac{1}{2}}$; $\cos b = z_2/(z_1^2 + z_2^2 + z_3^2)^{\frac{1}{2}}$; $\cos c = z_3/(z_1^2 + z_2^2 + z_3^2)^{\frac{1}{2}}$. In the alternative interpretation of $X = (x_1, x_2, x_3)$ as a general vector in V_3, these results may be phrased in the following form:

180 INNER-PRODUCT SPACES

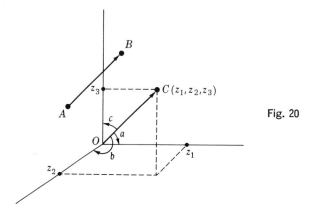

Fig. 20

The *direction angles a, b, c* of a vector $X = (x_1, x_2, x_3) \in V_3$ are a, b, c, where

$$\cos a = \frac{x_1}{|X|} = \frac{x_1}{(x_1^2 + x_2^2 + x_3^2)^{\frac{1}{2}}}$$

$$\cos b = \frac{x_2}{|X|} = \frac{x_2}{(x_1^2 + x_2^2 + x_3^2)^{\frac{1}{2}}}$$

$$\cos c = \frac{x_3}{|X|} = \frac{x_3}{(x_1^2 + x_2^2 + x_3^2)^{\frac{1}{2}}}$$

It should be clear that either the direction cosines or direction angles of a directed line segment make precise our previously intuitive idea of *direction:* Line vectors are either *identical* or *positive real multiples* of each other (directed line segments are either *identical, parallel,* or *positive real multiples* of each other) if and only if they have the same direction cosines (see Prob. 10).

Finally, it is also quite easy to give an algebraic expression that will characterize the *angle between* two arbitrary geometric vectors in 3-space. Figure 21 shows two such vectors AB and CD in coordinatized 3-space, and we are interested in the angle between them. We first replace AB and CD by the respectively equivalent vectors OP and OQ, emanating from the origin O. If we use the coordinates of the points A, B, C, D, P, Q, as indicated in Fig. 21, it is clear that $x_1 = b_1 - a_1$, $x_2 = b_2 - a_2$, $x_3 = b_3 - a_3$, and $y_1 = d_1 - c_1$, $y_2 = d_2 - c_2$, $y_3 = d_3 - c_3$. An application of the law of cosines to the triangle OPQ now yields $|PQ|^2 = |OP|^2 + |OQ|^2 - 2|OP|\,|OQ| \cos \theta$, where θ is the angle in question between OP and OQ (or AB and CD). But, since $|OP|^2 = x_1^2 + x_2^2 + x_3^2$ and $|OQ|^2 = y_1^2 + y_2^2 + y_3^2$, and $|PQ|^2 =$

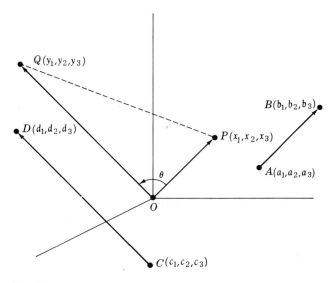

Fig. 21

$(x_1 - y_1)^2 + (x_2 - y_2)^2 + (x_3 - y_3)^2$, an easy reduction of the above equation leads to the following result:

$$\cos \theta = \frac{x_1 y_1 + x_2 y_2 + x_3 y_3}{|OP| \, |OQ|}$$

Again, if we regard $X = (x_1, x_2, x_3)$ and $Y = (y_1, y_2, y_3)$ as arbitrary elements of V_3, the result implies that the angle θ between X and Y is such that

$$\cos \theta = \frac{x_1 y_1 + x_2 y_2 + x_3 y_3}{|X| \, |Y|}$$

The numerator of the above expression for $\cos \theta$ is the important function to which we referred earlier, and this leads to the following definition.

Definition   If $X = (x_1, x_2, x_3)$ and $Y = (y_1, y_2, y_3)$ are vectors in V_3, the real number $x_1 y_1 + x_2 y_2 + x_3 y_3$ is called the *standard inner product* or *dot product* of X and Y and is designated (X, Y).

In *this section*, it will also be convenient to use the same symbolism to designate this type of function in the space of geometric vectors. That is, if AB and CD are geometric vectors in 3-space, which are represented in V_3 by X and Y, respectively, then (AB, CD) may be used to designate the same real number as (X, Y).

The standard inner product, which we have just defined, plays a very important role in the geometry of 3-space, because it is easy to see how to express both the length of a vector and the angle between two vectors in terms of it. Thus, if $X = (x_1, x_2, x_3)$, then $|X| = (x_1^2 + x_2^2 + x_3^2)^{\frac{1}{2}} = (X,X)^{\frac{1}{2}}$. Also, if $Y = (y_1, y_2, y_3)$ is another vector in V_3, and θ is the angle between X and Y, then

$$\cos \theta = \frac{x_1 y_1 + x_2 y_2 + x_3 y_3}{|X| \, |Y|} = \frac{(X,Y)}{|X| \, |Y|}$$

It is clear that this *inner product* is a mapping of ordered pairs of vectors in V_3 to **R**, that is, a *scalar* function; another type of mapping which is a *vector* function is described in Prob. 23. The following properties are immediate consequences of the definition, but they should be checked (Prob. 8) by the reader:

(1) $(X,Y) = (Y,X)$
(2) $(rX,Y) = r(X,Y)$, for any $r \in$ **R**
(3) $(X + Y, Z) = (X,Z) + (Y,Z)$
(4) $(X,X) > 0$, if $X \neq 0$

In the preceding statements, the vectors X, Y, Z are arbitrary elements of V_3, except as indicated otherwise. We may recall that in Sec. 1.3 we were unable to give an analytic characterization of orthogonal geometric vectors, but it is now easy to do this in V_3 in terms of the standard inner product:

If X and Y are line vectors or elements of V_3, they are orthogonal or perpendicular if $(X,Y) = 0$.

Inasmuch as we obtained this result from a formula that was derived from geometric considerations, it is clear that this characterization of *orthogonality* is consistent with geometric common sense.

It is easy to see that what we have called the *standard inner product* in V_3 and geometric 3-space can be *specialized* in an obvious way to V_2 and the geometric plane. A *generalization* to V_n and geometric n-space is also clear. However, instead of dwelling on such ramifications here, we shall include them in our general discussion in the following section. We shall close this section with two examples, both of which have a flavor that is distinctly geometric, and so in line with our introductory "metric" comments.

EXAMPLE 1

We recall from analytic geometry that the *equation* of a surface, and, in particular, of a plane, in 3-space is a symbolic statement of the

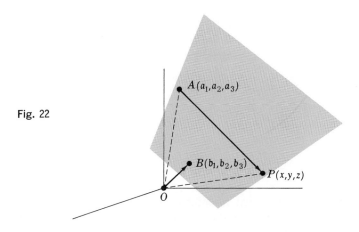

Fig. 22

relationship that must exist between the coordinates of an arbitrary point on the surface. Many of the problems of geometry that involve the notion of orthogonality, heretofore beyond our reach, can now be solved easily by the vector methods just developed. For example, we consider the following problem: *Determine the equation of the plane that is perpendicular to a given line and passes through a given point.* The situation is depicted in Fig. 22, in which $A(a_1,a_2,a_3)$ is the given point and OB is a segment of the line, with (b_1,b_2,b_3) the coordinate triple of B. (Since there is no loss in generality involved, we may assume that the given line passes through the origin.) If $P(x,y,z)$ is an arbitrary point on the plane, the condition of the problem is that the vector AP is to be orthogonal to the vector OB. Since $AP = OP - OA$, this means that $(AP, OB) = (OP - OA, OB) = 0$. If we now identify the line vectors OA, OB, OP, respectively, with the elements (a_1,a_2,a_3), (b_1,b_2,b_3), (x,y,z) of V_3, we see that the orthogonality condition becomes $(X, Y) = 0$, where $X = (x - a_1, y - a_2, z - a_3)$ and $Y = (b_1,b_2,b_3)$. An evaluation of this inner product gives $b_1x + b_2y + b_3z = a_1b_1 + a_2b_2 + a_3b_3$ as the desired equation.

EXAMPLE 2

Find a vector in V_3 with integral components that is orthogonal to the plane containing the points $A(1,2,-1)$, $B(-2,1,1)$, $C(3,-2,1)$.

Solution

The condition in the problem, as depicted in Fig. 23, requires that the unknown vector be orthogonal to any vector in the given plane, and, in particular, to the vectors AB and AC. There is no loss in generality in assuming (Prob. 9)

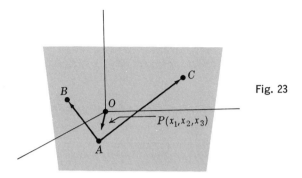

Fig. 23

that the unknown geometric vector is OP, where O is the origin and the coordinate triple of P is (x_1, x_2, x_3) with $x_1 = 1$. The requirement then is that $(OP, AB) = 0$ and $(OP, AC) = 0$. If we represent the line vectors AB and AC by $(-3, -1, 2)$ and $(2, -4, 2)$, respectively, these two conditions become equivalent to the following pair of equations:

$$-3 - x_2 + 2x_3 = 0$$
$$2 - 4x_2 + 2x_3 = 0$$

or, equivalently,

$$x_2 - 2x_3 = -3$$
$$4x_2 - 2x_3 = 2$$

On solving, we find that $x_2 = \frac{5}{3}$ and $x_3 = \frac{7}{3}$, so that $(1, x_2, x_3) = (1, \frac{5}{3}, \frac{7}{3})$. Since any scalar multiple of a vector has the same orthogonality properties as the original vector, we may select $(3, 5, 7)$ as a vector in V_3 with the desired properties. It is easy to check that this vector is orthogonal to both $(-3, -1, 2)$ and $(2, -4, 2)$, and so it must be orthogonal to the plane containing the points A, B, C.

Problems 4.1

1. Find the length of each of the following vectors in V_3: (a) $(1, 1, 2)$; (b) $(1, 1, 1)$; (c) $(-2, 4, 3)$.
2. Find the length of the line vector AB in 3-space, being given that the respective coordinates of A and B are (a) $(1, -1, 3)$, $(2, 1, -2)$; (b) $(-2, -1, 1)$, $(3, 1, 1)$; (c) $(2, 1, -3)$, $(-2, -1, 4)$.
3. Determine (X, Y), where X and Y are, respectively (a) $(1, -1, 1)$, $(2, 1, -2)$; (b) $(1, 0, 2)$, $(2, 3, 1)$.
4. Determine (X, Y), where X and Y are, respectively (a) $(2, -1, 1)$, $(1, 2, -2)$; (b) $(3, 1, 2)$, $(1, -2, 0)$.
5. Determine (AB, CD), where the coordinates of the points A, B, C, D are, respectively $(1, 1, -2)$, $(2, -1, 1)$, $(-3, 2, 2)$, $(3, 1, -4)$.

6. If $A(3,-1,1)$ and $B(3,2,z)$ are points in 3-space, determine the smallest positive integer z such that the line vector AB is orthogonal to the line vector from the origin to $(1,-2,3)$.

7. Review the distinction that we are making between line or geometric *vectors* and line *segments*.

8. Verify that the four properties, asserted to be possessed by the standard inner product in V_3, are possessed by this function.

9. In Example 2, explain why there is no loss in generality in the assumption that $x_1 = 1$.

10. Verify that line vectors are identical or positive real multiples of each other (directed line segments are identical, parallel, or positive real multiples of each other) if and only if they have the same direction cosines.

11. Use vector methods to determine the equation of the plane through the point $A(1,0,4)$, that is orthogonal to the line through the origin and the point $(2,3,4)$.

12. Use vector methods to determine the equation of the plane through the point $A(0,-1,1)$ that is orthogonal to the line through the origin and the point $(3,1,-3)$.

13. Let $A(1,2,1)$, $B(-2,1,-2)$, $C(2,1,4)$ be three coplanar points with the directed line segments AB and AC regarded as the unit vectors of a (skew) coordinate system of the plane. Find the coordinates of a point D, such that AB and AD may be regarded as the unit vectors of a *cartesian* coordinate system of the plane.

14. It is shown in texts on solid geometry that an equation such as $2x + y - 3z = 4$ is the equation of a plane in 3-space. Find, in this plane, a pair of orthogonal line segments and a pair that make an angle of 60° with each other.

15. If $X = (1,-1,2)$ and $Y = (2,1,-3)$ are given vectors of V_3, determine (a) two orthogonal vectors in $\mathcal{L}\{X,Y\}$; (b) a vector in V_3 that is orthogonal to both vectors determined in (a); (c) three vectors of unit length, respectively parallel to the vectors determined in (a) and (b).

16. Prove that any vector, that is orthogonal to each of vectors $X, Y \in V_3$, is also orthogonal to any vector in $\mathcal{L}\{X,Y\}$.

17. Prove that mutually orthogonal vectors in V_3 are linearly independent.

18. Use vector methods to prove that the diagonals of a rhombus are perpendicular to each other.
19. Use vector methods to prove that the diagonals of a rectangle are equal in length.
20. Use vector methods to prove that the midpoints of the sides of a quadrilateral are the vertices of a parallelogram.
21. Use vector methods to prove that the base angles of an isosceles triangle are equal to each other.
22. If X and Y are arbitrary vectors of V_3, prove the *parallelogram law* that $|X + Y| + |X - Y| = 2|X|^2 + 2|Y|^2$.
23. If $X = (x_1,x_2,x_3)$, $Y = (y_1,y_2,y_3)$ are given vectors of V_3, we define the *cross product* $X \times Y$ of X and Y to be the vector $Z = (z_1,z_2,z_3)$, where $z_1 = x_2y_3 - x_3y_2$, $z_2 = x_3y_1 - x_1y_3$, $z_3 = x_1y_2 - x_2y_1$. Regard X, Y, Z as geometric vectors and prove that the volume of the parallelepiped that is determined by them is $|(X, Y \times Z)|$. If T is a nonsingular linear operator on V_3, try to discover the relationship between the volumes of the parallelepipeds determined by X, Y, Z and TX, TY, TZ.

4.2 General Euclidean Spaces

In Sec. 4.1, we showed how the notion of the *standard inner product* can be used to play a key role in expressing the concepts of length and angle in the geometry of ordinary 3-space. As was suggested in that section, we shall now use the basic properties of this real-valued function as motivation for the desired generalization to any real vector space. For convenience, however, we shall reduce the set of four properties, as listed in Sec. 4.1 in the notation of coordinate vectors, to an equivalent three (Prob. 10).

Definition   Let V be any real vector space. An inner product on V is a real-valued function that maps each ordered pair of vectors $\alpha,\beta \in V$ onto a scalar $(\alpha,\beta) \in \mathbf{R}$ such that the following hold for $\alpha,\beta,\gamma \in V$ and $a,b \in \mathbf{R}$, arbitrary except as indicated:

(1) $(\alpha,\beta) = (\beta,\alpha)$
(2) $(a\alpha + b\beta, \gamma) = a(\alpha,\gamma) + b(\beta,\gamma)$
(3) $(\alpha,\alpha) > 0$ if $\alpha \neq 0$

Moreover, any real vector space on which an inner-product function has been defined is called a *euclidean* space.

It is immediate that the standard inner product, as defined in Sec. 4.1, makes V_3 into a euclidean space, and the specialization of this to V_1 and V_2 and its generalization to V_n (for any $n > 3$) are clear. Thus, if $X = (x_1,x_2)$ and $Y = (y_1,y_2)$ are elements of V_2, we may define $(X,Y) = x_1y_1 + x_2y_2$; if $X = (x_1,x_2, \ldots, x_n)$ and $Y = (y_1,y_2, \ldots, y_n)$ are elements of V_n, for any $n > 0$, we define $(X,Y) = x_1y_1 + x_2y_2 + \cdots + x_ny_n$. It is natural to refer to the latter as the *standard inner product* in V_n, and we leave to the reader (Prob. 10) the verification that it possesses the required properties. It is easy to see how one can make some of the other vector spaces, to which we have often referred throughout the text, into euclidean spaces, and we give a few examples.

EXAMPLE 1

Let V be the space of all continuous functions on the interval $[a,b]$, with $a \neq b$. Then, for arbitrary $f,g \in V$, we may define $(f,g) = \int_a^b f(x)g(x)\,dx$, and it is not difficult to see that this is a valid inner product (Prob. 11). Hence this space is euclidean.

EXAMPLE 2

An example very similar to Example 1 can be supplied by considering the space P_n of real polynomials of degree less than n (>0) in an indeterminate or real variable. In this space, with p and q arbitrary polynomials in P_n, we define $(p,q) = \int_a^b p(x)q(x)\,dx$, for any real and distinct numbers a, b. Essentially the same proof as that for Example 1 would then show that this space is euclidean.

It is clear from Example 2 that, by varying the numbers a, b, we may obtain different inner-product functions. Thus, there may be many different *euclidean* spaces with the same basic set of vectors. Our next example shows that even in the case of V_n, although the standard inner product is an important one, there are other possibilities.

EXAMPLE 3

Let $X = (x_1,x_2)$ and $Y = (y_1,y_2)$ be elements of V_2, and define $(X,Y) = x_1y_1 - x_2y_1 - x_1y_2 + 2x_2y_2$. Since $(X,X) = x_1^2 - 2x_1x_2 + x_2^2 + x_2^2 = (x_1 - x_2)^2 + x_2^2 > 0$, unless $X = 0$, we have verified Property 3 of an inner product, and it is easy to check (Prob. 12) the other two.

EXAMPLE 4

Let V and W be arbitrary real vector spaces, with T a nonsingular linear transformation from V to W. Then, if $(\ ,\ )$ designates an inner product in W, it is possible to define an inner product in V, designated $(\ ,\ )_T$, as follows: $(\alpha,\beta)_T = (T\alpha,T\beta)$, for arbitrary $\alpha,\beta \in V$. We leave the verification that this defines an inner product in V to the reader (Prob. 11) and describe a special case of this in Example 5.

EXAMPLE 5

Let $W = V$ be the space of continuous functions, as in Example 1, and let T be defined on V so that $(Tf)(x) = x^2 f(x)$, for any $f \in V$ and $x \in [a,b]$. The reader is asked to verify (Prob. 13) that T is nonsingular, and we then define the new inner product $(\ ,\ )_T$ in terms of the old inner product in Example 1. That is, in line with the general construction in Example 4, we make the following definition, for arbitrary $f,g \in V$:

$$(f,g)_T = \int_a^b [(Tf)(x)][(Tg)(x)]\,dx$$
$$= \int_a^b f(x)g(x)x^4\,dx$$

There are many other simple illustrations of euclidean spaces, but these will suffice for the present. There is a very important (but somewhat more complicated) example of this type of space whose elements coincide with those of the familiar space V_n. This example will provide a point of contact with quadratic forms in the next chapter, and so we postpone discussion of it until then.

It is clear that we can use the definitions of *length* and *angle*, as developed in Sec. 4.1 for ordinary 3-space, and apply these notions, at least *formally*, to the vectors of any euclidean space. Thus, if α and β are arbitrary vectors of a euclidean space, with $(\ ,\ )$ the indicated inner-product function, it is natural to define (see Prob. 31) the *length* or *norm* $|\alpha|$ of α and the *angle* θ between α and β as follows:

$$|\alpha| = (\alpha,\alpha)^{\frac{1}{2}}$$
$$\cos\theta = \frac{(\alpha,\beta)}{|\alpha|\,|\beta|}$$

The question arises at this point whether these definitions make "geometric sense," and our answer is a qualified affirmative. The case of V_2, with the standard inner product, is the familiar geometry of the plane, and is analogous to the development of V_3 in Sec. 4.1, while the geometry of V_1 is valid—though trivial. For these cases the answer is then clear. The geometry of n-space ($n > 3$) is a purely formal extension of the geometry of 3-space, in which we extend the *language* of this familiar space to apply to their formal extensions in V_n. Hence, for euclidean space V_n, with the standard inner product, the concepts of *length* and *angle* make at least *as much* "geometric sense" as any of the other "geometric" concepts attached to it. Whether it is reasonable to use this geometric language for an arbitrary euclidean space, however, is a question that cannot be answered conclusively, and there is likely no categorical answer. However, we are able to support this practice by showing that there are valid extensions of at least two of the basic theorems of ordinary euclidean geometry to any euclidean space: the *Pythagorean theorem* and the *triangle inequality*. We do this after an illustration of orthogonality in V_3.

EXAMPLE 6

Find all vectors in V_3 that are orthogonal to $(2,-1,1)$.

Solution
If (x,y,z) is a vector in V_3, our requirement is that $((x,y,z),(2,-1,1)) = 0$. We assume the standard inner product on V_3, so that this condition is equivalent to $2x - y + z = 0$. The desired vectors are then all solutions of this equation, and it is easy to see (Prob. 14) that they comprise a subspace of V_3 of dimension 2. If we write the equation in the form $x = y/2 - z/2$, we readily find $(1,2,0)$ and $(-1,0,2)$ as two linearly independent solutions, so that the solution space can be described as $\mathcal{L}\{(1,2,0),(-1,0,2)\}$.

If two subspaces are so related that every vector in one is orthogonal to every vector in the other, it is convenient to say that the *subspaces are orthogonal*. We may then conclude from Example 6 that $\mathcal{L}\{(2,-1,1)\}$ and $\mathcal{L}\{(1,2,0),(-1,0,2)\}$ are orthogonal subspaces of V_3.

We now proceed with the program of geometry, previously referred to, for a general euclidean space. In these discussions, it should be understood that the symbol (,) will be used to designate a general, but unspecified, inner-product function. Although we do not establish the existence of an inner-product function for an arbitrary vector space, the reader is referred to Prob. 15 (a special case of Example 4) for the existence of such a function for spaces of finite dimension. It is natural to extend the geometric notion of orthogonality in V_n to an arbitrary euclidean space V and say that α, $\beta \in V$ are *orthogonal* if $(\alpha,\beta) = 0$.

Theorem 2.1 (Pythagorean Theorem)
> If α and β are orthogonal vectors of a euclidean space V, than $|\alpha + \beta|^2 = |\alpha|^2 + |\beta|^2$.

Proof
The orthogonality assumption is that $(\alpha,\beta) = (\beta,\alpha) = 0$. But, by the definition of *length*, $|\alpha + \beta|^2 = (\alpha + \beta, \alpha + \beta) = (\alpha,\alpha) + (\alpha,\beta) + (\beta,\alpha) + (\beta,\beta) = (\alpha,\alpha) + (\beta,\beta) = |\alpha|^2 + |\beta|^2$, as desired. ∎

It would be easy to use an inductive argument to generalize the result in Theorem 2.1 to read as follows: $|\alpha_1 + \alpha_2 + \cdots + \alpha_n|^2 = |\alpha_1|^2 + |\alpha_2|^2 + \cdots + |\alpha_n|^2$, for any mutually orthogonal vectors $\alpha_1, \alpha_2, \ldots, \alpha_n$ of V.

In ordinary euclidean geometry, the *distance* between two points is the *length* of the line segment joining the two points. The following definition should then appear quite natural.

Definition   The distance between two vectors α and β of a euclidean space is $|\alpha - \beta|$.

If α and β happen to be orthogonal vectors, it should be clear that $|\alpha - \beta| = |\alpha + \beta|$. The characteristic properties of the *distance* function are listed in Prob. 27.

Before obtaining the triangle inequality in an arbitrary euclidean space, let us first consider the reasonableness of the definition of $\cos \theta$ as $(\alpha,\beta)/|\alpha| \, |\beta|$ in such a space. In particular, does there always exist an angle θ that satisfies this equation? To answer this question in the affirmative, we show that the inequality $-1 \leq (\alpha,\beta)/|\alpha| \, |\beta| \leq 1$ or, equivalently, $(\alpha,\beta)^2 \leq |\alpha|^2|\beta|^2$ always holds for vectors α, β in the space.

Lemma (Schwarz Inequality)
> For any two vectors α, β of a euclidean space V, $(\alpha,\beta)^2 \leq |\alpha|^2|\beta|^2$.

Proof
Let us consider the vector $\alpha - b\beta$, for any real number b. Since $(\alpha - b\beta, \alpha - b\beta) \geq 0$, we may conclude (after a little computation) that $b^2(\beta,\beta) - 2b(\alpha,\beta) + (\alpha,\alpha) \geq 0$. The quadratic polynomial on the left side of this inequality has a nonnegative value for any real number b, and so its discriminant must be nonpositive. Thus, $(\alpha,\beta)^2 - (\alpha,\alpha)(\beta,\beta) \leq 0$, whence $(\alpha,\beta)^2 \leq |\alpha|^2|\beta|^2$, as desired. ∎

The triangle inequality can now be obtained quite easily for the space V.

Theorem 2.2 (Triangle Inequality)

If α and β are arbitrary vectors of a euclidean space V, then $|\alpha + \beta| \leq |\alpha| + |\beta|$.

Proof

We know from the properties of an inner product that $|\alpha + \beta|^2 = (\alpha + \beta, \alpha + \beta) = (\alpha,\alpha) + 2(\alpha,\beta) + (\beta,\beta)$, and the Schwarz inequality implies that $(\alpha,\beta) \leq |\alpha| |\beta|$. Hence, $|\alpha + \beta|^2 \leq (\alpha,\alpha) + 2|\alpha| |\beta| + (\beta,\beta) = (|\alpha| + |\beta|)^2$, so that $|\alpha + \beta| \leq |\alpha| + |\beta|$. ∎

With the completion of the proofs of these two theorems, we repeat an earlier comment. Although the results just established prove nothing about the credibility of the concepts of *length* and *angle* in an arbitrary euclidean space, they make more acceptable and plausible the use of these geometric terms in this more general context.

We must point out that, *in the sequel*, whenever a *coordinate* space V_n is assumed to be euclidean, *we shall understand that the metric is based on the standard inner product*, unless the contrary is stated.

Problems 4.2

1. Find (X,Y), where (a) $X = (1,2,-1)$, $Y = (2,2,1)$; (b) $X = (1,1,1,2)$, $Y = (-1,2,-1,2)$.
2. Determine x if the indicated vectors are orthogonal in V_4: (a) $(1, x-1, x+1, 2)$, $(2x,3,1,x)$; (b) $(2, x+1, 3x, 1)$, $(x, 1-x, 3x, -7)$.
3. Determine the length of each of the following coordinate vectors: (a) $(1,2)$; (b) $(1,-1,1)$; (c) $(2,-1,3,1)$.
4. Find the cosine of the angle between each of the following pairs of coordinate vectors: (a) $(1,2)$, $(3,-1)$; (b) $(1,-1,1)$, $(3,-1,2)$; (c) $(2,-1,-2)$, $(1,2,1)$.
5. Find the distance between the designated coordinate vectors: (a) $(1,1)$, $(2,-3)$; (b) $(1,-1,1)$, $(3,-1,2)$; (c) $(2,-1,-2)$, $(1,2,1)$.
6. Determine the subspace of vectors in V_2 that are orthogonal to $(2,-1)$.
7. Determine the subspace of vectors in V_3 that are orthogonal to $(1,-1,2)$.
8. Let V be the function space of Example 1, with $a = 0$ and $b = 1$. Determine $|f|$, where (a) $f(x) = x^2$; (b) $f(x) = 2x + 1$.
9. Let p and q be polynomials in the space P_4 of Example 2, with $a = 0$ and $b = 1$. Determine (p,q), $|p|$, and $|q|$, where $p(x) = 1 + 2x + x^2$ and $q(x) = 3 - 2x^2 + x^3$.

10. Show that the four properties of a (standard) inner-product function, as listed in Sec. 4.1, are equivalent to the corresponding three listed in this section. Hence deduce that a standard inner product in V_n is an inner product.

11. Verify that the scalar-valued functions defined in Examples 1 and 4 are inner products.

12. Complete the checking requested in Example 3.

13. Explain why the nonsingular nature of T is necessary in Example 4, and prove that the transformation T in Example 5 is nonsingular.

14. Explain why the solutions of the equation $2x - y + z = 0$ comprise a subspace of V_3 of dimension 2.

15. If V is any vector space of dimension n, with basis $\{\alpha_i\}$, let T be the isomorphic mapping of V onto V_n defined by $T\alpha_i = E_i$, $i = 1, 2, \ldots, n$. Then use the construction in Example 4 to determine an inner product for V, and deduce the existence of an inner product for any finite-dimensional vector space.

16. Let $\{1,l,l^2,l^3\}$ be a basis for the space P_4 of Example 2, and use the construction in Prob. 15 to determine an inner product for this space. Then, if $p(t) = 1 + t + t^2$ and $q(t) = 2 - t^2 + t^3$, determine first $(p,q)_T$ from the function just constructed and then (p,q) from the function defined in Example 2.

17. With inner product defined as in Example 1, prove that f and g are orthogonal functions in the space of continuous functions on $[0,\pi]$, where $f(x) = \sin x$ and $g(x) = \cos x$.

18. Find a vector in V_3 that is orthogonal to both $(1,1,-2)$ and $(-2,4,1)$.

19. Prove that the only vector common to two orthogonal subspaces of a euclidean space is the zero vector. Hence conclude that the sum of orthogonal subspaces is always direct (see Sec. 1.9).

20. If $V_3 = V \oplus W$ for orthogonal subspaces V and W, find the (*orthogonal*) projection (see Sec. 1.9) of α on V, where (*a*) $\alpha = (1,2,1)$, $V = \mathcal{L}\{(1,0,1),(-1,1,2)\}$; (*b*) $\alpha = (2,1,3)$, $V = \mathcal{L}\{(1,0,0),(0,1,0)\}$.

21. Prove that $(\alpha,\beta)^2 = |\alpha| |\beta|$ if and only if α and β are linearly dependent.

22. Prove that the collection of vectors orthogonal to a given set of vectors in a euclidean space comprises a subspace.

23. Prove that the diagonals of a rectangle bisect each other.

24. If V is a euclidean space, prove that (a) $(0,\alpha) = 0$, for any $\alpha \in V$; (b) if $(\alpha,\beta) = 0$, for all $\beta \in V$, then $\alpha = 0$.
25. Prove that the sum of two inner-product functions on a vector space V is also an inner-product function on V. Does the set of all inner-product functions on V form a vector space?
26. Establish the validity of the parallelogram law in any euclidean space: $|\alpha + \beta|^2 + |\alpha - \beta|^2 = 2|\alpha|^2 + 2|\beta|^2$.
27. For convenience, let $d(\alpha,\beta)$ designate the *distance* between the vectors α, β in a euclidean space V. Verify that the distance function d has the following properties, for arbitrary vectors α, β, γ in V: *metric*
    (a) $d(\alpha,\beta) \geq 0$
    (b) $d(\alpha,\beta) = 0$ if and only if $\alpha = \beta$
    (c) $d(\alpha,\beta) = d(\beta,\alpha)$
    (d) $d(\alpha,\beta) \leq d(\alpha,\gamma) + d(\gamma,\beta)$
28. If $\{\alpha_1, \alpha_2, \ldots, \alpha_n\}$ is a basis for a euclidean space V, and $c_1, c_2, \ldots, c_n$ are arbitrary real numbers, prove that there exists a unique vector $\alpha \in V$ such that $(\alpha,\alpha_i) = c_i$, $i = 1, 2, \ldots, n$.
29. Let T be the linear operator on V_2, which "rotates" each vector through 90° and is defined by $T(x,y) = (-y,x)$. Note that $(X, TX) = 0$, for any $X \in V_2$, and determine all inner-product functions on V_2 such that any vector and its image under T are orthogonal.
30. If two inner products are defined on a vector space V, prove that they are the same if and only if the associated norms, for arbitrary vectors in V, coincide.
31. In a nongeometric context, the symbol $|\alpha|$ for the *norm* of a vector α is often replaced by $\|\alpha\|$, and its characteristic properties may be listed as: $\|\alpha\| \geq 0$, and $\|\alpha\| = 0$ if and only if $\alpha = 0$; $\|c\alpha\| = |c| \cdot \|\alpha\|$, for any scalar (real or complex) c; $\|\alpha + \beta\| \leq \|\alpha\| + \|\beta\|$. Verify that the following are alternative definitions for the norm of a vector $X = (x_1, x_2, \ldots, x_n) \in V_n$: (a) $\|X\| = \max_{1 \leq i \leq n} \{|x_i|\}$; (b) $\|X\| = \sum_{i=1}^{n} |x_i|$; (c) $\|X\| = \left[\sum_{i=1}^{n} |x_i|^p\right]^{1/p}$, for any positive integer p.

4.3 The Gram-Schmidt Process of Orthogonalization

To solve a problem in analytic geometry, it would be foolish to use other than mutually orthogonal axes, unless, for some reason, they are required by the nature of the problem. It is a "fact of life" that it is much easier to work with a set of cartesian axes, and it is likely that the reader is already persuaded of this fact! For similar reasons it is much easier, in general, to solve problems and obtain results in a vector space if there is available a basis of mutually orthogonal vectors. Such a basis is called an *orthogonal basis*, and it is the purpose of this section to outline a procedure for obtaining a basis of this kind from any given basis of the space. This procedure is often referred to as the *Gram-Schmidt process of orthogonalization*. It is clear that such a process makes sense only in a euclidean space, in which an inner product, and hence a notion of *orthogonality*, has been defined. The principal result of this section may be formally stated as the following theorem.

Theorem 3.1 *Every subspace W ($\neq \mathcal{O}$) of a finite-dimensional euclidean space V possesses an orthogonal basis.*

The proof usually given for this theorem proceeds by mathematical induction, and we shall also give this kind of proof. However, before giving the actual proof, we shall go through more than the minimal number of steps required for an inductive proof, so that the reader can get a better idea of the procedure. It is the considered opinion of the author that to many students there is an element of trickery in inductive proofs, the mathematician being somewhat in the position of a magician who "pulls rabbits out of a hat" or at least seems to be assuming the very thing that he claims to be proving! The basic idea of the process to be described is simple enough, if it is considered geometrically: The first vector of the original basis is retained as the first member of the new basis; then a vector orthogonal to this one is found in the subspace generated by the first two vectors of the original basis; this is followed by the determination of a vector orthogonal to the two vectors of the new basis and which lies in the subspace generated by the first three vectors of the original basis; etc. The Gram-Schmidt procedure provides a formula to replace the "etc.," and from which to obtain, one at a time, all members of the desired basis. But, as suggested above, let us go through a few of the steps.

We assume the presence of a basis $\{\alpha_1, \alpha_2, \ldots, \alpha_r\}$ of W, and, since basis vectors are never zero, we pick any one of these, say α_1, for the first member of the new basis. For convenience we rename this first vector β_1, so that $\alpha_1 = \beta_1$. We now hunt for a vector in $\mathcal{L}\{\alpha_1, \alpha_2\}$, or, equivalently, in $\mathcal{L}\{\beta_1, \alpha_2\}$, that is orthogonal to β_1. Thus, if $\beta_2 = a_1\beta_1 + a_2\alpha_2$, we must determine $a_1, a_2 \in \mathbf{R}$ such that $(\beta_2, \beta_1) = 0 = (a_1\beta_1 + a_2\alpha_2, \beta_1) = a_1(\beta_1, \beta_1) + a_2(\alpha_2, \beta_1)$, and it is clear that we may assume that one of the coefficients, say a_2, is 1 (Prob. 9). It follows that $a_1 = -(\alpha_2, \beta_1)/(\beta_1, \beta_1) = -(\alpha_2, \beta_1)/|\beta_1|^2$, and so $\beta_2 = a_2\alpha_2 + a_1\beta_1 = \alpha_2 - [(\alpha_2, \beta_1)/|\beta_1|^2]\beta_1$. We have now obtained two basis elements β_1 and β_2 that are orthogonal and lie in $\mathcal{L}\{\alpha_1,\alpha_2\}$. We seek the third vector in $\mathcal{L}\{\alpha_1,\alpha_2,\alpha_3\}$ that is orthogonal to both β_1 and β_2. This means that we must solve the equation $\beta_3 = b_1\beta_1 + b_2\beta_2 + b_3\alpha_3$ for $b_1,b_2,b_3 \in \mathbf{R}$, subject to the conditions that $(\beta_3,\beta_1) = (\beta_3,\beta_2) = 0$, with the understanding implied from our previous discussion that $\mathcal{L}\{\beta_1,\beta_2\} = \mathcal{L}\{\alpha_1,\alpha_2\}$. It is easy to see that the conditions just stated are equivalent to the following equations:

$$(\beta_3,\beta_1) = 0 = (b_1\beta_1 + b_2\beta_2 + b_3\alpha_3, \beta_1) = b_1(\beta_1,\beta_1) + b_3(\alpha_3,\beta_1)$$
$$(\beta_3,\beta_2) = 0 = (b_1\beta_1 + b_2\beta_2 + b_3\alpha_3, \beta_2) = b_2(\beta_2,\beta_2) + b_3(\alpha_3,\beta_2)$$

There is no loss in generality if we assume that $b_3 = 1$ in these equations, and so we easily obtain $b_1 = -(\alpha_3,\beta_1)/|\beta_1|^2$, $b_2 = -(\alpha_3,\beta_2)/|\beta_2|^2$. The desired vector is then

$$\beta_3 = b_3\alpha_3 + b_2\beta_2 + b_1\beta_1 = \alpha_3 - \frac{(\alpha_3,\beta_2)}{|\beta_2|^2}\beta_2 - \frac{(\alpha_3,\beta_1)}{|\beta_1|^2}\beta_1$$

and we now have obtained three vectors β_1, β_2, β_3 for the orthogonal basis. The next step would be to obtain a vector β_4, orthogonal to β_1, β_2, β_3, in $\mathcal{L}\{\alpha_1,\alpha_2,\alpha_3,\alpha_4\}$ or, equivalently, in $\mathcal{L}\{\beta_1,\beta_2,\beta_3,\alpha_4\}$. The process will terminate when the number of orthogonal vectors obtained is the same as the dimension of the subspace W. The formal proof of the theorem now proceeds without difficulty.

Proof
If r is the dimension of W, we make the inductive assumption that the theorem is true for all subspaces of V of dimension less than r, it being trivially true for one-dimensional subspaces. If $\{\alpha_1,\alpha_2, \ldots, \alpha_r\}$ is a basis for W, we consider the subspace $\mathcal{L}\{\alpha_1,\alpha_2, \ldots, \alpha_{r-1}\}$ of dimension $r-1$. As a result of our inductive assumption, we may assume the existence of vectors $\{\beta_1,\beta_2, \ldots, \beta_{r-1}\}$ in $\mathcal{L}\{\alpha_1,\alpha_2, \ldots, \alpha_{r-1}\}$ such that

(1) $\mathcal{L}\{\beta_1,\beta_2, \ldots, \beta_{r-1}\} = \mathcal{L}\{\alpha_1,\alpha_2, \ldots, \alpha_{r-1}\}$.
(2) $\beta_1,\beta_2, \ldots, \beta_{r-1}$ are mutually orthogonal.

We now define

$$\beta_r = \alpha_r - \frac{(\alpha_r,\beta_{r-1})}{|\beta_{r-1}|^2}\beta_{r-1} - \frac{(\alpha_r,\beta_{r-2})}{|\beta_{r-2}|^2}\beta_{r-2} - \cdots - \frac{(\alpha_r,\beta_1)}{|\beta_1|^2}\beta_1$$

and verify that $\{\beta_1,\beta_2, \ldots, \beta_r\}$ is an orthogonal basis for W. It is clear that $\beta_r \neq 0$, for, otherwise, $\alpha_r \in \mathcal{L}\{\beta_1,\beta_2, \ldots, \beta_{r-1}\}$, whence $\alpha_r \in \mathcal{L}\{\alpha_1,\alpha_2,$

..., α_{r-1}} by our induction hypothesis, contrary to the linear independence of basis vectors (Prob. 10). But, since $\beta_1, \beta_2, \ldots, \beta_{r-1}$ are mutually orthogonal,

$$(\beta_r, \beta_i) = (\alpha_r, \beta_i) - \frac{(\alpha_r, \beta_{r-1})}{|\beta_{r-1}|^2}(\beta_{r-1}, \beta_i) - \frac{(\alpha_r, \beta_{r-2})}{|\beta_{r-2}|^2}(\beta_{r-2}, \beta_i) - \cdots$$

$$- \frac{(\alpha_r, \beta_1)}{|\beta_1|^2}(\beta_1, \beta_i) = (\alpha_r, \beta_i) - (\alpha_r, \beta_i) = 0 \qquad \text{(Prob. 11)}$$

Hence $\{\beta_1, \beta_2, \ldots, \beta_r\}$ is a mutually orthogonal set of vectors, each of which is in W. The definition of β_r implies that $\mathcal{L}\{\beta_1, \beta_2, \ldots, \beta_r\} \subset \mathcal{L}\{\beta_1, \beta_2, \ldots, \beta_{r-1}, \alpha_r\}$, and so $\mathcal{L}\{\beta_1, \beta_2, \ldots, \beta_r\} \subset \mathcal{L}\{\alpha_1, \alpha_2, \ldots, \alpha_{r-1}, \alpha_r\}$, as a result of assumption (1). Moreover, $\alpha_r \in \mathcal{L}\{\beta_1, \beta_2, \ldots, \beta_r\}$, so that $\mathcal{L}\{\beta_1, \beta_2, \ldots, \beta_r\} \supset \mathcal{L}\{\alpha_1, \alpha_2, \ldots, \alpha_r\}$, also by assumption (1). On combining the two inclusions, we see that $\mathcal{L}\{\alpha_1, \alpha_2, \ldots, \alpha_r\} = \mathcal{L}\{\beta_1, \beta_2, \ldots, \beta_r\}$, and the proof is complete. ∎

EXAMPLE 1

Use the Gram-Schmidt process to transform $\{(1,0,1), (1,2,-2), (2,-1,1)\}$ into an orthogonal basis for V_3, assuming the standard inner product in this space.

Solution
For notational purposes, we depart from our usual symbolism for coordinate vectors and let $\alpha_1 = (1,0,1)$, $\alpha_2 = (1,2,-2)$, and $\alpha_3 = (2,-1,1)$, and pick $\alpha_1 = \beta_1$ for the first element of the orthogonal basis. Using the formula given in the proof of Theorem 3.1, we now obtain

$$\beta_2 = \alpha_2 - \left[\frac{(\alpha_2, \beta_1)}{|\beta_1|^2}\right]\beta_1 = (1,2,-2) - \left[\frac{(-1)}{2}\right](1,0,1)$$
$$= (1,2,-2) + (\tfrac{1}{2}, 0, \tfrac{1}{2}) = (\tfrac{3}{2}, 2, -\tfrac{3}{2})$$

In like manner,

$$\beta_3 = \alpha_3 - \left[\frac{(\alpha_3, \beta_2)}{|\beta_2|^2}\right]\beta_2 - \left[\frac{(\alpha_3, \beta_1)}{|\beta_1|^2}\right]\beta_1 = \alpha_3 - \left[\frac{(-\tfrac{1}{2})}{\tfrac{34}{4}}\right](\tfrac{3}{2}, 2, -\tfrac{3}{2})$$
$$- \left(\frac{2+1}{2}\right)(1,0,1) = (2,-1,1) + (\tfrac{1}{17})(\tfrac{3}{2}, 2, -\tfrac{3}{2}) - (\tfrac{3}{2})(1,0,1)$$
$$= (2,-1,1) + (\tfrac{3}{34}, \tfrac{2}{17}, -\tfrac{3}{34}) - (\tfrac{3}{2}, 0, \tfrac{3}{2}) = (\tfrac{10}{17}, -\tfrac{15}{17}, -\tfrac{10}{17})$$

SEC. 4.3 THE GRAM-SCHMIDT PROCESS OF ORTHOGONALIZATION

Since the dimension of the space V_3 is 3, this is the desired orthogonal basis, as obtained by the Gram-Schmidt process: $\{(1,0,1),(\frac{3}{2},2,-\frac{3}{2}),(\frac{10}{17},-\frac{15}{17},-\frac{10}{17})\}$. It is clear that any scalar multiples of these basis vectors have the same properties of orthogonality; and so we may, if we wish, take the vectors $\{(1,0,1),(3,4,-3),(10,-15,-10)\}$ or, even more simply, $\{(1,0,1),(3,4,-3),(2,-3,-2)\}$ to comprise the orthogonal basis.

A vector of unit length is called a *unit vector*, and it is evident that any nonzero vector α in a euclidean space can be transformed into a unit vector on division by $|\alpha|$. If each vector of an orthogonal basis is a unit vector, the resulting basis is said to be *orthonormal*. In the case of Example 1, the orthonormal basis that is associated with the orthogonal basis obtained there is easily seen to be $\{(1/\sqrt{2},0,1/\sqrt{2}),(3/\sqrt{34},4/\sqrt{34},-3/\sqrt{34}),(2/\sqrt{17},-3/\sqrt{17},-2/\sqrt{17})\}$. It should also be clear that another orthonormal basis for the space V_3 in that example is the familiar $\{E_i\}$ basis.

A word of comment about the Gram-Schmidt *process:* In actual practice, it makes no difference, as far as the final outcome is concerned, whether we normalize (i.e., transform into unit vectors) each of the orthogonal vectors as soon as it is found or do so after we have obtained a complete orthogonal basis. We did the latter in the above example, but if calculations are done on a machine, it is preferable to do the normalizing as each new vector is obtained, since fewer divisions are required by this procedure.

EXAMPLE 2

Determine an orthonormal basis for the euclidean polynomial space P_3, described in Example 2 of Sec. 4.2, with $a = -1$ and $b = 1$.

Solution

If t is the indeterminate used in the construction of the polynomial elements of P_3, it is clear that the set $\{1,t,t^2\}$ is a basis for the space, and we shall apply the Gram-Schmidt process to this basis. If we let $1 = \alpha_1$, $t = \alpha_2$, and $t^2 = \alpha_3$ for notational purposes, we first pick $\beta_1 = \alpha_1 = 1$ as the initial member of the new basis. Also, $\beta_2 = \alpha_2 - [(\alpha_2,\beta_1)/|\beta_1|^2]\beta_1$, where $(\alpha_2,\beta_1) = \int_{-1}^{1} x \, dx = 0$, and $|\beta_1|^2 = \int_{-1}^{1} dx = 2$, so that $\beta_2 = \alpha_2 = t$. (Since t is an indeterminate, we prefer to use the symbol x as the real variable of integration.) Finally, $\beta_3 = \alpha_3 - [(\alpha_3,\beta_2)/|\beta_2|^2]\beta_2 - [(\alpha_3,\beta_1)/|\beta_1|^2]\beta_1$, where $(\alpha_3,\beta_2) = \int_{-1}^{1} x^2(x) \, dx = \int_{-1}^{1} x^3 \, dx = 0$, and $(\alpha_3,\beta_1) = \int_{-1}^{1} x^2 \, dx = \frac{2}{3}$. Hence $\beta_3 = \alpha_3 - [(\frac{2}{3})/2]\beta_1 = t^2 - \frac{1}{3}$. The orthogonal basis, which we obtain from the original basis by the Gram-Schmidt method, is then $\{1, t, t^2 - \frac{1}{3}\}$. We leave it for the reader (Prob. 12) to verify that the associated orthonormal basis for P_3 is $\{1/\sqrt{2}, \sqrt{6}\,t/2, (3\sqrt{10}/4)(t^2 - \frac{1}{3})\}$.

Problems 4.3

1. Give some reasons why orthogonal axes are usually preferred over oblique axes in analytic geometry.
2. Is it implied by Theorem 3.1 that V, as well as W, also has an orthogonal basis?
3. In one form of an inductive proof of a proposition p_n, we prove it for $n = 1$ and then *assume* that p_k is true for an arbitrary positive integer k. Why is it not then the case that we *are* assuming the proposition to be true for all positive integers n?
4. Use the Gram-Schmidt process to transform $\{(-1,2),(1,1)\}$ into an orthogonal basis of V_2, with integral components.
5. Extend $\{(1,2,1),(-2,1,0)\}$ to an orthogonal basis of V_3, with integral components.
6. Extend $\{(2,3,-1),(1,-2,-4)\}$ to an orthogonal basis of V_3, with integral components.
7. Use the Gram-Schmidt process to find an orthogonal basis, with integral components, for $\mathcal{L}\{(2,1,1),(1,-2,1),(-2,1,1)\}$.
8. Use the Gram-Schmidt process to find an orthogonal basis, with integral components, for $\mathcal{L}\{(1,1,1,1),(-1,2,3,1),(3,-2,1,2)\}$.

9. Explain why we are permitted to assume that $a_2 = 1$ and $b_3 = 1$ in the discussion prior to the formal proof of Theorem 3.1.
10. Give the details of the argument leading to a contradiction in the formal proof of Theorem 3.1, from the tentative assumption that $\alpha_r \in \mathcal{L}\{\beta_1, \beta_2, \ldots, \beta_{r-1}\}$.
11. Fill in the details in the proof of Theorem 3.1 that $(\beta_r, \beta_i) = 0$, for any positive integer $i \leq r - 1$.
12. Verify that the final basis, determined in Example 2, is orthonormal.

13. Use the Gram-Schmidt process to extend $\{(1,2,-1),(-2,2,2)\}$ to an orthonormal basis of V_3 by first (a) normalizing each vector; (b) finding a nonnormal orthogonal basis.
14. Use the method of Example 2 to find an orthogonal basis for the space P_4 of real polynomials of degree less than 4. The polynomials obtained, *apart from multiplicative constants*, should be the first four *Legendre polynomials*, which form an orthogonal—but not an orthonormal—basis for the space.

15. Determine the orthonormal basis associated with the basis found in Prob. 14.
16. If the Gram-Schmidt process is applied to a set of vectors to produce a mutually orthogonal set, why must the given vectors have been linearly independent?
17. If $|\alpha| = |\beta|$, for vectors α, β in a euclidean space, prove that $\alpha - \beta$ and $\alpha + \beta$ must be orthogonal.
18. Prove the converse of the statement in Prob. 17.
19. Find a basis for the subspace of all vectors in V_4 that are orthogonal to $(1,1,2,1)$ and $(2,0,1,0)$.
20. In the space of real integrable functions on **R**, with the inner product defined as in Example 1 of Sec. 4.2 ($a = -1$, $b = 1$), find a polynomial function that is of degree 2 and orthogonal to f and g, where $f(x) = 2$ and $g(x) = x + 1$ for all $x \in$ **R**.
21. Let V be a euclidean space with $\{\alpha_1, \alpha_2, \ldots, \alpha_n\}$ an orthonormal basis of the space. Then, if we define a linear operator T as an *isometry* provided it preserves inner products [that is, $(T\alpha, T\beta) = (\alpha, \beta)$, for arbitrary $\alpha, \beta \in V$], prove that (a) a linear operator T is an isometry if and only if $\{T\alpha_1, T\alpha_2, \ldots, T\alpha_n\}$ is an orthonormal basis of V; (b) an isometry T is invertible and T^{-1} is also an isometry.
22. If $\alpha_1, \alpha_2, \ldots, \alpha_r$ are arbitrary vectors of a euclidean space V, we define the *gramian* matrix G to be $[a_{ij}]$, where $a_{ij} = (\alpha_i, \alpha_j)$, $i, j = 1, 2, \ldots, r$. Prove that $\alpha_1, \alpha_2, \ldots, \alpha_r$ are linearly dependent if and only if $\det G = 0$. If $\dim V = n$, select an orthonormal basis for V and use it to represent G as the product of an $r \times n$ and an $n \times r$ matrix. Then verify that $\det G \geq 0$.

4.4 Orthogonal Complements

In this section, we examine the general problem of the possible decomposition of a vector space into the direct sum of a *given* subspace and one that is orthogonal to it. The geometric analog of this in ordinary 3-space is the following: given a plane (or line) in 3-space, find a line (or plane) orthogonal to the plane (or line). It is well known from elementary geometry that such *orthogonal complements* always exist, and we shall see that an analogous situation prevails in any finite-dimensional euclidean space. It may be of interest to recall that, in Sec. 1.9, we discussed direct sums of subspaces, but now we are to impose the stronger condition of *orthogonality* on the subspaces. However, before we discuss the principal result of the section, there are two preliminary matters of importance on which we wish to comment.

It has already been pointed out that quite possibly more than one inner-product function can be defined on a given vector space, so that it may be "euclidean" in more than one way. However, *during any one study*, it is usually the case that only one inner product is considered to have been defined on a space; and, although this remark may seem trivial, its significance is often overlooked. An inner product is usually defined in terms of the *coordinates* of vectors, and these in turn depend on the *basis* chosen for the representation of the vectors of the space. Thus, if the vectors of a space are expressed in terms of a basis that is not the one used in the definition of the inner-product function, it is necessary to express the vectors in terms of the latter basis before their inner products can be computed. For example, the standard inner product of two vectors in V_n can be described verbally as the sum of the products of their respective coordinates, *relative to the* $\{E_i\}$ *basis*. If some other basis is used to represent the elements of V_n, due cognizance of this must be taken in the determination of any *standard* inner product. (See Prob. 8.) The matter of orthogonality is directly connected with the definition of inner product, and vectors that are orthogonal according to one definition are possibly not orthogonal relative to a different inner-product function. It is also interesting and important to observe that an inner-product function is completely determined by a basis that is orthonormal. Suppose $\{\gamma_1, \gamma_2, \ldots, \gamma_n\}$ is an orthonormal basis for a space V, with α and β arbitrary vectors of V. Then, since $\alpha = a_1\gamma_1 + a_2\gamma_2 + \cdots + a_n\gamma_n$ and $\beta = b_1\gamma_1 + b_2\gamma_2 + \cdots + b_n\gamma_n$, for real numbers $a_1, a_2, \ldots, a_n, b_1, b_2, \ldots, b_n$, the requirements of an inner product and an orthonormal basis demand that $(\alpha,\beta) = (a_1\gamma_1 + a_2\gamma_2 + \cdots + a_n\gamma_n, b_1\gamma_1 + b_2\gamma_2 + \cdots + b_n\gamma_n) = a_1b_1 + a_2b_2 + \cdots + a_nb_n$. However, although a euclidean space of finite dimension has an unambiguous *rule* for the formation of an inner product in the presence of an orthonormal basis, it should be recalled that the actual inner product of two vectors may vary with the inner product on which the orthonormal nature of the basis is dependent. Only if two orthonormal bases are related in a very special way, to be discussed in Chap. 5, will the *values* as well as the rule of formation of the associated inner products be the same (see Prob. 6).

The other matter of preliminary interest is the observation of the ease with which the coordinates of a vector can be obtained, *if an orthonormal basis is assumed for the space*. Thus, let α be any vector of a euclidean space, for which $\{\beta_1, \beta_2, \ldots, \beta_n\}$ is an orthonormal basis. Since $\alpha = a_1\beta_1 + a_2\beta_2 + \cdots + a_n\beta_n$, for real numbers $a_1, a_2, \ldots, a_n$, it follows that $(\alpha,\beta_i) = (a_1\beta_1 + a_2\beta_2 + \cdots + a_n\beta_n, \beta_i) = a_i(\beta_i,\beta_i) = a_i$, for $1 \leq i \leq n$. Hence, *the coordinates of a vector, relative to an orthonormal basis of the space, are the respective inner products of the vector with the basis elements*. In view of this result, it seems appropriate to speak of the coordinates of α as *orthogonal projections* of α on the respective basis vectors—a use of the word *projection* that is consistent with its use in elementary plane and solid geometry (see Probs. 15 and 16).

EXAMPLE 1

Determine the coordinates of $X = (2,1,2)$ in V_3, relative to the orthonormal basis $\{(1,0,0),(0,1/\sqrt{2},1/\sqrt{2}),(0,-1/\sqrt{2},1/\sqrt{2})\}$.

Solution
The coordinates a_1, a_2, a_3 of X are easily computed as follows: $a_1 = ((2,1,2),(1,0,0)) = 2 + 0 + 0 = 2$; $a_2 = ((2,1,2),(0,1/\sqrt{2},1/\sqrt{2})) = 0 + 1/\sqrt{2} + 2/\sqrt{2} = 3/\sqrt{2} = 3\sqrt{2}/2$; $a_3 = ((2,1,2),(0,-1/\sqrt{2},1/\sqrt{2})) = 0 - 1/\sqrt{2} + 2/\sqrt{2} = 1/\sqrt{2} = \sqrt{2}/2$. Thus $(2,3\sqrt{2}/2, \sqrt{2}/2)$ is the coordinate vector of $(2,1,2)$, relative to the given orthonormal basis, and so $(2,1,2) = 2(1,0,0) + (3\sqrt{2}/2)(0,1/\sqrt{2},1/\sqrt{2}) + (\sqrt{2}/2)(0,-1/\sqrt{2},1/\sqrt{2})$.

EXAMPLE 2

It is easy to extend (see Prob. 15 of Sec. 4.3) the result in Example 2 of Sec. 4.3 to obtain the set of *normalized* Legendre polynomials $P_0(t)$, $P_1(t), \ldots, P_n(t)$ of degrees 0, 1, ..., n, respectively. If we use these polynomials as an orthonormal basis for the polynomial space P_{n+1} (assuming the inner product defined in the earlier example), it is now a very simple matter to determine the coordinates of an arbitrary polynomial vector $P(t) \in P_{n+1}$. In classical language, this is the process of expressing a polynomial of degree n or less as a linear combination of the first $n + 1$ normalized Legendre polynomials. But if $P(t) = a_0 P_0(t) + a_1 P_1(t) + a_2 P_2(t) + \cdots + a_n P_n(t)$, for real numbers $a_0, a_1, \ldots, a_n$, we are now able to use the above result to compute $a_i = \int_{-1}^{1} P(x) P_i(x) \, dx$, $i = 0, 1, \ldots, n$. To illustrate, we already know from the earlier example referred to that $P_2(t) = (3\sqrt{10}/4)(t^2 - \frac{1}{3})$. Then, if $n = 4$, and we wish to express $P(t) = t^3 - 2t + 1$ in terms of the first four normalized Legendre polynomials, we find easily that $a_2 = \int_{-1}^{1} (x^3 - 2x + 1)(3\sqrt{10}/4)(x^2 - \frac{1}{3}) \, dx = -\sqrt{10}/8$. In like manner, the coefficients a_0, a_1, a_3 can be obtained, but we leave their computation to the reader (Prob. 13). When these are found, the expression of $P(t)$ in the desired form is immediate. It is known that $P_k(x) = (K/2^k k!)[(d^k/dx^k)(x^2 - 1)^k]$, for some normalizing constant K (dependent on k) which can be determined quite easily. The method of expressing any polynomial of degree n or less in terms of the first $n + 1$ Legendre polynomials is then quite like the case $n = 4$ just illustrated. [*Note:* The symbol t may be either an indeterminate or real variable, but in the formulas involving integration and differentiation the symbol x *must* be a real variable.]

We now come to the principal topic of this section: *orthogonal subspaces*. This is a very easy one to investigate, in view of our prior development of the Gram-Schmidt orthogonalization process. The principal result is stated in the following theorem.

Theorem 4.1 *If W is a subspace of a finite-dimensional euclidean vector space V, there exists a unique subspace $W^\perp$, orthogonal to W, such that $V = W \oplus W^\perp$. Each of W and $W^\perp$ is called the orthogonal complement of the other in V.*

Proof
Let $\{\alpha_1, \alpha_2, \ldots, \alpha_r\}$ be a basis for W and, by Theorem 8.2 (Chap. 1), this may be extended to a basis $\{\alpha_1, \alpha_2, \ldots, \alpha_r, \gamma_{r+1}, \gamma_{r+2}, \ldots, \gamma_n\}$ of V. These n basis vectors can now be normalized by the Gram-Schmidt process to produce an orthogonal basis $\{\beta_1, \beta_2, \ldots, \beta_n\}$ of V, and the details of the orthogonalization process assures that $\mathcal{L}\{\alpha_1, \alpha_2, \ldots, \alpha_r\} = \mathcal{L}\{\beta_1, \beta_2, \ldots, \beta_r\}$. If we let $W^\perp = \mathcal{L}\{\beta_{r+1}, \beta_{r+2}, \ldots, \beta_n\}$, it follows at once that $V = W + W^\perp$, and it is easy to see (Prob. 12), in fact, that $V = W \oplus W^\perp$. The uniqueness of the orthogonal complement $W^\perp$ is still to be established. We know that each vector in $W^\perp$ is orthogonal to every vector in W, but we now show that *every* vector in V that is orthogonal to W lies in $W^\perp$. To this end, suppose that α is such a vector, where $\alpha = \alpha_1 + \alpha_2$, with $\alpha_1 \in W$ and $\alpha_2 \in W^\perp$. Then $(\alpha_1, \alpha_1) = (\alpha - \alpha_2, \alpha_1) = (\alpha, \alpha_1) - (\alpha_2, \alpha_1) = 0 - 0 = 0$, and Property 3 of an inner-product function now requires that $\alpha_1 = 0$. Hence $\alpha = \alpha_2 \in W^\perp$, and so every vector in V that is orthogonal to all vectors in W lies in $W^\perp$, so that $W^\perp$ may be characterized as the unique (Prob. 11) subspace of V with this property. It is then appropriate to speak of *the* orthogonal complement of W, since it is uniquely determined by V and W.

EXAMPLE 3

Determine the subspace of V_3 that is the orthogonal complement of $\mathcal{L}\{(1,0,1),(1,2,-2)\}$.

Solution
It is a simple matter to extend the basis of the subspace to a basis of V_3. For example, such a basis might be $\{(1,0,1),(1,2,-2),(2,-1,1)\}$. By the

result in Example 1 of Sec. 4.3, an application of the Gram-Schmidt process to this basis yields the following orthogonal basis: $\{(1,0,1),(3,4,-3),(2,-3,-2)\}$. Moreover, we recall from the orthogonalization process that $\mathcal{L}\{(1,0,1), (1,2,-2)\} = \mathcal{L}\{(1,0,1),(3,4,-3)\}$; and, if we designate this subspace W_1, we see that $V_3 = W_1 \oplus W_2$, where $W_2 = \mathcal{L}\{(2,-3,-2)\}$. The subspace W_2 is then the desired orthogonal complement. In the language of geometry, this orthogonal complement is the one-dimensional subspace of line vectors that are orthogonal to the two-dimensional space (or plane) of vectors containing $(1,0,1)$ and $(1,2,-2)$.

Problems 4.4

1. Use geometric language to explain the meaning of the orthogonal complement of (a) $\mathcal{L}\{X\}$, for $X (\neq 0) \in V_3$; (b) $\mathcal{L}\{X,Y\}$, for X, Y linearly independent vectors of V_3.
2. Use geometric language to describe what is meant by the orthogonal complement of $\mathcal{L}\{X\}$, where X is a nonzero vector of V_2.
3. Explain why an orthogonal decomposition of a vector space is a more restrictive type than a mere direct sum of subspaces.
4. If X and Y are elements of V_n, is it true that the *only* definition of (X,Y) requires it to equal the sum of the products of the respective components of X and Y? Explain.
5. The space V_2 can be made euclidean by defining $(X,Y) = a_1b_1 + a_2b_2$, for arbitrary $X = a_1X_1 + a_2X_2$ and $Y = b_1X_1 + b_2X_2$, with $\{X_1,X_2\}$ any basis of V_2. Use $X = (2,1)$ and $Y = (1,-1)$ to compare the different values (X,Y), relative to each of the indicated bases: (a) $X_1 = E_1 = (1,0)$, $X_2 = E_2 = (0,1)$; (b) $X_1 = (2,3)$, $X_2 = (-1,-3)$; (c) $X_1 = (2,3)$, $X_2 = (-3,2)$.
6. Use the directions in Prob. 5 but with $X_1 = (\frac{1}{2},\sqrt{3}/2)$, $X_2 = (\sqrt{3}/2,-\frac{1}{2})$. Compare this result with that in Prob. 5a.
7. If X and Y are coordinate vectors, relative to an orthonormal basis associated with vectors of a three-dimensional vector space, determine (X,Y), where (a) $X = (3,1,3)$, $Y = (-1,1,2)$; (b) $X = (-2,2,1)$, $Y = (2,1,-3)$.
8. Find the standard inner product of X and Y, where $X = (1,2,1)$ and $Y = (-2,1,-3)$. Now regard X and Y as the respective coordinates of vectors in V_3, relative to the basis $\{(1,1,-1),(1,0,1),(0,2,1)\}$, and determine the (standard) inner product of these two vectors. Compare the two results.
9. Assume the standard inner product in V_3 and determine the subspace that is orthogonal to (a) $\mathcal{L}\{(1,1,-1)\}$; (b) $\mathcal{L}\{(1,1,2), (0,1,2)\}$.

10. Fill in the details of the argument, outlined in the text, that an inner product is completely determined by an orthonormal basis of the space.
11. With reference to the proof of uniqueness in Theorem 4.1, amplify the argument that the "characteristic" property of $W^\perp$ establishes its uniqueness.
12. In the proof of Theorem 4.1, explain why the sum of W and $W^\perp$ is direct.
13. Complete Example 2 with a determination of a_0, a_1, a_3.
14. Let $\{P_0(t), P_1(t), P_2(t), P_3(t)\}$ be the orthonormal basis of Legendre polynomials of the polynomial space P_4, as discussed in Example 2. If $p = 2P_0 + P_2 - 3P_3$ and $q = 4P_1 + P_2 - 2P_3$, find an expression for (p,q), (assuming the inner-product function of the example) but do not carry out any integration.
15. Use a cartesian coordinate system in the plane to prove the assertion that our present use of the word *projection* is consistent with its use in plane geometry.
16. Use the directions, analogous to those in Prob. 15, for solid geometry.

17. Use the method given in this section to find the coordinates of the vector $(1,2) \in V_2$, relative to an orthonormal basis that includes $\left(\dfrac{2}{\sqrt{5}}, -\dfrac{1}{\sqrt{5}}\right)$ as one of its members.
18. Determine the orthogonal projection of $(1,-2,0,0)$ on $\mathcal{L}\{(0,1,2,0),(1,0,0,1)\}$.
19. Show that the condition that W_1 and W_2 are orthogonal complements in V_n implies that each is the solution space of a system of homogeneous linear equations. [*Hint:* Assume a basis for one of the subspaces, and apply the orthogonality condition to an arbitrary vector in the other.]
20. Use the result in Prob. 19 to explain how $\mathcal{L}\{(1,1,0,1),(0,1,1,0)\}$ can be described as the solution space of the following system of equations:

$$2x_1 + x_2 - x_3 - 3x_4 = 0$$
$$x_1 - x_4 = 0$$

Is it possible to describe every subspace of V_n in such a way?

21. Assume an orthonormal basis $\{\alpha_1, \alpha_2, \ldots, \alpha_n\}$ for a vector space V, and establish Bessel's inequality for $\alpha \in V$:

$$\sum_{i=1}^{n} |(\alpha, \alpha_i)|^2 \leq |\alpha|^2$$

22. Assume an orthonormal basis $\{\alpha_1, \alpha_2, \ldots, \alpha_n\}$ for a vector space V and establish Parseval's identity for $\alpha, \beta \in V$:

$$(\alpha, \beta) = \sum_{i=1}^{n} (\alpha, \alpha_i)(\beta, \alpha_i)$$

23. Give geometric interpretations in 3-space of the results in Probs. 21 and 22.

24. Use vector methods to prove that line segments joining the midpoints of adjacent sides of any quadrilateral form a parallelogram. What can be said about the parallelogram if the quadrilateral is a square?

25. A vector space V is defined as $\mathcal{L}\{1, \cos x, \sin x, \cos 2x, \sin 2x, \ldots, \cos nx, \sin nx\}$, for a real variable x. If we define $(f, g) = \int_0^{2\pi} f(x)g(x)\, dx$, for any $f, g \in V$, prove that the given generators form an orthogonal basis of V.

26. Verify that $\{1/\sqrt{2\pi}, 1/\sqrt{\pi} \cos x, 1/\sqrt{\pi} \sin x, \ldots, 1/\sqrt{\pi} \cos nx, 1/\sqrt{\pi} \sin nx\}$ is an orthonormal basis for the space V in Prob. 25.

27. If

$$\begin{bmatrix} 1 & 0 & -1 \\ 2 & 1 & 2 \\ 3 & 0 & -2 \end{bmatrix}$$

is the matrix of a linear operator T on V_3, determine (X, Y) and (TX, TY), where (a) $X = (1, 2, -3)$, $Y = (2, 1, 0)$; (b) $X = (1, 0, -2)$, $Y = (0, -3, -2)$; (c) $X = (1, 1, 1)$, $Y = (2, -3, 2)$.

28. Let V be the euclidean space of continuous functions, as described in Example 1 of Sec. 4.2 with $a = -1$ and $b = 1$, with W the subspace of "odd" functions h such that $h(-x) = -h(x)$. Determine the orthogonal complement of W, noting that V is not of finite dimension. {Hint: $f(x) = G(x) + H(x)$, where $G(x) = [f(x) + f(-x)]/2$ and where $H(x) = [f(x) - f(-x)]/2$.}

206   INNER-PRODUCT SPACES

29. Let $\{\alpha_1, \alpha_2, \ldots, \alpha_n\}$ be an orthonormal basis for a euclidean space V. Then, if $A = [a_{ij}]$ is the matrix, relative to the given basis, of a linear operator T on V, prove that $a_{ij} = (T\alpha_j, \alpha_i)$, where $i, j = 1, 2, \ldots, n$.
30. Regard the system of $n \times n$ matrices as a vector space of dimension n^2, and use the analogy of the standard inner product in V_n to define an inner product (A,B) for any two matrices A, B in the space. (a) Verify that $(A,B) = \text{tr}(AB')$, recalling Prob. 20 of Sec. 2.10; (b) determine the orthogonal complement of the subspace of diagonal matrices.

4.5 Orthogonal Transformations

A study of elementary geometry consists, in large part, of a study of the invariants of geometric figures under the *rigid motions* of the space that contains them. It is not difficult to see (Prob. 1) that these rigid motions are composites of certain *translations* and *rotations* in the space. It is clear that the *translations* of a vector space V (that is, the transformations that map each vector $\alpha \in V$ onto $\alpha + \alpha_0$, for some fixed $\alpha_0 \in V$) are not linear, and we have previously encountered mappings of this type. However, the geometric *rotations*, as discussed in Sec. 2.1, are linear, and they also have the special property that is characteristic of the transformations to be studied in this section: They are *distance-preserving*. The notion of *distance* is dependent, of course, on what inner-product function has been defined in the space, but, regardless of the definition, we shall denote the inner product by (,). We now give the precise definition of the type of linear operator currently under study.

Definition   *A linear operator T on a euclidean space V is said to be orthogonal if $(T\alpha, T\alpha) = (\alpha, \alpha)$, for any $\alpha \in V$.*

That is, a linear operator is orthogonal if it preserves the length of every vector in the space. It is an elementary matter to verify that this is equivalent to the requirement that it preserve inner products (Prob. 10), and so an orthogonal operator also preserves the "angle" between any two vectors of the space (Prob. 10). In Prob. 21 of Sec. 4.3, we have already called an orthogonal operator an *isometry* of the space.

Our first interesting observation is that the set S of orthogonal operators on a vector space V is a subgroup of the full linear group of V. Let T be such an operator. Since a nonzero vector cannot be transformed by T into the zero vector, the null space of T must consist of the zero vector alone. It follows from the definition in Sec. 2.6 that T is nonsingular, so that Theorem

6.4 (Chap. 2) implies that S is a *subset* of the full linear group of V. If $T_1, T_2 \in S$, neither T_1 nor T_2 can change the length of any vector in V, and so neither can the product T_1T_2. Hence S is closed under multiplication. Finally, for any $T \in S$, T^{-1} must exist, and if T^{-1} were to change the length of a vector $\alpha \in V$, the length of $T(T^{-1}\alpha) = \alpha$ would be different from that of $T\alpha$. But this would contradict the defining property of T as an orthogonal operator, and so T^{-1} must preserve the length of any vector in V, whence $T^{-1} \in S$. Hence S is a subgroup of the full linear group and is known as the *orthogonal group* of V *with respect to the inner product in* V.

Not very much can be said about the matrices that represent orthogonal operators, in general, in view of the very important role played by the bases with respect to which the various matrix representations of an operator are to be made. However, if an *orthonormal* basis is used for a space, a very interesting result can be obtained.

Theorem 5.1 *If* $\{\alpha_1, \alpha_2, \ldots, \alpha_n\}$ *is an orthonormal basis for a euclidean space* V, *and* T *is an orthogonal operator whose matrix with respect to this basis is* A, *then the following is true:*

(i) *The columns of* A, *as vectors of* V_n, *are orthonormal with respect to the standard inner product.*

(ii) $A^{-1} = A'$, *so that the rows of* A *are also orthonormal vectors of* V_n.

(iii) $\det A = \pm 1$. *Moreover,* T *is said to be a rotation if* $\det A = 1$, *and a reflection if* $\det A = -1$.

Proof

(i) If $A = [a_{ij}]$, we know that $T\alpha_i = a_{1i}\alpha_1 + a_{2i}\alpha_2 + \cdots + a_{ni}\alpha_n$, for $i = 1, 2, \ldots, n$. Hence $1 = (\alpha_i, \alpha_i) = (T\alpha_i, T\alpha_i) = (a_{1i}\alpha_1 + a_{2i}\alpha_2 + \cdots + a_{ni}\alpha_n, a_{1i}\alpha_1 + a_{2i}\alpha_2 + \cdots + a_{ni}\alpha_n) = a_{1i}^2 + a_{2i}^2 + \cdots + a_{ni}^2$. But this implies that the standard inner product of $(a_{1i}, a_{2i}, \ldots, a_{ni})$ with itself is 1, so that the length of each column vector of A is 1. Moreover, since α_i and α_j ($i \neq j$) are orthogonal, so also are $T\alpha_i$ and $T\alpha_j$, whence $0 = (a_{1i}\alpha_1 + a_{2i}\alpha_2 + \cdots + a_{ni}\alpha_n, a_{1j}\alpha_1 + a_{2j}\alpha_2 + \cdots + a_{nj}\alpha_j) = a_{1i}a_{1j} + a_{2i}a_{2j} + \cdots + a_{ni}a_{nj}$. Hence the columns of A are orthogonal vectors of V_n, and this, when combined with the preceding result, proves that they are orthonormal.

(ii) If the product $A'A$ is computed, the element at the intersection of the ith row and jth column of the product matrix is the standard inner product of the ith and jth columns of A, regarded as elements of V_n. In view of statement (i), this shows that $A' = A^{-1}$, and so the rows of A are also orthonormal vectors of V_n.

(iii) Since $A^{-1} = A'$, $1 = \det I = \det A'A = (\det A')(\det A) = (\det A)^2$, from which we can conclude that $\det A = \pm 1$. ∎

The following definition will not seem unnatural in the light of the preceding result.

Definition A nonsingular matrix A, such that $A^{-1} = A'$, is called an *orthogonal matrix*.

It would be a simple matter to verify that the set of orthogonal $n \times n$ matrices constitutes a group that is isomorphic to the group of orthogonal operators on an n-dimensional euclidean space (Prob. 11). However, it must be understood that each choice of orthonormal basis for the space determines a different isomorphic mapping. It *should not* be implied from Theorem 5.1 that every orthogonal matrix represents an orthogonal operator, and this fact puts a slight "crimp" into our customary isomorphic treatment of linear transformations and their representing matrices. However, we do have the following result which asserts, in a sense, that "all is well" if we stick to orthonormal bases.

Theorem 5.2 Any orthogonal $n \times n$ matrix A represents an orthogonal linear operator T on an n-dimensional vector space V, provided the representation is with respect to an orthonormal basis of V.

Proof
Let $\{\alpha_1, \alpha_2, \ldots, \alpha_n\}$ be an orthonormal basis, with $\alpha = a_1\alpha_1 + a_2\alpha_2 + \cdots + a_n\alpha_n$ an arbitrary vector of V, and $a_1, a_2, \ldots, a_n \in \mathbf{R}$. Then if $T\alpha = b_1\alpha_1 + b_2\alpha_2 + \cdots + b_n\alpha_n$, for $b_1, b_2, \ldots, b_n \in \mathbf{R}$, we know (see Sec. 2.8) that $[b_1 b_2 \cdots b_n]' = A[a_1 a_2 \cdots a_n]'$. But the $\{\alpha_i\}$ basis is orthonormal, so that $(T\alpha, T\alpha) = b_1^2 + b_2^2 + \cdots + b_n^2 = [b_1 b_2 \cdots b_n][b_1 b_2 \cdots b_n]'$. If we now use the orthogonal nature of A, we see that $(T\alpha, T\alpha) = [a_1 a_2 \cdots a_n]A'A[a_1 a_2 \cdots a_n]' = [a_1 a_2 \cdots a_n][a_1 a_2 \cdots a_n]' = a_1^2 + a_2^2 + \cdots + a_n^2 = (\alpha, \alpha)$. Hence the operator T is orthogonal. ∎

Before giving some examples of orthogonal operators, it may be well to point out that it is possible for a mapping of a euclidean space to be distance-preserving without being linear. We have already referred to a distance-preserving mapping of a vector space as a *rigid motion* of the space. In the context of the present section, a *rigid motion* of a euclidean space V is a mapping T of V such that $|T\alpha - T\beta| = |\alpha - \beta|$, for any $\alpha, \beta \in V$. Of course, every orthogonal operator on V is a rigid motion, but, as we have just remarked, there exist rigid motions that are not even linear. An elementary example of this latter type of mapping is a *translation*, mentioned at the beginning of this section. It was asserted there, however, that any rigid motion is the composite of a translation and an orthogonal operator.

EXAMPLE 1

Let T be an orthogonal operator on a one-dimensional euclidean space V, generated by the single vector α. Then, necessarily, $T\alpha = \lambda\alpha$, for some real number λ, and, since $(T\alpha, T\alpha) = (\alpha, \alpha)$, it follows that $\lambda^2(\alpha, \alpha) = (\alpha, \alpha)$. Hence $\lambda = \pm 1$. There are then only two possibilities for an orthogonal operator T on a one-dimensional space: Either $T\beta = \beta$ or $T\beta = -\beta$, for each $\beta \in V$. The one is the identity operator, and the other, in a geometrical sense, reverses the direction of each vector.

EXAMPLE 2

We now consider the possibilities for an orthogonal operator T on a two-dimensional euclidean space V. If $A = \begin{bmatrix} a & c \\ b & d \end{bmatrix}$ is the matrix of T, relative to an orthonormal basis, there are two cases to be considered: (i) $\det A = 1$; (ii) $\det A = -1$.

(i) If $\det A = 1$, then $ad - bc = 1$, and the condition of orthogonality requires that $\begin{bmatrix} a & c \\ b & d \end{bmatrix}^{-1} = \begin{bmatrix} a & b \\ c & d \end{bmatrix}$. But an easy application of the cofactor rule for computing inverses shows that $\begin{bmatrix} a & c \\ b & d \end{bmatrix}^{-1} = \begin{bmatrix} d & -c \\ -b & a \end{bmatrix}$, whence $\begin{bmatrix} a & b \\ c & d \end{bmatrix} = \begin{bmatrix} d & -c \\ -b & a \end{bmatrix}$. Hence $d = a$ and $c = -b$, so that $A = \begin{bmatrix} a & -b \\ b & a \end{bmatrix}$, with $a^2 + b^2 = 1$. There always exists a real number θ such that $a = \cos\theta$ and $b = \sin\theta$ with $a^2 + b^2 = 1$; and so a typical matrix A for this case has the form $\begin{bmatrix} \cos\theta & -\sin\theta \\ \sin\theta & \cos\theta \end{bmatrix}$. But this implies that an arbitrary vector in V, with coordinate pair (x,y), is mapped by T onto a vector whose coordinate pair is (x',y'), where

$$\begin{bmatrix} \cos\theta & -\sin\theta \\ \sin\theta & \cos\theta \end{bmatrix} \begin{bmatrix} x \\ y \end{bmatrix} = \begin{bmatrix} x' \\ y' \end{bmatrix}$$

Thus (see Example 2 of Sec. 2.7), in the geometry of ordinary 2-space, an orthogonal operator of this type represents a rotation of the plane through an angle of θ rad.

(ii) If $\det A = -1$, then $ad - bc = -1$, and, this time, we are led to the matrix equality $\begin{bmatrix} a & b \\ c & d \end{bmatrix} = \begin{bmatrix} -d & c \\ b & -a \end{bmatrix}$. Hence $d = -a$ and

210 INNER-PRODUCT SPACES

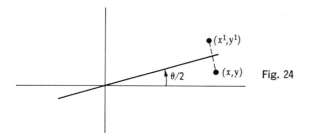

Fig. 24

$c = b$, so that $A = \begin{bmatrix} a & b \\ b & -a \end{bmatrix}$, with $a^2 + b^2 = 1$. Again, it is possible to replace a by $\cos \theta$ and b by $\sin \theta$, for some real number θ, so that

$$A = \begin{bmatrix} \cos \theta & \sin \theta \\ \sin \theta & -\cos \theta \end{bmatrix}$$

In this case, the effect of T is to map a vector of V, with coordinate pair (x,y), onto the vector with coordinate pair (x',y'), where

$$\begin{bmatrix} \cos \theta & \sin \theta \\ \sin \theta & -\cos \theta \end{bmatrix} \begin{bmatrix} x \\ y \end{bmatrix} = \begin{bmatrix} x' \\ y' \end{bmatrix}$$

We leave it as an exercise for the reader to verify, with the help of Fig. 24, that this type of operator in ordinary 2-space reflects the plane in a line through the origin that makes an angle of $\theta/2$ rad with the x axis (Prob. 13).

These results from geometry then provide a reason for the names given to the two types of orthogonal operators in the statement of Theorem 5.1.

Problems 4.5

1. Give an intuitive explanation of why any rigid motion in ordinary 3-space must be a composite of translations and rotations (see Prob. 14).
2. Review the reason why a translation, which is not the identity mapping of a vector space, is not linear.

3. Find the inverse of the orthogonal matrix A, where

(a) $$A = \begin{bmatrix} 1 & 0 & 0 \\ 0 & -\sqrt{3}/2 & -\tfrac{1}{2} \\ 0 & -\tfrac{1}{2} & \sqrt{3}/2 \end{bmatrix}$$

(b) $$A = \begin{bmatrix} \cos\theta & -\sin\theta & 0 \\ \sin\theta & \cos\theta & 0 \\ 0 & 0 & -1 \end{bmatrix}$$

4. Explain why an orthogonal operator must map an orthonormal basis onto a basis that is also orthonormal.

5. Verify that

$$\begin{bmatrix} \tfrac{2}{3} & -\tfrac{2}{3} & -\tfrac{1}{3} \\ \tfrac{1}{3} & \tfrac{2}{3} & -\tfrac{2}{3} \\ \tfrac{2}{3} & \tfrac{1}{3} & \tfrac{2}{3} \end{bmatrix}$$

is an orthogonal matrix.

6. Let the matrix in Prob. 5 represent an orthogonal operator T on V_3, relative to the $\{E_i\}$ basis, and verify that T preserves the length of the vector (3,0,4).

7. Let the matrix in Prob. 5 represent an operator T on V_3, relative to the basis $\{(1,0,1),(1,1,0),(0,0,1)\}$, and determine the length of T(3,0,4). Explain the difference from the result in Prob. 6.

8. Verify that the operator T, as defined in Prob. 6, preserves the angle between the vectors (3,0,4) and (2,2,1).

9. Find the matrix of the operator T in Prob. 6, relative to the basis $\{(0,1/\sqrt{2},1/\sqrt{2}),(0,-1/\sqrt{2},1/\sqrt{2}),(1,0,0)\}$.

10. Prove that a linear operator preserves lengths of vectors if and only if it preserves inner products of vectors. Then deduce that an orthogonal operator also preserves the angle between any two vectors in the space on which it operates. [*Hint:* Let $\alpha - \beta = \gamma$, and use the fact that $(T\gamma,T\gamma) = (\gamma,\gamma)$.]

11. Verify that the set of orthogonal $n \times n$ matrices forms a multiplicative group and that this orthogonal group is isomorphic to the group of orthogonal operators on an n-dimensional vector space.

12. In Example 2, point out where we used condition (i) that det $A = 1$ and condition (ii) that det $A = -1$.

13. Verify that the orthogonal operator in case (ii) of Example 2 is a reflection in the line through the origin that makes an angle of $\theta/2$ rad with the x axis.
14. Prove that any rigid motion in 2-space that leaves the origin fixed is an orthogonal operator. Then conclude that any rigid motion in this space is a translation followed by an orthogonal operator and so is either a translation followed by a rotation or a translation followed by a reflection and a rotation (see Prob. 13).
15. Let T be a linear operator on a vector space V, with $\{\alpha_1, \alpha_2, \ldots, \alpha_n\}$ an orthonormal basis of V. Then prove that T is orthogonal if $(T\alpha_i, T\alpha_j) = \delta_{ij}$, where δ_{ij} is the Kronecker delta, and deduce the converse of the result in Prob. 4.
16. Refer to Prob. 15, and prove that the matrix of transition from one orthonormal basis of a vector space to another is orthogonal.
17. Use the $\{E_i\}$ basis of V_2 to give a matrix representation of the orthogonal operator that rotates each vector through an angle of (a) 2 rad; (b) 30°; (c) 60°.
18. Find a symmetric orthogonal matrix whose first column is $(\frac{3}{5}, 0, \frac{4}{5})$.
19. Determine necessary and sufficient conditions for the matrix $\begin{bmatrix} x+y & y-x \\ x-y & y+x \end{bmatrix}$ to be orthogonal.
20. What is the matrix representation, relative to the $\{E_i\}$ basis of V_2, of the operator that reflects each point of the plane in the diagonal line bisecting the first and third quadrants? Verify that this matrix is orthogonal (cf. Example 2).
21. If A is the matrix of an orthogonal operator on a vector space, prove that $\det A = \pm 1$, regardless of whether the basis used in the space is orthonormal. Moreover, for any given orthogonal operator, prove that the determinants of the representing matrices are either all 1 or all -1.
22. If $\{\alpha_1, \alpha_2, \alpha_3\}$ is an orthonormal basis for a euclidean space V, find the associated matrix of an orthogonal operator that maps α_2 onto $\frac{2}{3}\alpha_1 - \frac{1}{3}\alpha_2 + \frac{2}{3}\alpha_3$.
23. Let $A = [a_{ij}]$ be an $n \times n$ skew matrix (that is, $a_{ij} = -a_{ji}$, $i \neq j$; $a_{ii} = 0$, for $i, j = 1, 2, \ldots, n$). Then prove that $I_n + A$ is nonsingular and that $(I_n - A)(I_n + A)^{-1}$ is orthogonal.
24. If $X (\neq 0) \in V_n$, regarded as a column vector, prove that the matrix $I_n - \dfrac{2}{(X,X)} XX'$ is orthogonal and symmetric.

25. Let X_i be the ith row of an $n \times n$ matrix A, with $AA' = B = [b_{ij}]$, where $b_{ij} = (X_i, X_j)$. Then, if the rows of A are mutually orthogonal, prove that det $B = (|X_1| |X_2| \cdots |X_n|)^2$.

26. Let T be a linear transformation from a euclidean space V into a euclidean space W. Then T is said to be a *euclidean-space isomorphism* of V onto W if it is a vector-space isomorphism such that inner products are preserved. Prove that V and W are isomorphic as euclidean spaces if and only if they have the same (finite) dimension.

27. Let V be the space of real continuous functions on the interval $[0,1]$, with $(f,g) = \int_0^1 f(x)g(x)x^2\,dx$, while W consists of the same vectors but with $(f,g) = \int_0^1 f(x)g(x)\,dx$, for arbitrary functions f, g of the respective spaces. Then check that the mapping T: $V \to W$, such that $(Tf)(x) = xf(x)$, is linear and preserves inner products, but explain why T is not a euclidean-space isomorphism of V onto W. [See Problem 26.]

4.6 Linear Functionals and Adjoints

The basic idea of a linear function or functional on a vector space was introduced in Chap. 2, along with a few of the elementary properties of the *dual space* of all such functionals. It is the purpose of this section to make a further study of functionals on a euclidean space V, emphasizing their close relationship with the inner product defined on V. We shall then see how this association leads to the notion of an *adjoint* operator T* uniquely determined by any linear operator T on a finite-dimensional euclidean space V.

For any fixed $\beta \in V$, it is clear that we can define a linear functional f_β on V, such that $f_\beta(\alpha) = (\alpha,\beta)$, for any $\alpha \in V$ (Prob. 9). The following theorem may be somewhat surprising, however, in its assertion that every linear functional on V may be considered to arise in this way.

Theorem 6.1 *For every linear functional f, defined on a finite-dimensional euclidean space V, there exists a unique vector $\beta \in V$, such that $f(\alpha) = (\alpha,\beta)$, for all $\alpha \in V$.*

Proof
Let $\{\alpha_1, \alpha_2, \ldots, \alpha_n\}$ be an orthonormal basis of V. We now choose the vector β so that $\beta = f(\alpha_1)\alpha_1 + f(\alpha_2)\alpha_2 + \cdots + f(\alpha_n)\alpha_n$, and define $f_\beta(\alpha) = (\alpha,\beta)$, for any $\alpha \in V$. The properties of an inner product show that $f_\beta(\alpha_i) = [\alpha_i, f(\alpha_1)\alpha_1 + f(\alpha_2)\alpha_2 + \cdots + f(\alpha_n)\alpha_n] = f(\alpha_i)$, $i = 1, 2, \ldots, n$, since the basis is orthonormal. But two linear functionals that agree on a basis must be identical, and so $f_\beta = f$, as asserted. In order to establish the

uniqueness of β, suppose that $(\alpha,\beta) = (\alpha,\gamma)$, for all $\alpha \in V$, and some fixed $\gamma \in V$. Then $(\alpha, \beta - \gamma) = (\alpha,\beta) - (\alpha,\gamma) = 0$, and (Prob. 10) we must conclude that $\gamma = \beta$. ∎

EXAMPLE 1

Let $Y = (1,2,3)$ be a vector of V_3. Then, for arbitrary $X = (x_1,x_2,x_3) \in V_3$, we can define $f_Y(X) = (X,Y) = x_1 + 2x_2 + 3x_3$. This is the linear function on V_3, determined by Y. In particular, if $X = (-1,0,1)$, then $f_Y(X) = -1 + 0 + 3 = 2$.

EXAMPLE 2

Let $\{X_1, X_2, X_3\}$ be an orthonormal basis of V_3, where $X_1 = \left(\frac{1}{\sqrt{2}}, \frac{1}{\sqrt{2}}, 0\right)$, $X_2 = \left(-\frac{1}{\sqrt{2}}, \frac{1}{\sqrt{2}}, 0\right)$, $X_3 = (0,0,1)$. Then, if f is a linear functional on V_3, with $f(X_1) = 2$, $f(X_2) = -1$, $f(X_3) = 1$, determine the vector Y such that $f(X) = (X,Y)$, for all $X \in V_3$.

Solution

The proof of Theorem 6.1 gives the method for constructing the vector Y. Thus $Y = f(X_1)X_1 + f(X_2)X_2 + f(X_3)X_3 = 2X_1 - X_2 + X_3 = \left(\frac{3}{\sqrt{2}}, \frac{1}{\sqrt{2}}, 1\right)$.

The preceding discussion would be quite irrelevant for a space in which no inner product has been defined. Our ability to "capture" all linear functionals on a euclidean space in the way outlined was quite critically dependent on the existence of an inner product in the space. It may be recalled, however, that in the case of an arbitrary vector space any linear functional can be expressed in terms of the basis that is dual to any given basis of the space. We now turn to the idea of the *adjoint* of a linear operator T on a euclidean space V, and we shall see that the adjoint has a point of contact with the transpose operator, as defined in Chap. 2.

Theorem 6.2 If T is a linear operator on a finite-dimensional euclidean space V, there exists a unique linear operator T* on V, called the adjoint of T, such that $(T\alpha,\beta) = (\alpha,T^*\beta)$, for all $\alpha,\beta \in V$.

Proof

If β is a fixed vector in V, it is easy to see (Prob. 9) that a linear functional f on V can be defined by requiring that $f(\alpha) = (T\alpha,\beta)$, for any $\alpha \in V$. But we know by Theorem 6.1 that there exists in V a vector β' such that $(T\alpha,\beta) =$

(α,β'); we use the association between β and β' to define the adjoint T* so that $T^*\beta = \beta'$, for any $\beta \in V$. Thus, T* is an operator associated with T on V in such a way that $(T\alpha,\beta) = (\alpha,T^*\beta)$, for arbitrary $\alpha,\beta \in V$. Moreover, the uniqueness of β', for a given β, ensures the uniqueness of T*, for a given T. There remains the simple verification that T* is linear. But, for arbitrary $\alpha,\beta,\gamma \in V$, and real numbers a and b, we know that $[\alpha,T^*(a\beta + b\gamma)] = (T\alpha, a\beta + b\gamma) = (T\alpha,a\beta) + (T\alpha,b\gamma) = a(T\alpha,\beta) + b(T\alpha,\gamma) = a(\alpha,T^*\beta) + b(\alpha,T^*\gamma) = (\alpha, aT^*\beta + bT^*\gamma)$. Since this equality is true for any $\alpha,\beta,\gamma \in V$, and so *a fortiori* for any $\alpha \in V$, we can assert that $T^*(a\beta + b\gamma) = aT^*\beta + bT^*\gamma$, for any $\beta,\gamma \in V$ and $a,b \in \mathbf{R}$. Hence T* is linear. ∎

Theorem 6.2 defines the adjoint as an operator that is uniquely associated with any linear operator on a finite-dimensional vector space. Although our principal interest in this book is with spaces of finite dimension, the concept of an adjoint is meaningful in some other spaces, and so it is appropriate to make the following definition in a more general context.

Definition   Let T be a linear operator on a euclidean space V. Then, if there exists a linear operator T* on V such that $(T\alpha,\beta) = (\alpha,T^*\beta)$, for arbitrary $\alpha,\beta \in V$, the operator T* is called the adjoint of T.

If V is not of finite dimension, the adjoint of a linear operator on V does not necessarily exist; in some cases it does and in some cases it does not exist. It is also worth noting that the nature of T*, even when V is finite-dimensional, depends not only on T but also on the inner product that has been defined on the space. We now present two examples which will help to familiarize the reader with the concept of an adjoint operator.

EXAMPLE 3

If T is the linear operator defined on V_3 so that $T(a,b,c) = (a + b, b, a + b + c)$, for arbitrary $(a,b,c) \in V_3$ determine T*.

Solution
If $(x,y,z) \in V_3$, the definition of T* requires that $(T(a,b,c),(x,y,z)) = ((a,b,c),T^*(x,y,z))$. But then $((a,b,c),T^*(x,y,z)) = ((a + b, b, a + b + c), (x,y,z)) = x(a + b) + yb + z(a + b + c) = a(x + z) + b(x + y + z) + cz = ((a,b,c),(x + z, x + y + z, z))$. Since (a,b,c) and (x,y,z) are arbitrary elements of V_3, we may conclude from this that $T^*(x,y,z) = (x + z, x + y + z, z)$.

EXAMPLE 4

Let V be the space of all real polynomials of degree 2 or less in an indeterminate or real variable t, with an inner product (f,g) defined

216 INNER-PRODUCT SPACES

on V, so that $(f,g) = \int_0^1 f(x)g(x)\,dx$, for any $f,g \in V$. Then, if D is the ordinary differentiation operator, so that $Df = f'$, determine D^*. (See Note in Example 2 of Sec. 4.4.)

Solution
We proceed, as in Example 3, by first noting that $(f,D^*g) = (Df,g) = \int_0^1 f'(x)g(x)\,dx = \int_0^1 g(x)[f'(x)\,dx] = [f(x)g(x)]_0^1 - \int_0^1 f(x)g'(x)\,dx = f(1)g(1) - f(0)g(0) - (f,Dg)$. We now use Theorem 6.1 to determine vectors h_1, h_2 such that $f(1) = (f,h_1)$ and $f(0) = (f,h_2)$, for any $f \in V$ (see Example 3 of Sec. 2.4). In order to apply the theorem, we must first find an orthonormal basis for V. We leave it to the reader to verify (Prob. 11) that, if we apply the Gram-Schmidt orthogonalization process to the basis $\{1,l,l^2\}$, we obtain the orthonormal basis $\{\alpha_1,\alpha_2,\alpha_3\}$, where $\alpha_1 = 1$, $\alpha_2 = 2\sqrt{3}\,t - \sqrt{3}$, $\alpha_3 = 6\sqrt{5}\,l^2 - 6\sqrt{5}\,l + \sqrt{5}$. It then follows from the proof of Theorem 6.1 that $h_1 = \alpha_1(1)\alpha_1 + \alpha_2(1)\alpha_2 + \alpha_3(1)\alpha_3$ and so $h_1(l) = 1 + \sqrt{3}\,(2\sqrt{3}\,t - \sqrt{3}) + \sqrt{5}\,(6\sqrt{5}\,l^2 - 6\sqrt{5}\,l + \sqrt{5}) = 30l^2 - 24l + 3$. In like manner we use $f(0) = (f,h_2)$ to find that $h_2(l) = 30l^2 - 36l + 9$. Hence $(f,D^*g) = f(1)g(1) - f(0)g(0) - (f,Dg) = g(1)(f,h_1) - g(0)(f,h_2) - (f,Dg) = [f,\, h_1 g(1) - h_2 g(0) - Dg]$; and, since f and g are arbitrary elements of V, we may conclude that $D^*g = g(1)h_1 - g(0)h_2 - Dg$. If we replace g by f as the arbitrary polynomial in V, we have shown that $D^*f(l) = (30l^2 - 24l + 3)f(1) - (30l^2 - 36l + 9)f(0) - f'(l)$. It is clear that, in the special case where $f(1) = f(0) = 0$, $D^*f(l) = -f'(l)$.

EXAMPLE 5

In this example we shall study the adjoint of a linear operator T on the euclidean space V_n, from the point of view of its representing matrix *relative to an orthonormal basis* of V. If X is an arbitrary element of V_n, regarded here as a column matrix, then $TX = AX$ where A is the matrix that represents T. But now, recalling that the rule for computing an inner product in the presence of an orthonormal basis is invariant, we see that $(TX,Y) = (AX,Y) = Y'(AX) = (A'Y)'X = (X,A'Y)$, for any $Y \in V_n$, whence $T^*Y = A'Y$.

The important observation from Example 5 is that the matrix representing T* is A', where A is the matrix that represents T—both relative to the same orthonormal basis of the space. The next theorem shows in effect that there is nothing unusual about this result.

SEC. 4.6 LINEAR FUNCTIONALS AND ADJOINTS

Theorem 6.3   Let T be a linear operator on a finite-dimensional euclidean space V. Then, if the matrix of T, relative to an orthonormal basis of V, is A, the matrix of T*, relative to the same basis, is A'.

Proof
We let $\{\alpha_1, \alpha_2, \ldots, \alpha_n\}$ be the orthonormal basis of V, relative to which $A = [a_{ij}]$ is the matrix representation of T. Since the basis is orthonormal (see Sec. 4.4), an arbitrary vector $\alpha \in V$ can be expressed in the form $\alpha = (\alpha, \alpha_1)\alpha_1 + (\alpha, \alpha_2)\alpha_2 + \cdots + (\alpha, \alpha_n)\alpha_n$. But the definition of A requires that $T\alpha_j = a_{1j}\alpha_1 + a_{2j}\alpha_2 + \cdots + a_{nj}\alpha_n$, and, by the remark just made, $T\alpha_j = (T\alpha_j, \alpha_1)\alpha_1 + (T\alpha_j, \alpha_2)\alpha_2 + \cdots + (T\alpha_j, \alpha_n)\alpha_n$. The uniqueness of the coordinates of a vector, relative to any basis, now allows us to assert that $a_{ij} = (T\alpha_j, \alpha_i)$. On the other hand, if $B = [b_{ij}]$ is the matrix of T*, relative to the $\{\alpha_i\}$ basis, a similar argument shows that $b_{ij} = (T^*\alpha_j, \alpha_i) = (\alpha_i, T^*\alpha_j) = (T\alpha_i, \alpha_j) = a_{ji}$. Hence $B = A'$, as asserted in the theorem. ∎

It is possible to use this result as an alternative method to describe the adjoint of an operator. For instance, in Example 4, the differentiation operator D is represented, relative to the orthonormal basis used in the example, by the matrix

$$\begin{bmatrix} 0 & 2\sqrt{3} & 0 \\ 0 & 0 & 6\sqrt{5}/\sqrt{3} \\ 0 & 0 & 0 \end{bmatrix}$$

It then is a consequence of Theorem 6.3 that D* is represented, relative to the same basis, by

$$\begin{bmatrix} 0 & 0 & 0 \\ 2\sqrt{3} & 0 & 0 \\ 0 & 6\sqrt{5}/\sqrt{3} & 0 \end{bmatrix}$$

The reader is asked in Prob. 12 to make a new determination of the descriptive result in Example 4, making use of this matrix representation of D*.

It is implied by Theorem 6.3 that, if a linear operator T is defined on a euclidean space V, the matrix that represents the adjoint T*, *relative to an orthonormal basis of V*, and the matrix that represents the transpose T', relative to the dual basis, are the same. But it should not be concluded from this that T* and T' are then identical operators! Many *different* things can be represented by the same matrix; this is true even for different transformations of the *same* vector space (relative to different bases). However, in the case of T* and T', it should be recalled that, although T* operates on the same space V as T, the transpose operator T' operates on the dual space $\hat{V}$ of linear functionals on V. The connection between the two operators may be discovered, however, with another use of Theorem 6.1, which we leave to the reader (Prob. 27).

We now introduce our final notion in this section with an example.

EXAMPLE 6

Let V be the vector space of real polynomials, with inner product defined so that $(f,g) = \int_0^1 f(x)g(x)\,dx$, for any $f,g \in V$. If T_h is the mapping defined on V for any fixed $h \in V$, such that $T_h f = hf$, for all $f \in V$, it is not difficult to see that this "multiplication by h" mapping is a linear operator on V (Prob. 13). But $(T_h f, g) = (hf, g) = \int_0^1 [h(x)f(x)]g(x)\,dx = \int_0^1 f(x)[h(x)g(x)]\,dx = (f, hg)$, so that $T^* g = hg = Tg$, for any $g \in V$. This example is illustrative of two points: Although we asserted earlier (without proof) that an adjoint of a linear operator on a space of infinite dimension does not *necessarily* exist, it is still *possible* for one to exist on such a space; an operator may be *self-adjoint*, as described in the following definition.

Definition   *A linear operator T is self-adjoint if* $T = T^*$.

It is an immediate consequence of Theorem 6.3 that *the matrix of a self-adjoint linear operator on a finite-dimensional euclidean space, relative to any orthonormal basis of the space, is symmetric.*

The final theorem of this section summarizes some of the most important elementary properties of adjoints, the proofs of which we leave to the reader (Prob. 14).

Theorem 6.4   *If T, T_1, T_2 are linear operators on a finite-dimensional euclidean space, and $c \in \mathbf{R}$, then*

(a) $(T_1 + T_2)^* = T_1^* + T_2^*$ (b) $(cT)^* = cT^*$
(c) $(T_1 T_2)^* = T_2^* T_1^*$ (d) $(T^*)^* = T$

It may be observed from the properties listed in Theorem 6.4 that "taking the adjoint" is an operation much like that of "taking the conjugate" of a complex number. It would take us too far afield to discuss this matter here, but it is in fact the presence of these properties that makes adjoint operators so important on vector spaces over the field **C** of complex numbers. The discussion in this section has perhaps not made clear the importance of adjoints in the context of vector spaces, except that it is an "interesting" operator associated with each operator of a finite-dimensional space. However, we shall see in Chap. 6 how adjoints play a role in the normalization of the matrix representations of certain important linear operators.

Problems 4.6

1. If $Y = (-1,2,1) \in V_3$, use the method of Example 1 to define the linear functional f_Y on V, and determine $f_Y(X)$, where $X = (1,-1,1)$. Interchange the roles of X and Y and determine $f_X(Y)$.
2. A linear functional f is defined on V_4, so that $f(E_1) = f(E_3) = 2$ and $f(E_2) = f(E_4) = -2$, with $\{E_i\}$ the familiar basis of the space. Use Theorem 6.1 to determine a vector $Y \in V_4$ such that $f(X) = (X,Y)$, for any $X \in V_4$.
3. A linear functional f is defined on V_2 such that $f(X_1) = 1$ and $f(X_2) = 3$, where $X_1 = (1/\sqrt{2}, -1/\sqrt{2})$ and $X_2 = (-1/\sqrt{2}, -1/\sqrt{2})$ are the members of an orthonormal basis of the space. Use Theorem 6.1 to determine a vector $Y \in V_2$ such that $f(X) = (X,Y)$, for any $X \in V_2$.
4. If T is a linear operator on a euclidean space V, as defined below for each $\alpha \in V$, determine the adjoint T*: (a) $T\alpha = 2\alpha$; (b) $T\alpha = -\alpha$; (c) $T\alpha = 0$.
5. Let T be a linear operator, defined on V_2 in terms of the familiar basis $\{E_1, E_2\}$ as indicated, and determine T* in each case: (a) $TE_1 = (-1,2)$, $TE_2 = (1,-3)$; (b) $TE_1 = (1,3)$, $TE_2 = (-2,4)$.
6. A linear operator T on a euclidean space is represented, relative to an orthonormal basis, by the indicated matrix. Use Theorem 6.3 to describe T*α for an arbitrary α in the space.

(a) $\begin{bmatrix} 2 & -1 \\ 3 & 2 \end{bmatrix}$ (b) $\begin{bmatrix} 2 & 1 & -1 \\ 1 & 1 & 2 \\ -1 & 2 & 1 \end{bmatrix}$

(c) $\begin{bmatrix} 2 & 0 & 1 & 1 \\ 1 & 2 & 0 & -1 \\ 1 & 2 & 3 & -1 \\ 3 & 0 & 0 & 2 \end{bmatrix}$

7. If T is the rotation operator in the plane, defined so that $T(x,y) = (x \cos\theta - y \sin\theta, x \sin\theta + y \cos\theta)$, verify that T* is the rotation that is inverse to T.
8. Prove that any linear operator T on a finite-dimensional vector space can be expressed in the form $T = T_1 + T_2$, where $T_1^* = T_1$ and $T_2^* = -T_2$. [*Hint:* (T + T*)/2 is self-adjoint.]

9. Let α, β be arbitrary vectors of a euclidean space V. Then prove that (a) $f_\beta(\alpha) = (\alpha,\beta)$, for fixed β, defines f_β as a linear functional on V; (b) $f(\alpha) = (T\alpha,\beta)$, for fixed β and a fixed linear operator T on V, defines f as a linear functional on V.

10. With reference to the last sentence of the proof of Theorem 6.1, explain why we must conclude that $\gamma = \beta$.

11. Use the Gram-Schmidt construction to obtain the orthonormal basis used in Example 4.

12. Use the matrix method, based on Theorem 6.3, to obtain the result in Example 4.

13. Verify that the mapping T_h, as defined in Example 6, is a linear operator on the space V.

14. Prove Theorem 6.4.

15. Use the result in Theorem 6.3 to describe T* where (a) T is the operator in Prob. 5; (b) T is the operator in Prob. 7; (c) T is the operator in Example 3.

16. Describe T* if T is the indicated operator on ordinary 3-space:
    (a) T is a rotation about the y axis through an angle of $\pi/4$.
    (b) T is a projection onto the xy plane.
    (c) T is a reflection in the yz plane.

17. If T is the indicated operator defined on V_2, compare the adjoints T*, according as the standard inner product or the one defined by $[(x_1,y_1),(x_2,y_2)] = x_1 x_2 + (x_1 + y_1)(x_2 + y_2)$ is used: (a) $T(x,y) = (x + 2y, x - y)$; (b) $T(x,y) = (x + y, 2x - y)$.

18. Let T be defined as in Prob. 17, and let X and Y designate arbitrary elements of V_2 as column matrices. Then compare the adjoints T*, according as the standard inner product or the one defined by $(X,Y) = (AX,Y)$, where $A = \begin{bmatrix} 5 & -3 \\ -3 & 2 \end{bmatrix}$, is used. (We *assume* here that the latter is an inner product, a matter to be discussed in the following section.)

19. If T_1 and T_2 are self-adjoint operators, show that $T_1 T_2$ is self-adjoint if and only if $T_1 T_2 = T_2 T_1$.

20. Let V be the space of $n \times n$ matrices, with $(A,B) = \text{tr}(AB')$ defining the inner-product function for arbitrary matrices $A,B \in V$. If P is any nonsingular $n \times n$ matrix, prove that $T_P(A) = P^{-1}AP$ defines a linear operator T_P on V, and describe T_P^*.

21. Let W be the subspace of V in Example 6, consisting of the polynomials with degrees not exceeding 3. Determine $g \in W$,

such that $(f,g) = f(a)$, for all $f \in W$ and any fixed real number a.

22. With V the space of polynomials in Example 4, determine the adjoint of each of the following operators on V, as defined: (a) $\mathsf{T}_1 f(t) = t[Df(t)]$; (b) $\mathsf{T}_2 f(t) = (D + 2\mathsf{I})f(t)$; (c) $\mathsf{T}_3 f(t) = (1/t)\int_0^1 f(x)\, dx$. [*Caution:* I in (b) is the identity operator.]

23. Use the same orthonormal basis as in Example 4 to represent both T and T* as matrices, for each of the cases in Prob. 22.

24. With $\mathsf{T} = D$, as in Example 4, and $\mathsf{T}_1, \mathsf{T}_2, \mathsf{T}_3$ as in Prob. 22, describe (a) TT^*; (b) $(\mathsf{T}_1\mathsf{T}_2)^*$; (c) $(\mathsf{T}_2\mathsf{T}_3)^*$; (d) $(\mathsf{T}+\mathsf{T}_2)^*$.

25. If T is a linear operator on the euclidean space V_n, prove that the range of T* is the orthogonal complement of the null space of T. Then extend this result to any finite-dimensional euclidean space V.

26. If T is a linear operator on a finite-dimensional vector space, prove that T and T* have the same rank. Also prove, if T is invertible, that $(\mathsf{T}^*)^{-1} = (\mathsf{T}^{-1})^*$.

27. Let T' and T* be, respectively, the transpose and adjoint of a linear operator T on a finite-dimensional euclidean space V. Then, if ϕ is the mapping of $\hat{V}$ into V, as defined in Theorem 6.1 by $\phi(f) = \beta$, prove that ϕ is one-to-one and linear and that $\phi \mathsf{T}' = \mathsf{T}^* \phi$.

4.7 Inner Products and Positive Operators

The material of this section provides a nice blend of the theories of linear operators, inner products, adjoints, and matrices, along with a point of contact with the topic of the next chapter: quadratic forms. Let V be a vector space, which has been made euclidean by the definition of an inner product $(\ ,\ )$. If T is a linear operator on V, the reader was asked in Prob. 9 of Sec. 4.6 to verify that $(\mathsf{T}-,\beta)$, for any fixed $\beta \in V$, defines a linear functional on V. Moreover, it was shown in Theorem 6.1 that every linear functional—and, in particular, this one—can be expressed in the form $(\ ,\gamma)$, for some fixed vector $\gamma \in V$. (We have replaced the β of the theorem by γ, in order to avoid confusion with the β mentioned previously.) The question arises, however, as to whether $(\mathsf{T}\alpha,\beta)$ might itself define an inner product on V, for arbitrary vectors α, β in the space. It is clear, of course, that if $\mathsf{T} = \mathsf{I}$ is the identity operator, we have merely recovered the original inner product defined on V. But it is the search for other operators T, such that $(\mathsf{T}\alpha,\beta)$ is an inner product on the space, that motivates our present study.

The definition of $(\ ,\ )$ leads to the linearity of $(\mathsf{T}\alpha,\beta)$, relative to α, without any difficulty (Prob. 11). Hence, if $(T\alpha,\beta)$ is to define an inner product of α and β, a check of the definition in Sec. 4.2 shows that the following conditions must be met:

(1) $\quad\quad\quad\quad\quad\quad (T\alpha,\beta) = (T\beta,\alpha)$
(2) $\quad\quad\quad\quad\quad\quad (T\alpha,\alpha) > 0 \quad$ if $\alpha \neq 0$

Since $(T\beta,\alpha) = (\alpha,T\beta)$, the first condition implies that T must be *self-adjoint*, and we *define* a linear operator T to be *positive* if it satisfies *both* conditions. This means that, by definition, a linear operator T is positive if and only if $(T-,-)$ defines an inner product in the space on which T operates. Our first theorem asserts that *all* inner products on a finite-dimensional euclidean space can be obtained in this way.

Theorem 7.1 *If V is a finite-dimensional euclidean space, any inner product in V can be expressed in the form $(T-,-)$ for some unique positive linear operator T on the space.*

Proof
If we designate an arbitrary inner-product function in V by $p(-,-)$, we know from the basic properties of such a function that $p(-,\beta)$ is a linear functional for any fixed $\beta \in V$. It is then a consequence of Theorem 6.1 that $p(-,\beta) = (-,\beta')$, for some fixed $\beta' \in V$. We now think of β as varying over the whole space V, and we define T as the function that maps β onto β': that is, $T\beta = \beta'$, for all $\beta \in V$. (Cf. Theorem 6.2.) This means that $p(\alpha,\beta) = (\alpha,\beta') = (\alpha,T\beta)$, for arbitrary $\alpha,\beta \in V$. The commutative property of $p(\alpha,\beta)$ leads to the deduction that $p(\alpha,\beta) = p(\beta,\alpha) = (\beta,T\alpha) = (T\alpha,\beta)$, so that $p(-,-)$ has the formal appearance of the desired inner product. The linearity of T follows from the fact that $(T(a\alpha + b\beta),\gamma) = p(a\alpha + b\beta,\gamma) = a(p(\alpha,\gamma)) + b(p(\beta,\gamma)) = a(T\alpha,\gamma) + b(T\beta,\gamma) = (a(T\alpha),\gamma) + (b(T\beta),\gamma) = (a(T\alpha) + b(T\beta),\gamma)$, for any $\alpha,\beta,\gamma \in V$, so that $T(a\alpha + b\beta) = a(T\alpha) + b(T\beta)$, as desired. The fact that T is positive is a consequence of our assumption that $p(\alpha,\beta)$ is an inner product. Finally, to show that T is unique, let us suppose that $(T\alpha,\beta) = (T_1\alpha,\beta)$ and so $[(T - T_1)\alpha,\beta] = 0$, for arbitrary $\alpha,\beta \in V$. For a fixed α, this implies that $(T - T_1)\alpha$ is orthogonal to every vector in V; and, since α can be selected at random in V, we conclude (Prob. 12) that $T - T_1 = 0$ and $T_1 = T$. The proof of the theorem is now complete. ∎

The preceding theorem tells us that all inner products on a finite-dimensional euclidean space can be described in terms of positive linear operators on the space. In our next result, some light is shed on the structure of these positive operators.

SEC. 4.7 INNER PRODUCTS AND POSITIVE OPERATORS

Theorem 7.2 Let T be a linear operator on a finite-dimensional euclidean space V. Then T is positive if and only if $T = U^*U$, for some invertible linear operator U on V.

Proof
If $T = U^*U$, as described in the theorem, then $T^* = (U^*U)^* = U^*U = T$, so that T is self-adjoint. Moreover, $(T\alpha,\alpha) = (U^*U\alpha,\alpha) = (U\alpha,U\alpha) \geq 0$, from the basic properties of an adjoint and an inner-product function. Since U is invertible, the null space of U is $\mathcal{O}$, so that $U\alpha \neq 0$, if $\alpha \neq 0$. Hence $(T\alpha,\alpha) = (U\alpha,U\alpha) > 0$, if $\alpha \neq 0$, from which we conclude that T is positive. Conversely, suppose that T is a positive linear operator, so that $(T-,-)$ defines an inner product in V. Let $\{\alpha_1,\alpha_2, \ldots, \alpha_n\}$ and $\{\beta_1,\beta_2, \ldots, \beta_n\}$ be bases of the space that are orthonormal with respect to the basic inner product $(-,-)$ and the inner product $(T-,-)$, respectively. This implies that $(\alpha_i,\alpha_j) = \delta_{ij} = (T\beta_i,\beta_j)$, for $i, j = 1, 2, \ldots, n$. If we now define the mapping U by $U\beta_i = \alpha_i$, so that U maps the $\{\beta_i\}$ basis onto the $\{\alpha_i\}$ basis, it is clear [corollary to Theorem 6.3 (Chap. 2)] that U is an invertible linear operator. Moreover, $(T\beta_i,\beta_j) = (\alpha_i,\alpha_j) = (U\beta_i,U\beta_j)$. Hence if $\alpha = \sum_{i=1}^{n} x_i\beta_i$ and $\beta = \sum_{j=1}^{n} y_j\beta_j$ are arbitrary vectors of V, it follows that

$$(T\alpha,\beta) = \left(T\sum_{i=1}^{n} x_i\beta_i, \sum_{j=1}^{n} y_j\beta_j\right) = \sum_{i=1}^{n}\sum_{j=1}^{n} x_iy_j(T\beta_i,\beta_j) = \sum_{i=1}^{n}\sum_{j=1}^{n} x_iy_j(U\beta_i,U\beta_j)$$

$$= \left[\sum_{i=1}^{n} x_i(U\beta_i), \sum_{j=1}^{n} y_j(U\beta_j)\right] = (U\alpha,U\beta) = (U^*U\alpha,\beta)$$

Since this holds for arbitrary $\alpha,\beta \in V$, we conclude that $T = U^*U$, as the theorem asserts. ∎

It is possible to gain more insight into the nature of positive linear operators if we look at their associated matrices *relative to some orthonormal basis* $\{\alpha_1,\alpha_2, \ldots, \alpha_n\}$. We recall, from Sec. 4.6, that, under such an assumption, if the matrix of a linear operator T is A, the matrix of T^* is A'. It follows that the "matrix" condition, which is equivalent to the property of self-adjointness of T, is that A is symmetric. In order to translate the other condition on a positive linear operator into matrix language, we consider an arbitrary $\alpha \in V$, where $\alpha = x_1\alpha_1 + x_2\alpha_2 + \cdots + x_n\alpha_n$. We now let $A = A' = [a_{ij}]$, and this condition becomes

$$(T\alpha,\alpha) = \left[\sum_{i=1}^{n} x_i(T\alpha_i), \sum_{j=1}^{n} x_j\alpha_j\right] = \sum_{i=1}^{n}\sum_{j=1}^{n} x_ix_j(T\alpha_i,\alpha_j)$$

$$= \sum_{i=1}^{n}\sum_{j=1}^{n} a_{ji}x_ix_j = \sum_{i=1}^{n}\sum_{j=1}^{n} a_{ij}x_ix_j > 0$$

provided $\alpha \neq 0$. If we think of the coordinate vector $(x_1, x_2, \ldots, x_n)$ of α as the column matrix X', where $X = [x_1 \; x_2 \; \cdots \; x_n]$, the concept of a positive linear operator can now be reformulated as follows:

> A linear operator T on a finite-dimensional euclidean space V is positive if and only if the matrix A of T, relative to an orthonormal basis, satisfies the following conditions:
> (1) A is symmetric.
> (2) $X'AX > 0$, for any nonzero coordinate vector X, regarded as a column matrix.

Any polynomial $X'AX$, regardless of the nature of A, is known as a *quadratic form* in $x_1, x_2, \ldots, x_n$, but there is clearly (Prob. 13) no loss of generality in assuming that A is symmetric. However, if A satisfies *both* of the above conditions, the matrix and the quadratic form are said to be *positive* (definite), and the matrix is sometimes said to *represent* the form. We shall discuss quadratic forms very briefly in Chap. 5, and the mere definition will be sufficient here. It is clear from the foregoing that it is possible to describe a positive linear operator in terms of either a positive (definite) matrix or a positive (definite) quadratic form. We reemphasize, however, that *the association is strongly dependent on the assumption of an orthonormal basis* in the space under study.

Although it may seem natural to associate a positive matrix with a positive linear operator, it should be clear that this association is by choice and not of necessity. For a matrix A is positive provided it satisfies the above two conditions; that is, A is symmetric and the quadratic form $X'AX$ that it represents is positive. The following result is an immediate consequence of Theorem 7.2 and the familiar representation, via an orthonormal basis, of linear operators by matrices. It also provides a convenient method for the construction of positive matrices.

Corollary   A matrix A is positive if and only if there exists a nonsingular matrix Q such that $A = Q'Q$.

A special case of this type of decomposition occurred earlier when $A = I$, in which case $Q' = Q^{-1}$. A matrix Q of this kind was then said to be *orthogonal*.

We restate the principal result of this section in parallel operator-matrix terms. Thus, any inner product on a finite-dimensional euclidean space V can be defined by $(T\alpha, \beta)$, for $\alpha, \beta \in V$ and some positive linear operator T; or, alternatively, by the standard inner product $(AX, Y) = Y'AX$ in the associated coordinate space, where A is a positive matrix and X and Y are arbitrary coordinate vectors (written as column matrices) relative to some orthonormal basis of V. It is sometimes convenient to refer to such an inner product as a

SEC. 4.7 INNER PRODUCTS AND POSITIVE OPERATORS

T-inner product or an A-inner product, according as T or A is used in the definition.

EXAMPLE 1

It is clear that the identity matrix of any order is positive and so may be used to define either a positive quadratic form or an inner product on any space of the proper dimension. If $I = I_n$, then $Y'IX = Y'X = [y_1 \ y_2 \ \cdots \ y_n][x_1 \ x_2 \ \cdots \ x_n]' = x_1y_1 + x_2y_2 + \cdots + x_ny_n$, and the quadratic form represented by I is $X'X = x_1^2 + x_2^2 + \cdots + x_n^2$. The quadratic form induced by I is clearly positive, and the associated inner product in V_n is the standard one. If X and Y are considered the coordinate vectors (relative to an orthonormal basis) of elements in some other euclidean space V, it is clear that the operator T defined on V by the matrix I is the identity.

EXAMPLE 2

In order to construct a nontrivial example of an A-inner product, we may select a nonsingular matrix Q at random and then apply the corollary above. Thus, if we let

$$Q = \begin{bmatrix} 1 & 0 & 1 \\ 0 & 1 & 2 \\ 1 & 1 & 1 \end{bmatrix}$$

we find easily that

$$A = \begin{bmatrix} 1 & 0 & 1 \\ 0 & 1 & 1 \\ 1 & 2 & 1 \end{bmatrix} \begin{bmatrix} 1 & 0 & 1 \\ 0 & 1 & 2 \\ 1 & 1 & 1 \end{bmatrix} = \begin{bmatrix} 2 & 1 & 2 \\ 1 & 2 & 3 \\ 2 & 3 & 6 \end{bmatrix}$$

If we let $X = [x_1 \ x_2 \ x_3]'$ and $Y = [y_1 \ y_2 \ y_3]'$, the A-inner product of X and Y is $(AX, Y) = 2x_1y_1 + x_1y_2 + 2x_1y_3 + x_2y_1 + 2x_2y_2 + 3x_2y_3 + 2x_3y_1 + 3x_3y_2 + 6x_3y_3$; the quadratic form represented by A is $2x_1^2 + 2x_2^2 + 6x_3^2 + 2x_1x_2 + 4x_1x_3 + 6x_2x_3$. Again, if X and Y are the coordinates (relative to an orthonormal basis) of two vectors α and β of any three-dimensional euclidean space, this result may be used to define a T-inner product $(T\alpha, \beta)$ in the space, since we have constructed A to be positive.

EXAMPLE 3

The matrix $A = \begin{bmatrix} 2 & 3 \\ 3 & -4 \end{bmatrix}$ is symmetric, but let us consider the associated quadratic form $X'AX$, for an arbitrary $X = [x_1 \; x_2]' \in V_2$. In this case,

$$[x_1 \; x_2] \begin{bmatrix} 2 & 3 \\ 3 & -4 \end{bmatrix} \begin{bmatrix} x_1 \\ x_2 \end{bmatrix} = 2x_1^2 + 6x_1 x_2 - 4x_2^2$$

but if we let $x_1 = 1$ and $x_2 = -1$, we find that $X'AX = -8$. Hence, although A is clearly a symmetric matrix, it is not positive and cannot represent a positive quadratic form nor be used to define a positive linear operator or inner product.

EXAMPLE 4

As a final example, let us consider the conditions under which a rotation operator in V_2 is positive. We assume the $\{E_i\}$ basis and the standard inner product in the space, the rotation operator T being defined by $T(x,y) = (x \cos \theta - y \sin \theta, x \sin \theta + y \cos \theta)$, for arbitrary $(x,y) \in V_2$. Since $TE_1 = (\cos \theta, \sin \theta)$ and since $TE_2 = (-\sin \theta, \cos \theta)$ the matrix associated with T, relative to the orthonormal $\{E_i\}$ basis, is

$$A = \begin{bmatrix} \cos \theta & -\sin \theta \\ \sin \theta & \cos \theta \end{bmatrix}$$

One of the requirements for T to be positive is that $A = A'$, and it is immediate that $\sin \theta = 0$, whence $\theta = n\pi$, for an integer n. But now

$$A = \begin{bmatrix} \cos \theta & 0 \\ 0 & \cos \theta \end{bmatrix} = (\cos \theta) I_2$$

and the quadratic form represented by A is $(x_1^2 + x_2^2)(\cos \theta)$. It is clear that this form is positive if and only if $\cos \theta > 0$, and this occurs if and only if $(4k-1)\pi/2 < \theta < (4k+1)\pi/2$. The necessary and sufficient condition on θ for "rotation through an angle θ" to be a positive operator is then that θ be a multiple of π and satisfy this inequality; that is, $\theta = 2n\pi$ for any integer n.

We close this section with a statement of omission. Although we have shown how to construct positive matrices—and so positive operators and inner products—we have not given a general test for determining whether a given matrix is positive. In Example 3, by a simple act of discovery we were able to decide that the matrix there is not positive, but it is clear that this method might be difficult if the matrix has large order. However, we leave this general matter to Chap. 5, when a criterion will be given for testing the positive nature of an arbitrary *symmetric* matrix.

Problems 4.7

1. Explain why an A-inner product is a generalization of the idea of the standard inner product in V_n.
2. With T a linear operator, and $X = (1,-1,1)$ and $Y = (-2,1,0)$ vectors in V_3, compute (TX,Y), where $T(x,y,z)$ is defined to be (a) $(x,\ x-y,\ y+z)$; (b) $(x,\ 0,\ 2x+y)$; (c) $(x-3z,\ x+2y,\ x+2y-3z)$.
3. Decide whether the following matrix can be the matrix A of an A-inner product or of a positive (symmetric) quadratic form:

   (a) $\begin{bmatrix} 2 & 3 \\ 3 & 5 \end{bmatrix}$ (b) $\begin{bmatrix} 2 & 1 & 1 \\ 1 & -1 & 2 \\ 3 & 2 & 1 \end{bmatrix}$

4. (a) Use the corollary to find a positive matrix A of order 3.
   (b) Write the quadratic form represented by the matrix A.
   (c) Determine the A-inner product of $(1,2,-1)$ and $(-2,3,2)$, regarded as elements of V_3.
5. If $X' = [x_1\ x_2\ x_3]$ and $Y' = [y_1\ y_2\ y_3]$, express $Y'AX$ in the form of a polynomial, where

   (a) $A = \begin{bmatrix} 2 & 1 & 1 \\ 1 & -1 & 1 \\ 2 & 0 & 1 \end{bmatrix}$ (b) $A = \begin{bmatrix} 2 & 0 & 1 \\ -1 & 1 & 1 \\ 0 & 2 & 3 \end{bmatrix}$

6. With X as in Prob. 5, express the quadratic form $X'AX$, where

   (a) $A = \begin{bmatrix} 2 & 0 & 4 \\ 0 & 3 & 2 \\ 4 & 2 & -1 \end{bmatrix}$ (b) $A = \begin{bmatrix} 1 & 1 & 0 \\ 1 & -2 & 2 \\ 0 & 2 & 1 \end{bmatrix}$

7. With respect to the quadratic form $3x_1^2 + 3x_1x_2 + 4x_2^2$, find (a) the matrix of the form, as it stands; (b) the symmetric matrix of the form.

8. Apply the directions in Prob. 7 to the form $x^2 - 2xy + xz - 4yz + y^2 - 6z^2$.
9. Apply the directions in Prob. 7 to the form $3x_1^2 + 4x_1x_2 - 5x_2x_3 + x_3^2$.
10. Explain the nature of the T-inner product, where T is the rotation operator in V_2, defined in Example 4.

11. If T is a linear operator on a vector space V, prove that $(T\alpha,\beta)$ is a linear function of $\alpha \in V$, for any fixed $\beta \in V$.
12. With reference to the end of the proof of Theorem 7.1, explain why we may conclude that $T - T_1 = 0$.
13. Explain why there is no loss in generality in assuming that the matrix A of a quadratic form $X'AX$ is symmetric.
14. If A is a positive matrix, verify *directly* that (AX,Y) defines an inner product for arbitrary vectors X and Y in a euclidean coordinate space of appropriate dimension.
15. Discover a 2×2 matrix A such that $X'AX > 0$, for arbitrary $X \in V_2$, but such that A is not positive.
16. Show that the differentiation operator D in the polynomial space P_n is not positive, assuming $(f,g) = \int_0^1 f(x)g(x)\,dx$, for $f,g \in P_n$.

17. Prove that the sum of two positive linear operators is positive.
18. Prove that the product of two positive linear operators is positive if and only if they commute.
19. If T is a positive linear operator on V_2, describe the possibilities for TE_1.
20. If A is a symmetric matrix, show that a real number c exists such that $cI + A$ is positive.
21. Let $X = (x_1, x_2, \ldots, x_n)$ and $Y = (y_1, y_2, \ldots, y_n)$ be the coordinate vectors of arbitrary vectors α and β, respectively, in a space V. Then, if the coordinate space is made euclidean with an inner product $(\ ,\ )$, show that $(X,Y) = Y'GX$, where G is the gramian matrix (see Prob. 22 of Sec. 4.3) of X and Y, with X and Y regarded as column matrices. That is, verify that the gramian matrix of any ordered basis of a space completely determines an inner-product function on the space.
22. If we define two coordinate vectors to be A-orthogonal whenever their A-inner product is 0, outline the Gram-Schmidt method for determining an A-orthogonal set of coordinate vectors.

23. Refer to Prob. 22, and use the $\{E_i\}$ basis of V_2 to determine a basis that is A-orthogonal, where $A = \begin{bmatrix} 1 & 2 \\ 2 & 5 \end{bmatrix}$. Also determine a matrix Q such that $A = Q'Q$, and so check that A is positive.

24. Assume the matrix

$$A = \begin{bmatrix} \frac{3}{2} & \frac{1}{2} & -1 \\ \frac{1}{2} & \frac{3}{2} & -1 \\ -1 & -1 & 3 \end{bmatrix}$$

is positive, and use the Gram-Schmidt method (see Prob. 22) to determine an A-orthogonal basis from the $\{E_i\}$ basis of V_3.

25. If an A-inner product is understood in V_n, explain how to define the related concept of *distance* in the space.

26. Use the definition given in Prob. 25, with $A = \begin{bmatrix} 2 & -1 \\ -1 & 5 \end{bmatrix}$, and find the distance ("A-distance") between $X = (1,2)$ and $Y = (3,-1)$ in V_2.

27. Prove that the gramian matrix of any ordered basis of a finite-dimensional vector space is positive (see Prob. 21).

4.8 Simple Applications of the Distance Function

It is a familiar fact of geometry that the shortest distance from a point to a line not containing it is the length of the unique perpendicular drawn from the point to the line. This situation is shown in Fig. 25, in which the shortest distance from the point P to the horizontal axis is the length of the perpendicular segment PM. If A is any point on the axis, we may regard AP as a line

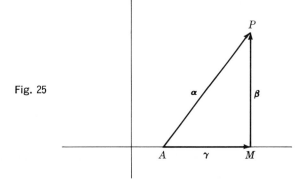

Fig. 25

vector in the plane, MP as the perpendicular vector drawn from the horizontal axis to P, with AM the *orthogonal projection* of AP in the one-dimensional space of line vectors lying on this axis. In the labeling in the diagram, the projection of α on the horizontal axis is γ, and $\beta = \alpha - \gamma$ is orthogonal to the space of horizontal line vectors.

It is easy to extend this geometric analogy to a general euclidean space V of finite dimension. Thus, if W is any proper subspace of V, we know from Theorem 4.1 that it is always possible to determine the "orthogonal projection" $\gamma \in W$ of an arbitrary vector $\alpha \in V$, such that $\beta = \alpha - \gamma$ is in the space $W^\perp$. Moreover, as in the geometric case, this particular β is of minimum length amongst all vectors β such that $\alpha = \beta + \gamma$, for some $\gamma \in W$. For, let $\gamma_1 \in W$, with $\gamma_1 \neq \gamma$. Then $\gamma - \gamma_1 \in W$ and is orthogonal to $\beta = \alpha - \gamma$ (Prob. 7). But, by Theorem 2.1 (Pythagorean theorem), it follows that $|\alpha - \gamma|^2 + |\gamma - \gamma_1|^2 = |\alpha - \gamma + \gamma - \gamma_1|^2 = |\alpha - \gamma_1|^2$, whence $|\alpha - \gamma_1| - |\alpha - \gamma|$. That is, $\beta = \alpha - \gamma$, with β in $W^\perp$, is of minimal length for all vectors of this form (i.e. with $\gamma \in W$ and α any fixed vector in V).

If we assign a basis to W, it is easy to compute the orthogonal projection γ of α. For, if $\{\alpha_1, \alpha_2, \ldots, \alpha_r\}$ is such a basis, we can write $\gamma = a_1\alpha_1 + a_2\alpha_2 + \cdots + a_r\alpha_r$, for real numbers $a_1, a_2, \ldots, a_r$. Since $\alpha - \gamma$ is in the subspace $W^\perp$, we know that $(\alpha - \gamma, \alpha_i) = 0$, or $(\alpha, \alpha_i) = (\gamma, \alpha_i)$, for $i = 1, 2, \ldots, r$. It then follows that $(\alpha, \alpha_i) = (a_1\alpha_1 + a_2\alpha_2 + \cdots + a_r\alpha_r, \alpha_i)$
$= a_1(\alpha_1, \alpha_i) + a_2(\alpha_2, \alpha_i) + \cdots + a_r(\alpha_r, \alpha_i)$, $i = 1, 2, \ldots, r$, and this is a system of r linear equations in r unknowns $a_1, a_2, \ldots, a_r$. If the basis $\{\alpha_i\}$ is orthonormal, the system is replaced by the very simple system

$$a_i = (\alpha, \alpha_i) \qquad i = 1, 2, \ldots, r$$

Since an orthonormal basis exists for any given finite-dimensional vector space, it is always possible to determine γ from a simple system like this. On the other hand, if an arbitrary basis $\{\alpha_i\}$ is used for W, the system of equations that determines γ is somewhat more complicated. To find the coefficient unknowns $a_1, a_2, \ldots, a_r$ in this case, we may use any convenient method of solving, but the uniqueness of the solution implies that the following determinant must be nonzero:

$$\begin{vmatrix} (\alpha_1,\alpha_1) & (\alpha_1,\alpha_1) & \cdots & (\alpha_1,\alpha_1) \\ (\alpha_2,\alpha_1) & (\alpha_2,\alpha_2) & \cdots & (\alpha_2,\alpha_r) \\ \vdots & \vdots & & \vdots \\ (\alpha_r,\alpha_1) & (\alpha_r,\alpha_2) & \cdots & (\alpha_r,\alpha_r) \end{vmatrix}$$

This is the *Gram* determinant, or determinant of the *gramian* matrix, as encountered previously in problems (Prob. 22 of Sec. 4.3; Prob. 21 of Sec. 4.7).

EXAMPLE 1

It is possible to use the linear-algebra method just described to determine the Fourier or Euler coefficients, which are probably very familiar to the student of elementary analysis. A study of Fourier series is a study of the representation of functions of a certain class by series involving sine and cosine terms. In our present oversimplified study, we consider the euclidean space V of all continuous functions defined on the interval $[0, 2\pi]$, in which the inner product of two such functions f and g is defined so that $(f,g) = \int_0^{2\pi} f(x)g(x)\,dx$, and the length of f is $\left\{\int_0^{2\pi} [f(x)]^2\,dx\right\}^{\frac{1}{2}}$. It is our desire to find the trigonometric polynomial $p(x) = a_0/2 + a_1 \cos x + b_1 \sin x + \cdots + a_n \cos nx + b_n \sin nx$ of "degree" n, which differs from a given $f \in V$ by as little as possible. We measure the *nearness* of $p(x)$ to $f(x)$ by the integral $\int_0^{2\pi} [f(x) - p(x)]^2\,dx$, which is the square of the distance from $f(x)$ to $p(x)$. The polynomials $p(x)$, as just defined, form a subspace W of V, and the problem here is to determine the length of the "perpendicular" from $f(x)$ to W.

Although the space V is not of finite dimension, it is clear that the subspace W is generated by the set $\{1, \cos x, \sin x, \ldots, \cos nx, \sin nx\}$. That the vectors in this generating set are independent is a problem in analysis, and we accept this fact (without proof), which is tantamount to asserting that the set constitutes a basis for W. It is now an easy matter to verify that $\alpha_0 = 1/\sqrt{2\pi}$, $\alpha_1 = (\cos x)/\sqrt{\pi}$, $\alpha_2 = (\sin x)/\sqrt{\pi}$, $\ldots$, $\alpha_{2n-1} = (\cos nx)/\sqrt{\pi}$, $\alpha_{2n} = (\sin nx)/\sqrt{\pi}$ constitute an orthonormal (Prob. 8) basis of W, which is observed to have dimension $2n + 1$. The desired polynomial $m(x)$ of minimum length is then determined immediately from the above results. To be more precise, $m(x) = c_0\alpha_0 + c_1\alpha_1 + c_2\alpha_2 + \cdots + c_{2n}\alpha_{2n}$, where $c_i = (f, \alpha_i)$, $i = 0, 1, \ldots, 2n$. In particular, this means that $c_0 = (1/\sqrt{2\pi}) \int_0^{2\pi} f(x)\,dx$, $c_1 = (1/\sqrt{\pi}) \int_0^{2\pi} f(x) \cos x\,dx$, $c_2 = (1/\sqrt{\pi}) \int_0^{2\pi} f(x) \sin x\,dx$, $\ldots$, $c_{2n-1} = (1/\sqrt{\pi}) \int_0^{2\pi} f(x) \cos nx\,dx$, $c_{2n+1} = (1/\sqrt{\pi}) \int_0^{2\pi} f(x) \sin nx\,dx$. The polynomial $m(x)$ can be otherwise expressed in the form $m(x) = a_0/2 + \sum_{k=1}^{n} (a_k \cos kx + b_k \sin kx)$, where $a_0 = (1/\pi) \int_0^{2\pi} f(x)\,dx$, $a_k = (1/\pi) \int_0^{2\pi} (fx) \cos kx\,dx$, $b_k = (1/\pi) \int_0^{2\pi} f(x) \sin kx\,dx$. The numbers $a_0, a_1, b_1, \ldots, a_n, b_n$ are the *Fourier* or *Euler coefficients*, used to represent the function f in the form of a trigonometric series. It might be assumed that the "fit" of $m(x)$ to $f(x)$ would become arbitrarily close, as n becomes infinite, but see Prob. 16.

EXAMPLE 2

We can also apply the linear-algebra procedure of this section to obtain what is essentially the *method of least squares* of statistics. In the general case, we are interested in determining an equation $y = a_1 x_1 + a_2 x_2 + \cdots + a_r x_r$, which is the "closest fit" to a set of data, usually obtained experimentally, consisting of values of $x_1, x_2, \ldots, x_r$ and y. If n such sets of data are available, and the kth set of values is $\{x_{1k}, x_{2k}, \ldots, x_{rk}; y_k\}$, the equations that we wish to approximate are the following:

$$x_{11}a_1 + x_{21}a_2 + \cdots + x_{r1}a_r = y_1$$
$$x_{12}a_1 + x_{22}a_2 + \cdots + x_{r2}a_r = y_2$$
$$\cdots \cdots \cdots \cdots \cdots \cdots \cdots \cdots$$
$$x_{1n}a_1 + x_{2n}a_2 + \cdots + x_{rn}a_r = y_n$$

These equations are to be subjected to the method of least squares, which means that we must minimize the following quantity:

$$\sum_{k=1}^{n} (a_1 x_{1k} + a_2 x_{2k} + \cdots + a_k x_{rk} - y_k)^2$$

We now let $X_1 = (x_{11}, x_{12}, \ldots, x_{1n})$, $X_2 = (x_{21}, x_{22}, \ldots, x_{2n})$, $\ldots, X_r = (x_{r1}, x_{r2}, \ldots, x_{rn})$ and assume (see Prob. 9) that these vectors constitute a basis for a subspace of V_n. In addition, we let $Y = (y_1, y_2, \ldots, y_n)$ be another vector in the subspace. With the standard inner product assumed in V_n, the above summation designates the square of the distance from Y to $a_1 X_1 + a_2 X_2 + \cdots + a_r X_r$, and our problem is to choose $a_1, a_2, \ldots, a_r$ in such a way that this distance is a minimum. The discussions in this section show that the desired numbers are the solutions for $a_1, a_2, \ldots, a_r$ of the following system of equations:

$$a_1(X_1, X_1) + a_2(X_2, X_1) + \cdots + a_r(X_r, X_1) = (Y, X_1)$$
$$a_1(X_1, X_2) + a_2(X_2, X_2) + \cdots + a_r(X_r, X_2) = (Y, X_2)$$
$$\cdots \cdots \cdots \cdots \cdots \cdots \cdots \cdots \cdots \cdots \cdots \cdots$$
$$a_1(X_1, X_r) + a_2(X_2, X_r) + \cdots + a_r(X_r, X_r) = (Y, X_r)$$

where $(Y,X_k) = \sum_{i=1}^{n} x_{ki}y_i$ and $(X_i,X_k) = \sum_{j=1}^{n} x_{ij}x_{kj}$. These are known as the *normal equations*, associated with the original system of equations. It may be noted that the coefficient matrix of this normal system is the transpose of the gramian matrix of $x_1, x_2, \ldots, x_n$ (see Prob. 22 of Sec. 4.3).

A special case of the least-squares method occurs when the original system of equations contains only one unknown m. In this case, the desired equation is linear of the form $y = mx$, and the n equations of the system may be expressed as $mx_1 = y_1$, $mx_2 = y_2, \ldots, mx_n = y_n$. The least-squares solution, for this system, is $m = (X,Y)/(X,X)$, where $X = (x_1,x_2,\ldots,x_n)$ and $Y = (y_1,y_2,\ldots,y_n)$, and so $m = \left[\sum_{k=1}^{n} x_k y_k\right] / \left[\sum_{k=1}^{n} x_k^2\right]$. In this very simple case, m is the slope of the *line of best fit* to the points (x_1,y_1), (x_2,y_2), $\ldots$, (x_n,y_n) in the cartesian plane, *and which passes through the origin*.

EXAMPLE 3

To illustrate the procedure for the simple case in Example 2, let us find the equation of the line through the origin which is "as close as possible" to the points $(-5,2)$, $(0,4)$, $(-4,4)$, $(5,8)$ in the cartesian plane. In this case, the normal equations are $m(-5) = 2$, $m(0) = 4$, $m(-4) = 4$, $m(5) = 8$. We now let $X = (-5,0,-4,5)$ and $Y = (2,4,4,8)$, so that the least-squares solution for m is $m = (X,Y)/(X,X) = (-10 + 0 - 16 + 40)/(25 + 0 + 16 + 25) = \frac{7}{33}$. The desired equation is then $y = 7x/33$ or $7x - 33y = 0$.

Problems 4.8

1. Find the Gram determinant of the vectors $X_1, X_2, X_3 \in V_3$, where $X_1 = (1,1,1)$, $X_2 = (-1,2,0)$, $X_3 = (2,-1,1)$.
2. Find the Gram determinant of the vectors f_1, f_2, f_3 in the space V of Example 1, where $f_1(x) = x$, $f_2(x) = -x^2$, $f_3(x) = 2$.
3. Refer to Example 3 to find the equation of the line through the origin and closest to the points $(2,-1)$, $(4,4)$, $(-2,1)$, $(3,1)$.
4. Apply the direction in Prob. 3 to the points $(1,-1)$, $(0,1)$, $(2,-2)$, $(1,1)$, $(2,-3)$.
5. Use the methods of analytic geometry to find the distance from the point $(1,1,1)$ to the plane in 3-space whose equation is $2x - 3y + z = 2$. [See Prob. 18.]

6. Use the methods of analytic geometry to find the distance from the point $(-2,3)$ to the line in 2-space whose equation is $3x + y = 2$. [See Problem 19.]

7. With reference to the discussion in the first part of this section, explain why $\gamma - \gamma_1$ is orthogonal to $\alpha - \gamma$.

8. Review the proof that the vectors $\{\alpha_i\}$ in Example 1 are orthonormal.

9. Is it reasonable to assume, as we did in Example 2, that the vectors $X_1, X_2, \ldots, X_r$ are linearly independent? Explain.

10. In the discussion just prior to the display of the Gram determinant, explain why we may conclude that the system of equations in $a_1, a_2, \ldots, a_r$ has a unique solution, even without the assumption of an orthonormal basis.

11. It is desired to determine the functional equation $y = a_1 x_1 + a_2 x_2 + a_3 x_3$ from the following four sets of experimental data for $(x_1, x_2, x_3; y)$: $(1, -1, 2; 3)$, $(-1, 2, 1; 2)$, $(-1, 2, 1; -2)$, $(1, -1, 1; 3)$. Find the normal equations, from which it is possible to determine a_1, a_2, a_3 by the method of least squares.

12. Apply the directions given in Prob. 11 to the following sets of data: $(2, 1, 4; 1)$, $(-1, 2, 3; 4)$, $(2, -3, -1; 3)$, $(0, 1, -2; 5)$.

13. It is desired to find the functional equation $y = ax_1 + bx_2$, from the following experimental data for $(x_1, x_2; y)$: $(2, 1; 1)$, $(-1, 2; -1)$, $(3, 1; -3)$. Determine the equation by the method outlined in Example 3.

14. If $f(x) = x$, for $x \in [0, 2\pi]$, refer to Example 1 to find the Fourier polynomials of the second degree in the Fourier approximation of f.

15. Apply the directions given in Prob. 14 to the function f, defined on $[0, 2\pi]$ by $f(x) = x^2$.

16. Have we shown in Example 1 that it is possible to approximate $f(x)$ arbitrarily close, provided a sufficiently large number of terms of the polynomial $m(x)$ are taken? Explain your answer (see Prob. 17).

17. Refer to a book on Fourier series, and look up the statement and significance of the so-called Dirichlet conditions for a function f, defined on the interval $[0, 2\pi]$.

18. Use vector methods to solve Prob. 5, and compare answers.

19. Use vector methods to solve Prob. 6, and compare answers.

Selected Readings

Grimm, C. A.: A Vector Solution of Simultaneous Linear Equations, *Amer. Math. Monthly*, **67** (3): 263–265 (March, 1960).

Reiersol, O: Rational Orthogonal Matrices, *Amer. Math. Monthly*, **70** (1): 63–65 (January, 1963).

Robinson, D. W.: On the Generalized Inverse of an Arbitrary Linear Transformation, *Amer. Math. Monthly*, **69** (5): 412–416 (May, 1962).

Walsh, Bertram: The Scarcity of Cross Products on Euclidean Spaces, *Amer. Math. Monthly*, **74** (2): 188–194 (February, 1967).

5

BILINEAR AND QUADRATIC FORMS

5.1 Bilinear Functions and Forms

During the course of this book, we have often had occasion to refer to *linear functions* or *functionals* on a vector space V as linear mappings of the vectors of V to scalars in the real field **R**. We have also encountered examples of *bilinear functions* on $V \times V$ as mappings of *pairs* of vectors from V to real numbers, the mappings being linear in both arguments. In Chap. 4, we were very much concerned with inner-product functions, which are of this kind: the defining properties of an inner product (,) in a vector space V require that (α,β) be a real number, with the mapping linear in both α and β. Another example arose in our development of determinants in Chap. 3: for any $X, Y \in V_2$, we know that det (X,Y) is a real number with the det function linear in both X and Y. However, although the idea is not new to us, we shall define the concept of a bilinear function in somewhat more general and explicit terms.

Definition *Let V and W be any two vector spaces, the same or distinct. A bilinear function f on $V \times W$ is a mapping that assigns to each pair $(\alpha,\beta) \in V \times W$ a real number $f(\alpha,\beta)$ and such that the following properties hold for arbitrary $\alpha_1, \alpha_2 \in$*

236

V, $\beta_1, \beta_2, \in W$, $a, b \in \mathbf{R}$:
(1) $f(a\alpha_1 + b\alpha_2, \beta) = af(\alpha_1, \beta) + bf(\alpha_2, \beta)$
(2) $f(\alpha, a\beta_1 + b\beta_2) = af(\alpha, \beta_1) + bf(\alpha, \beta_2)$

It is now our intention to imitate some of the procedures in our axiomatic theory of determinants, in order to develop a general expression or *form* for the value $f(\alpha, \beta)$ of such a bilinear function. Since our present discussions will be centered on coordinate spaces, and matrices will play an important role, it will often be convenient to regard our coordinate vectors as column matrices, as we have frequently done in preceding discussions.

To this end, let $\{\alpha_1, \alpha_2, \ldots, \alpha_m\}$ and $\{\beta_1, \beta_2, \ldots, \beta_n\}$ be bases of the respective spaces V and W, and we shall determine an expression for $f(\alpha, \beta)$, with f a general bilinear function on $V \times W$, and $\alpha \in V$, $\beta \in W$. Then we may write $\alpha = x_1\alpha_1 + x_2\alpha_2 + \cdots + x_m\alpha_m$ and $\beta = y_1\beta_1 + y_2\beta_2 + \cdots + y_n\beta_n$, so that the coordinates of α and β (relative to the given bases) are $X = (x_1, x_2, \ldots, x_m)$ and $Y = (y_1, y_2, \ldots, y_n)$, respectively. The bilinearity of f requires that $f(\alpha, \beta) = f(x_1\alpha_1 + x_2\alpha_2 + \cdots + x_m\alpha_m, \beta) = x_1 f(\alpha_1, \beta) + x_2 f(\alpha_2, \beta) + \cdots + x_m f(\alpha_m, \beta)$. But $f(\alpha_i, \beta) = f(\alpha_i, y_1\beta_1 + y_2\beta_2 + \cdots + y_n\beta_n) = y_1 f(\alpha_i, \beta_1) + y_2 f(\alpha_i, \beta_2) + \cdots + y_n f(\alpha_i, \beta_n)$, for $i = 1, 2, \ldots, m$. The expression for $f(\alpha, \beta)$ then becomes $\sum_{i=1}^{m} \sum_{j=1}^{n} x_i f(\alpha_i, \beta_j) y_j = \sum_{i=1}^{m} \sum_{j=1}^{n} a_{ij} x_i y_j$, where $a_{ij} = f(\alpha_i, \beta_j)$, $i = 1, 2, \ldots, m$ and $j = 1, 2, \ldots, n$.

It is to be noted that the value $f(\alpha, \beta)$ of the bilinear function f is completely determined by the mn scalars $a_{ij} = f(\alpha_i, \alpha_j)$. The $m \times n$ matrix $A = [a_{ij}]$ is then uniquely associated with f, and it is an easy exercise in matrix multiplication to verify that $f(\alpha, \beta) = X'AY$, where the coordinate vectors X and Y are regarded as column matrices. The polynomial expression $X'AY$ will often be denoted by $B(X, Y)$ and is called a *bilinear form* in the $m + n$ variables $x_1, x_2, \ldots, x_m, y_1, y_2, \ldots, y_n$. A bilinear form is then identified with the *general* value of a bilinear function and is said to *represent* the function. If X and Y are the coordinate vectors of α and β, respectively, as in the preceding development, it is clear that $f(\alpha, \beta) = B(X, Y)$, but it is nonetheless convenient to have the notation $B(X, Y)$ to denote the general value of f as a polynomial in the coordinates. (It should be pointed out, however, that all writers do not make this distinction.) The matrix A, which is uniquely associated (relative to the given bases) with a bilinear function or form, is called the *matrix of the bilinear function* or the *matrix of the bilinear form*, as desired, and may be said to *represent* either the function f or the form $B(X, Y)$. It is of computational interest to observe that the entry a_{ij} in the matrix A is the coefficient of $x_i y_j$ in the form $X'AY$.

EXAMPLE 1

Let us determine the matrix of the form $x_1y_1 + 2x_1y_2 + 5x_1y_3 - 2x_2y_1 + x_2y_3 - 6x_3y_2 + 6x_3y_3$. We may regard the form as $B(X,Y) = X'AY$, where X and Y are the coordinate vectors (x_1,x_2,x_3) and (y_1,y_2,y_3), regarded as column matrices. But if we use the observation made just prior to this example, we see that $a_{11} = 1$, $a_{12} = 2$, $a_{13} = 5$, $a_{21} = -2$, $a_{22} = 0$, $a_{23} = 1$, $a_{31} = 0$, $a_{32} = -6$, $a_{33} = 6$, and so the matrix A of the form is

$$\begin{bmatrix} 1 & 2 & 5 \\ -2 & 0 & 1 \\ 0 & -6 & 6 \end{bmatrix}$$

An elementary matrix computation will now check that $X'AY$ is the given form (Prob. 10).

EXAMPLE 2

Let V be the space of all continuous real functions on **R**. Then, if $K(x,y)$ is a given continuous function of two real variables x and y, we can define $H(g,h) = \int_a^b \int_a^b K(x,y)g(x)h(y)\, dx\, dy$, for arbitrary $g,h \in V$ and any real numbers a, b. This function H is a mapping of $V \times V$ to **R**, and the properties of such a definite integral may be used to verify (Prob. 11) that H is bilinear. As a special case, we may let $K(x,y) = 1$, so that $H(g,h) = \int_a^b \int_a^b g(x)h(y)\, dx\, dy = \left[\int_a^b g(x)\, dx\right]\left[\int_a^b h(y)\, dy\right]$. In this case, $H(g,h)$ is the product of the two linear functions whose values are $\int_a^b g(x)\, dx$ and $\int_a^b g(y)\, dy$. The proof that the product of two linear functions is always a bilinear function is left as an exercise for the reader (Prob. 12).

Before we proceed to the important matter of the effect on a bilinear form of a change of basis for the spaces involved, let us consider a preliminary example.

EXAMPLE 3

Let a bilinear function $f(\alpha,\beta)$ be defined by the form $B(X,Y) = 2x_1y_1 - 3x_2y_3 + x_3y_3$, with $X = (x_1,x_2,x_3)$ and $Y = (y_1,y_2,y_3)$ the coordinate vectors of α and β, relative to some basis of the three-dimensional space V. Find the matrix of the form resulting when the basis is changed to vectors whose coordinates, relative to the original basis, are $(1,1,1)$, $(-2,1,1)$, and $(2,1,0)$.

Solution

The observation made just prior to Example 1 leads directly to the matrix of the new form. If the new basis elements are γ_1, γ_2, γ_3, whose coordinates are in the order listed above, it follows that the matrix of $f(\alpha,\beta)$, relative to this basis, is $[b_{ij}]$, where $b_{ij} = f(\gamma_i,\gamma_j)$. On using the given form for these basis vectors in pairs, we find that $b_{11} = 0$, $b_{12} = -6$, $b_{13} = 4$, $b_{21} = -6$, $b_{22} = 6$, $b_{23} = -8$, $b_{31} = 1$, $b_{32} = -11$, $b_{33} = 8$. Hence the matrix of the new form, which is merely a different representation of $f(\alpha,\beta)$, due to the different choice of basis for V, is

$$\begin{bmatrix} 0 & -6 & 4 \\ -6 & 6 & -8 \\ 1 & -11 & 8 \end{bmatrix}$$

If the new coordinate vectors of α and β are $X^1 = (x_1', x_2', x_3')$ and $Y^1 = (y_1', y_2', y_3')$, respectively the new form can be written directly from this matrix: $B(X^1, Y^1) = -6x_1'y_2' + 4x_1'y_3' - 6x_2'y_1' + 6x_2'y_2' - 8x_2'y_3' + x_3'y_1' - 11x_3'y_2' + 8x_3'y_3'$.

The construction in Example 3 now leads to the following very basic definition.

Definition   *Two bilinear forms are equivalent if they represent the same bilinear function, relative to specific bases of the spaces involved.*

With reference to Example 3, the given form and the final form are equivalent in the sense that they represent the same bilinear function f on $V \times V$ (see Prob. 13). As the example indicates, equivalent forms may be quite different in appearance, but we shall see later that they possess a numerical invariant.

We now state and prove the important *change of basis* theorem, as related to bilinear functions or forms.

Theorem 1.1   *Let $\{\alpha_1, \alpha_2, \ldots, \alpha_m\}$ and $\{\beta_1, \beta_2, \ldots, \beta_n\}$ be respective bases of vector spaces V and W, with A the matrix of a bilinear function $f(\alpha,\beta)$, for $\alpha \in V$ and $\beta \in W$. Then, if P and Q are respective matrices of transition from the given bases in V and W, the matrix of the new form representing f (and so equivalent to the original) is $P'AQ$.*

Proof

If X and Y are the coordinate vectors of α and β, relative to the $\{\alpha_i\}$ and $\{\beta_i\}$ bases and expressed as column matrices, we know that $f(\alpha,\beta) = X'AY$.

240 BILINEAR AND QUADRATIC FORMS

But, if $\tilde{X}$ and $\tilde{Y}$ are the new coordinate vectors of α and β, respectively, after the changes of bases, it follows from Theorem 9.1 (Chap. 2) that $X = P\tilde{X}$ and $Y = Q\tilde{Y}$. Hence $B(\alpha,\beta) = X'AY = (P\tilde{X})'A(Q\tilde{Y}) = \tilde{X}'(P'AQ)\tilde{Y}$, from which we observe that the matrix of the form, relative to the new bases, is $P'AQ$, as asserted. ∎

EXAMPLE 4

We can use the result of Theorem 1.1 to obtain our earlier result in Example 3. From the information given in that example,

$$P = Q = \begin{bmatrix} 1 & -2 & 2 \\ 1 & 1 & 1 \\ 1 & 1 & 0 \end{bmatrix}$$

so that

$$P' = \begin{bmatrix} 1 & 1 & 1 \\ -2 & 1 & 1 \\ 2 & 1 & 0 \end{bmatrix}$$

The matrix of the given form is

$$A = \begin{bmatrix} 2 & 0 & 0 \\ 0 & 0 & -3 \\ 0 & 0 & 1 \end{bmatrix}$$

and so the matrix of the form that results from the change of basis of V is

$$P'AQ = P'AP = \begin{bmatrix} 1 & 1 & 1 \\ -2 & 1 & 1 \\ 2 & 1 & 0 \end{bmatrix} \begin{bmatrix} 2 & 0 & 0 \\ 0 & 0 & -3 \\ 0 & 0 & 1 \end{bmatrix} \begin{bmatrix} 1 & -2 & 2 \\ 1 & 1 & 1 \\ 1 & 1 & 0 \end{bmatrix}$$

$$= \begin{bmatrix} 1 & 1 & 1 \\ -2 & 1 & 1 \\ 2 & 1 & 0 \end{bmatrix} \begin{bmatrix} 2 & -4 & 4 \\ -3 & -3 & 0 \\ 1 & 1 & 0 \end{bmatrix} = \begin{bmatrix} 0 & -6 & 4 \\ -6 & 6 & -8 \\ 1 & -11 & 8 \end{bmatrix}$$

This is in agreement with the earlier result.

Definition   *A matrix B is said to be equivalent to a matrix A if there exist nonsingular matrices P and Q such that $B = PAQ$.*

SEC. 5.1 BILINEAR FUNCTIONS AND FORMS 241

If we recall that transition matrices are nonsingular (see Sec. 2.9), as are also their transposes, the corollary below follows immediately.

Corollary *Two bilinear forms are equivalent if and only if their respective matrices are equivalent (Prob. 14).*

The very important lemma, which we now state, allows the transfer of the concept of *rank* from any representing matrix of a bilinear function or form to the function or form itself.

Lemma *Equivalent matrices have the same rank.*

Proof
It is possible to establish this result in many ways. We shall use what is perhaps the most direct approach but suggest two others to the reader in Probs. 19 and 20. If $P = [p_{ij}]$ is a nonsingular $m \times m$ matrix and $A = [a_{jk}]$ is any $m \times n$ matrix, we shall prove that the row spaces of PA and A are the same. To this end, let $PA = C = [c_{ik}]$, so that $c_{ik} = \sum_{j=1}^{m} p_{ij} a_{jk}$. If we let Y_i designate the ith row vector of PA, we see that $Y_i = (c_{i1}, c_{i2}, \ldots, c_{in})$; and, if $X_j = (a_{j1}, a_{j2}, \ldots, a_{jn})$ is the jth row vector of A, it follows that $Y_i = \sum_{j=1}^{m} p_{ij} x_j$. But this shows that the rows of PA are linear combinations of the rows of A, which implies that the row space of PA is contained in the row space of A. Since P is nonsingular, $A = P^{-1}(PA)$, and a similar argument will show that the row space of A is contained in the row space of PA. Thus the two spaces are identical, and in particular the row ranks—and hence (see Sec. 2.10) the ranks—are the same. It should be clear (Prob. 15) that the effect of Q as a *right multiplier of A* on the *column* space of A is subject to a discussion which is analogous to that just given for the row spaces of A and PA. Hence the column spaces of A and AQ are the same and this, when combined with the earlier result, assures the equality of the ranks of A, PA, and PAQ. ∎

As suggested earlier, it is now appropriate to make the following definition.

Definition *The rank of a bilinear function or bilinear form is the rank of any of its representing matrices.*

This is the numerical invariant that can be attached to any bilinear form or function. The importance of this invariant is made clear in the final theorem in this section, which describes a *canonical* or standard bilinear form.

Theorem 1.2 Let V and W be vector spaces of finite dimension. Then, any bilinear form of rank r on $V \times W$ is equivalent to $x_1 y_1 + x_2 y_2 + \cdots + x_r y_r$, where $(x_1, x_2, \ldots, x_r, \ldots, x_m)$ and $(y_1, y_2, \ldots, y_r, \ldots, y_n)$ are the coordinates (relative to appropriate bases) of arbitrary vectors in V and W, respectively.

Proof
In view of Theorem 1.1, our assertion can be phrased in matrix-theoretic terms: If A is the matrix of the given form, there exist nonsingular matrices P and Q such that $PAQ = \begin{bmatrix} I_r & 0 \\ 0 & 0 \end{bmatrix}$, where I_r is the identity matrix of order r and each O designates a matrix of zeros of size consistent with that of the $m \times n$ matrix A. We have seen in Sec. 3.2 that there exists a nonsingular matrix P such that PA is in Hermite normal or *row-echelon* form. If the discussions of that section are now paralleled by analogous discussions pertaining to *columns* of a matrix, it follows (Prob. 16) that there exists a nonsingular matrix Q such that $(PA)Q = PAQ$ is in what we may call *column-echelon* form, this having a meaning analogous to that for rows. Moreover, it should be noted (Prob. 17) that the elementary matrices involved in the construction of Q do not alter the submatrix of PA whose diagonal entries are the "pivotal" 1s in this row-echelon matrix. But then PAQ is in both row-echelon and column-echelon form, which means that its upper left portion must consist of an identity matrix and all other of its entries are O. Inasmuch as A and PAQ are equivalent matrices, the lemma assures that the rank of PAQ is the same as the rank of A, and it is clear that the rank of PAQ is the number of 1s in the identity submatrix. We have then shown that $PAQ = \begin{bmatrix} I_r & 0 \\ 0 & 0 \end{bmatrix}$, as asserted. If the bases of V and W are desired, relative to which the value of the bilinear function has the simple form of the theorem, it follows from Theorem 1.1 that, although Q is the basis transition matrix in W, the matrix of transition of the original basis in V is P'. ∎

As a result of Theorem 1.2, any bilinear form on $V \times W$ can be assumed, without loss of generality, to have the simple representation established in the theorem. The vector spaces V and W are quite arbitrary and are often (in fact usually) the same space, but a word of caution must be injected here: It may not be possible to find a *single* transition matrix for the given basis of V to produce the simple form. The reason for this is that, although two distinct transition matrices P and Q are available in the case of two spaces V and W, only one such matrix P is available if V and W are the same—unless one uses two distinct transition matrices in a very unnatural way to be associated with the general vectors in the bilinear form. If only one transition matrix P is used for the space V, the matrix A of a bilinear form on $V \times V$

is transformed into $P'AP$—a matrix that may be said to be *congruent* to A. Since congruent matrices are clearly a more specialized class than equivalent matrices, it is not surprising that it may not be possible for $P'AP$ to be in as simple a form as PAQ, for *any* nonsingular matrices P and Q. From the point of view of elementary operations, as discussed in Sec. 3.2, the matrix PAQ may be considered the result of applying any sequence of elementary row and elementary column operations to the matrix A. However, in the congruence transformation of A to $P'AP$, it is necessary to apply the *same* sequence of row operations as column operations; this is clearly more restrictive than the other case. Thus, *unless we allow the simultaneous use of two distinct bases for V*, we are not, in general, able to represent a bilinear function on $V \times V$ in the simple form obtained in Theorem 1.2.

It is likely that the single most important result in this section is that *equivalent matrices represent the same bilinear function*, provided appropriate bases are used for the spaces. It is an elementary matter to verify (Prob. 18) that *equivalence* is an *equivalence relation* in the set of all $m \times n$ matrices and is one of the three important equivalence relations that arise in matrix theory. The other two are *similarity*, which was discussed in Sec. 2.9, and *congruence*, which was mentioned above and which will be discussed more fully in the following section. It is the important characteristic of any equivalence relation that it determines disjoint classes of equivalent elements, and the *simplest* member of a class is usually considered the *canonical* member of the class. In the case of equivalence, as just discussed, the invariant that is common to all matrices of a given equivalence class is the *rank;* if the rank is r, the *canonical* matrix of this class is $\begin{bmatrix} I_r & 0 \\ 0 & 0 \end{bmatrix}$, as obtained above. We shall see later that a similar observation, but one which is not quite so simple, may be made for the equivalence classes under both congruence and similarity.

Problems 5.1

1. Decide which of the following functions f, as defined below on $V_2 \times V_2$, are bilinear, with $X = (x_1, x_2)$ and $Y = (y_1, y_2)$: (a) $f(X,Y) = (x_1 + y_1)^2 - (x_1 - y_1)^2$; (b) $f(X,Y) = 2$; (c) $f(X,Y) = x_1 y_2 - x_2 y_1$; (d) $f(x,y) = (x_1 + y_1)^2 - x_2 y_2$.

2. Explain why the bilinear function f, defined on $V_2 \times V_2$ by $f(X,Y) = \det(X,Y)$, is not capable of generalization to spaces V_n for $n > 2$.

3. Find the matrix of each of the following bilinear forms: (a) $2x_1 y_1 - 3x_1 y_3 + 2x_2 y_2 - 5x_2 y_3 + 4x_3 y_1 + x_3 y_3$; (b) $4x_1 y_1 + 2x_2 y_2 + x_3 y_3$; (c) $3x_1 y_1 + x_1 y_4 - 2x_2 y_3 + x_4 y_2 - 2x_4 y_4$; (d) $x_1 y_1 + x_2 y_2 + x_3 y_3 + x_4 y_4$.

4. Express each of the following bilinear forms as a polynomial:

(a) $[x_1 \; x_2 \; x_3] \begin{bmatrix} 2 & -1 & 1 \\ 0 & 3 & 0 \\ -1 & 0 & 1 \end{bmatrix} \begin{bmatrix} y_1 \\ y_2 \\ y_3 \end{bmatrix}$

(b) $[x \; y \; z] \begin{bmatrix} 2 & 1 & -1 \\ 0 & 2 & 0 \\ 1 & -3 & 1 \end{bmatrix} \begin{bmatrix} x' \\ y' \\ z' \end{bmatrix}$

5. Find the bilinear form whose matrix is

$$\begin{bmatrix} 1 & 1 & 0 & 1 \\ 2 & 0 & 1 & -1 \\ 1 & 0 & 2 & -1 \\ -1 & 1 & 1 & 0 \end{bmatrix}$$

6. A bilinear function f is defined on $V_2 \times V_2$ by $f(X,Y) = 2x_1y_1 + x_2y_1 - 3x_2y_2$, where $X = (x_1,x_2)$ and $Y = (y_1,y_2)$. Determine the value of f when (a) $X = (1,1)$, $Y = (-2,2)$; (b) $X = (2,-1)$, $Y = (1,-2)$.

7. The matrix A of a bilinear form is transformed into an equivalent matrix PAQ, with nonsingular matrices P and Q. If the forms are considered to represent a bilinear function on $V \times W$, for vector spaces V and W, express the new basis elements of V and W in terms of the original, given that

$$P = \begin{bmatrix} 1 & -1 & 2 \\ 1 & 1 & 1 \\ 2 & 0 & 1 \end{bmatrix} \quad \text{and} \quad Q = \begin{bmatrix} 2 & 1 & 0 \\ 1 & 0 & 1 \\ 1 & 1 & 1 \end{bmatrix}$$

8. If I_3 is the matrix of a bilinear form on $V \times W$, find the equivalent form after the bases of V and W have been changed by transition matrices P and Q, respectively, where

$$P = \begin{bmatrix} 1 & -1 & 1 \\ 2 & 1 & 2 \\ 3 & 2 & 1 \end{bmatrix} \quad \text{and} \quad Q = \begin{bmatrix} 2 & 1 & 1 \\ 1 & -1 & 2 \\ 3 & 1 & 0 \end{bmatrix}$$

9. A bilinear form on $V \times V$ is defined as $2x_1y_1 + x_2y_3 - 3x_2y_2 + x_3y_3$. If the matrix P in Prob. 8 is the matrix of transition to a new basis for V, determine the equivalent form relative to the new basis.

10. Check the matrix product, as requested in Example 1.

11. Prove that the function H in Example 2 is bilinear.
12. Prove that the product (*not* the composite!) of two linear functions on a vector space V is bilinear on $V \times V$.
13. With f as in Example 3, use the given form to determine $f(\alpha,\beta)$, where the coordinate vectors of α and β are $(1,-1,0)$ and $(2,1,-1)$, respectively. Now use the new form derived in Example 3 to check that the same values are obtained for $f(\alpha,\beta)$ in the new reference system.
14. The "if" portion of the corollary to Theorem 1.1 follows directly from the theorem. Supply the argument for the "only if" part.
15. Verify that the effect of a matrix Q as a right multiplier of a matrix A on the column space of A is analogous to its effect as a left multiplier on the row space of A (see Sec. 3.2).
16. Outline the discussion, to parallel that in Sec. 3.2, for the reduction of a matrix to column-echelon form with the use of elementary matrices as right multipliers.
17. Verify that an elementary matrix, as a right multiplier of a matrix in row-echelon form, does not disturb the row-echelon nature of this matrix.
18. Verify that matrix equivalence is an equivalence relation in the set of all $m \times n$ matrices.

19. Use an argument based on elementary matrices to prove the lemma in this section. [*Hint:* Refer to Theorem 2.1 (Chap. 3), and consider the effect of each type of elementary matrix.]
20. Use an argument based on the isomorphism between systems of linear transformations and matrices to prove the lemma in this section.
21. Let

$$P = \begin{bmatrix} 1 & -1 & 1 \\ 2 & 1 & 2 \\ 3 & 2 & 1 \end{bmatrix} \quad \text{and} \quad Q = \begin{bmatrix} 2 & 1 & 1 \\ 1 & -1 & 2 \\ 3 & 1 & 0 \end{bmatrix}$$

be basis transition matrices, associated with vector spaces V and W, respectively, with A the matrix of a bilinear form on $V \times W$. If $PAQ = I_3$, find the form relative to the original reference frame.

22. The general value of a bilinear function f on $V \times V$ is the form $2x_1y_1 - 3x_1y_3 + x_2y_2 - 4x_3y_3$, relative to a certain basis of V. If the coordinate vectors of new basis elements of V are

$\{(1,-1,1),(0,1,2),(2,1,1)\}$, determine the equivalent form that represents f in the new reference system for V.

23. If f is a bilinear function on $V \times V$, for any vector space V, show that $\tilde{f}$ is also bilinear if $\tilde{f}(\alpha,\beta) = f(\alpha,\beta) + f(\beta,\alpha)$, for arbitrary $\alpha,\beta \in V$.

24. If f is a bilinear function on $V \times V$, for a vector space V, show that $f(-,\beta)$, for any fixed $\beta \in V$, is a linear function on V and so an element of the dual space V^*.

Note Problems 25 to 30 make use of the following definition. A bilinear form $B(X,Y)$, for coordinate vectors $X \in V_m$ and $Y \in V_n$, is said to be *nondegenerate* or *nonsingular* if these conditions are satisfied:
(i) For each $Y \neq 0$, there exists Y such that $B(X,Y) \neq 0$.
(ii) For each $Y \neq 0$, there exists X such that $B(X,Y) \neq 0$.

25. Prove that either of the two conditions in the definition of a nondegenerate form implies the other.

26. Show that a bilinear form $B(X,Y)$, with $X,Y \in V_n$, is nondegenerate if and only if the rank of the form is n. Hence verify that a bilinear form is nondegenerate if and only if its matrix is nonsingular.

27. If $X,Y \in V_n$, prove that the standard inner product (X,Y) is a nondegenerate form.

28. Try to find a degenerate bilinear form $B(X,Y)$, with X and Y arbitrary coordinate vectors in V_2.

29. If $F(X,Y)$ and $G(X,Y)$ are bilinear forms, with $X,Y \in V_n$ and $G(X,Y)$ nonsingular, prove the existence of linear operators T_1 and T_2 on V_n such that $F(X,Y) = G(T_1X,Y) = G(X,T_2Y)$.

5.2 Quadratic Forms

The topic of this section is not completely new to the reader, because *positive* quadratic forms were used in Sec. 4.7 to define inner products on a euclidean space. However, we shall introduce them again here as important special cases of bilinear forms, which we have just discussed. If V and W are finite-dimensional vector spaces, we have referred to the *value* $f(\alpha,\beta)$ of a bilinear function f on $V \times W$, for general vectors $\alpha \in V$ and $\beta \in W$, as a *bilinear form*. The bilinear form is, of course, a polynomial in the coordinates of α and β, relative to some basis of each space. Moreover, if X and Y are the coordinate vectors of α and β, respectively, we have sometimes found it convenient to write $f(\alpha,\beta)$ as $B(X,Y)$ or $X'AY$, with A the matrix of the form. If we now assume that V and W are the same space, and let $\alpha = \beta$ and $X = Y$, the general value of the *bilinear* function f becomes $f(\alpha,\alpha) =$

$B(X,X) = X'AX$. Inasmuch as only one space V and one general vector $\alpha \in V$ are now involved, it is customary to call the function and its associated forms *quadratic*. The following definition makes the concept precise.

Definition Let V be any n-dimensional vector space, with α a general vector in V and $X = (x_1, x_2, \ldots, x_n)$ its coordinate vector in V_n, relative to some basis of V. Then a function h, defined on V so that $h(\alpha) = \sum_{i,j} a_{ij} x_i x_j$, with $a_{ij} \in \mathbf{R}$, is called a quadratic function on V, and the polynomial expression for the value $h(\alpha)$ is its associated quadratic form denoted by $Q(X)$.

Although a quadratic form, as a polynomial in $x_1, x_2, \ldots, x_n$, can be expressed without any prior reference to a quadratic function, there is no loss in generality—and usually a gain in understanding—if the two are associated. If we consider the polynomial expression for a quadratic form, it is clear that we may distribute the coefficients of $x_i x_j$ and $x_j x_i$ so that $a_{ij} = a_{ji}$, for any distinct $i, j = 1, 2, \ldots, n$. (For example, the terms $2x_1 x_2 + 4x_2 x_1$ may be replaced by $3x_1 x_2 + 3x_2 x_1$.) But, if we now write $Q(X) = \sum_{i,j} a_{ij} x_i x_j$ in matrix form as $X'AX$, this implies that the matrix A of the form is symmetric. In the sequel, *we shall always assume that the matrix of any quadratic form is symmetric*, unless explicitly stated otherwise; and we have just noted that there is no loss in generality in making this assumption. It should be observed that this simple situation does not exist for a general bilinear form $X'AY$, but, if $A = A'$, the bilinear form and its associated bilinear function are also said to be *symmetric*. We add, as a passing remark, that, if $A' = -A$, the bilinear form $X'AY$ and associated function are *skew-symmetric*; but, in view of our assumption stated above, nonzero quadratic forms of this sort need not be considered.

We have observed that a quadratic form $Q(X) = X'AX$ may be formally obtained from the symmetric bilinear form $B(X,Y) = X'AY$ by simply letting $Y = X$ in the latter expression. Under these circumstances, the bilinear form $X'AY$ is said to be *polar* to the quadratic form $X'AX$, and our first theorem asserts that the polar bilinear form is uniquely determined by the quadratic form.

Theorem 2.1 *If $Q(X)$ is any quadratic form, there exists a unique polar bilinear form $B(X,Y)$ such that $B(X,X) = Q(X)$.*

Proof
Since it is clear that a polar bilinear form $B(X,Y)$ can always be defined from any quadratic form $Q(X) = X'AX$ in the manner described, the burden of

the proof is to establish uniqueness. However, we know that $B(X + Y, X + Y) = B(X,X) + B(X,Y) + B(Y,X) + B(Y,Y)$, and the symmetry of a polar form allows $B(X,Y) = B(Y,X)$. Hence $B(X,Y) = \frac{1}{2}[B(X + Y, X + Y) - B(X,X) - B(Y,Y)] = \frac{1}{2}[Q(X + Y) - Q(X) - Q(Y)]$. Since the bilinear form is then completely determined by the quadratic form, the uniqueness of the former is assured. (See also Prob. 12.) ∎

It should be observed that the interesting polar relationship in Theorem 2.1 would not exist if the matrix of a polar form was not required to be symmetric: many distinct bilinear forms $X'AY$ could give rise to the same quadratic form by letting $X = Y$. For example, the quadratic form $2x_1^2 + 4x_1x_2 - 3x_2^2$ can be expressed in matrix form in many ways (if we discard symmetry of the matrix), such as $[x_1 \ x_2] \begin{bmatrix} 2 & 3 \\ 1 & -3 \end{bmatrix} \begin{bmatrix} x_1 \\ x_2 \end{bmatrix}$ and $[x_1 \ x_2] \begin{bmatrix} 2 & 4 \\ 0 & -3 \end{bmatrix} \begin{bmatrix} x_1 \\ x_2 \end{bmatrix}$; and the given quadratic form results by letting $(x_1,x_2) = (y_1,y_2)$ in either of the bilinear forms $[x_1 \ x_2] \begin{bmatrix} 2 & 3 \\ 1 & -3 \end{bmatrix} \begin{bmatrix} y_1 \\ y_2 \end{bmatrix}$ or $[x_1 \ x_2] \begin{bmatrix} 2 & 4 \\ 0 & -3 \end{bmatrix} \begin{bmatrix} y_1 \\ y_2 \end{bmatrix}$. Hence either of these forms [and there are infinitely many more (see Prob. 13)] could be considered polar to the original quadratic form *if* we did not require that the matrix of a polar form be symmetric. We may note that, if $B(X,Y)$ is an arbitrary (not necessarily symmetric) bilinear form on a space V, the form $[B(X,Y) + B(Y,X)]/2$ is symmetric (Prob. 14). Moreover, both $B(X,Y)$ and this associated symmetric form give rise to the *same* quadratic form if we let $Y = X$ (Prob. 15). Thus we see that the requirement that the matrix of a polar form be symmetric does not result in any loss of generality.

The following important *change of basis* theorem for quadratic forms will be seen to be a direct consequence of Theorem 1.1.

Theorem 2.2 Let A be the matrix of a quadratic form, regarded as the general value of a quadratic function on a vector space V. Then, if P is the matrix of transition to a new basis of V, the matrix of the new form that results from the given form is $P'AP$.

Proof
As we indicated earlier, we obtain this result from Theorem 1.1 by assuming that $W = V$ and letting $Y = X$: for, by Theorem 1.2, any quadratic form $Q(X)$ can be associated with its unique polar bilinear form $B(X,Y)$, to which Theorem 1.1 can be applied. But then, under the change of basis asserted in the statement of the present theorem, the matrix A of $B(X,Y)$ is changed into $P'AP$ (noting here that $Q = P$ in Theorem 1.1). Since $Q(X) = B(X,X)$, we have shown that the matrix of $Q(X)$, in the new frame of reference, is $P'AP$, thus completing the proof. ∎

The observant reader will have noted that the matrix theory associated with bilinear functions on $V \times V$, with a single basis assumed for V, is the same as that of quadratic functions on V. In either case, if the basis of V is changed by a transition matrix P, the matrix A of an associated form is transformed into $P'AP$. And *this* is true whether or not the matrix of either form is symmetric. The idea of *congruence*, as applied to two matrices, was suggested in Sec. 5.1, but the following definition is more precise.

Definition   If matrices A and B are so related that $B = P'AP$, for a nonsingular matrix P, then A and B are said to be *congruent*.

By analogy with bilinear forms on $V \times W$, two quadratic or bilinear forms on V are said to be *equivalent* if they represent the same quadratic or bilinear function. It then follows from Theorem 2.2 that *quadratic or bilinear forms on V are represented by matrices that are congruent*. Although there are parallel bilinear and quadratic theories, our present interest is in the quadratic case.

EXAMPLE 1

Determine the quadratic form equivalent to $2x^2 - 6xy + y^2 - 2xz + 3z^2$, caused by a change of basis of the underlying vector space, the matrix of transition being

$$P = \begin{bmatrix} 1 & 0 & 2 \\ 0 & 1 & -2 \\ 0 & 2 & -1 \end{bmatrix}$$

Solution
The (symmetric) matrix A of the given form is easily seen to be

$$\begin{bmatrix} 2 & -3 & -1 \\ -3 & 1 & 0 \\ -1 & 0 & 3 \end{bmatrix}$$

It is then an immediate consequence of Theorem 2.2 that the matrix of the desired equivalent form is

$$P'AP = \begin{bmatrix} 1 & 0 & 0 \\ 0 & 1 & 2 \\ 2 & -2 & -1 \end{bmatrix} \begin{bmatrix} 2 & -3 & -1 \\ -3 & 1 & 0 \\ -1 & 0 & 3 \end{bmatrix} \begin{bmatrix} 1 & 0 & 2 \\ 0 & 1 & -2 \\ 0 & 2 & -1 \end{bmatrix}$$

$$= \begin{bmatrix} 1 & 0 & 0 \\ 0 & 1 & 2 \\ 2 & -2 & -1 \end{bmatrix} \begin{bmatrix} 2 & -5 & 11 \\ -3 & 1 & -8 \\ -1 & 6 & -5 \end{bmatrix} = \begin{bmatrix} 2 & -5 & 11 \\ -5 & 13 & -18 \\ 11 & -18 & 43 \end{bmatrix}$$

250 BILINEAR AND QUADRATIC FORMS

If (x',y',z') is the general coordinate vector in the new frame of reference, the new quadratic form, equivalent to the original, is then $2x'^2 - 10x'y' + 13y'^2 + 22x'z' - 36y'z' + 43z'^2$.

EXAMPLE 2

If $X = (x,y,z)$ and $X' = (x',y',z')$ are the general coordinate vectors, associated as in Example 1, check that the two forms in that example give the same value when $X = (1,1,1)$.

Solution
The value of the original form, for $X = (1,1,1)$, is easily seen to be $2 - 6 + 1 - 2 + 3 = -2$. On computing P^{-1} (and reviewing Sec. 2.9), we find that $X' = (\frac{5}{3},\frac{1}{3},-\frac{1}{3})$. Then, if we put $x' = \frac{5}{3}$, $y' = \frac{1}{3}$, and $z' = -\frac{1}{3}$ in the new form, an elementary arithmetic computation shows that the value is also -2.

As was remarked in Sec. 5.1, the relation of congruence is an equivalence relation in the set of all $n \times n$ matrices (Prob. 16), and it is clear that *congruence is a special case of equivalence*. This equivalence relation then partitions the set of all $n \times n$ matrices into equivalence classes of congruent matrices. In the study of equivalence in Sec. 5.1, we discovered that *rank* was the characteristic invariant of each of the equivalence classes and that the class of matrices of rank r has a *canonical* representative of the form $\begin{bmatrix} I_r & 0 \\ 0 & 0 \end{bmatrix}$. We now raise the query of the possible existence of a comparable invariant characterization of the classes of congruent matrices, and of a canonical matrix that may be a simple representation of each class. We shall give an affirmative answer to both parts of this query in Sec. 5.3 and actually exhibit the invariants and canonical matrices for the relation of congruence.

Problems 5.2

1. If the standard inner product (X,Y) in the space V_n is considered a bilinear form, verify that the quadratic form $Q(X)$ to which it is polar designates the square of the length of X.

2. Review the reason why a vector space is euclidean if and only if a positive quadratic form has been defined on it.

3. Write each of the following quadratic forms as a polynomial:

   (a) $[x \ y \ z] \begin{bmatrix} 2 & -1 & 1 \\ -1 & 2 & 0 \\ 1 & 0 & 1 \end{bmatrix} \begin{bmatrix} x \\ y \\ z \end{bmatrix}$

   (b) $[x_1 \ x_2 \ x_3] \begin{bmatrix} 1 & 2 & 4 \\ 2 & 2 & -2 \\ 4 & -2 & 0 \end{bmatrix} \begin{bmatrix} x_1 \\ x_2 \\ x_3 \end{bmatrix}$

4. Express each of the following quadratic forms as a matrix product $X'AX$, with A a symmetric matrix and X an appropriate coordinate vector: (a) $2x^2 - 3xy + y^2 - 4yz - 4z^2$; (b) $x_1^2 + 4x_1x_2 - 3x_2^2 + x_3^2$.
5. Express each quadratic form in Prob. 4 as $X'AX$, with A not a symmetric matrix.
6. Use an appropriate symbolism to express the bilinear form that is polar to each of the quadratic forms in (a) Prob. 3; (b) Prob. 4.
7. Verify that any matrix that is congruent to a symmetric matrix is a symmetric matrix.
8. If
$$P = \begin{bmatrix} 2 & 1 & 2 \\ -1 & 1 & 1 \\ 0 & 1 & -3 \end{bmatrix}$$
find the matrix $P'AP$ congruent to A, where

(a) $$A = \begin{bmatrix} 2 & -1 & 1 \\ 1 & 0 & -1 \\ 0 & 2 & 0 \end{bmatrix}$$
(b) $$A = \begin{bmatrix} 1 & 0 & 1 \\ 2 & 1 & -1 \\ 0 & 2 & 3 \end{bmatrix}$$

9. If $x_1^2 - 2x_1x_2 + 3x_2^2 + 4x_3^2$ is a given quadratic form, use the formula developed in the proof of Theorem 2.1 to find the associated polar bilinear form.
10. Apply the directions in Prob. 9 to the quadratic form $2x_1^2 + 4x_1x_2 + x_2x_3 - 3x_3^2$.
11. If
$$P = \begin{bmatrix} 1 & 2 & -1 \\ 1 & 1 & 2 \\ 0 & 1 & -1 \end{bmatrix}$$

is the basis transition matrix for a vector space V, determine the quadratic form (regarded as the general value of a quadratic function on V) that is equivalent, relative to the new basis, to (a) $x^2 - 3xy + yz - y^2 + 4z^2$; (b) $3x^2 - 2xz + 4yz + 2z^2$.

12. If $B(X,Y)$ is a symmetric bilinear form and $Q(X) = B(X,X)$ is the quadratic form to which it is polar, prove the following polarization identity: $4B(X,Y) = Q(X + Y) - Q(X - Y)$.
13. If the matrix of a polar bilinear form were not required to be symmetric, explain why a quadratic form could have infinitely many distinct polar bilinear forms associated with it.

14. If $B(X,Y)$ is any bilinear form, on a vector space V, prove that $[B(X,Y) + B(Y,X)]/2$ is a symmetric form on V.
15. Verify that both $B(X,Y)$ and the associated symmetric form in Prob. 14 give rise to the same quadratic form if Y is replaced by X.
16. Prove that congruence is an equivalence relation in the set of $n \times n$ matrices.
17. Let $x^2 - 4xy - 2xz + z^2$ be a given quadratic form. If $x' = x - y$, $y' = x + y + z$, $z' = y - z$ are equations describing a change of variables, use matrix methods to determine the equivalent form in the new variables. Check your answer by direct substitutions.
18. Apply the directions in Prob. 17 to the quadratic form $3x^2 - 2xy + 4yz + y^2 + 3z^2$.
19. Check that the two forms in Prob. 17 and the two forms in Prob. 18 represent the same functions, by using the "test" coordinate vector $(x,y,z) = (2,0,-1)$ in both cases. (See Example 2.)
20. Check that the congruent forms, obtained in Prob. 11, represent the same quadratic functions, by using $(x,y,z) = (1,-1,1)$ as a "test" vector. (See Example 2.)
21. Prove that any quadratic form in x and y is equivalent to a form consisting of the sum of two squares with real coefficients.
22. Prove that the set of all quadratic functions on V comprises a subspace of the space of all real functions on V.
23. With reference to Prob. 22, try to find an isomorphic mapping of the subspace of quadratic functions onto the subspace of bilinear functions, without any reference to a basis of V.
24. What would be an appropriate definition of a nondegenerate quadratic form? (See Note prior to Prob. 25 of Sec. 5.1.) Is the polar form, associated with a nondegenerate quadratic form, necessarily nondegenerate?

5.3 Diagonal Quadratic Forms Under Congruence

As a prelude to our more serious search for a matrix in simple form to represent a class of equivalent matrices under congruence, let us first consider the question of the existence of such a *diagonal* representative. The principal result in this section affirms the existence of a diagonal matrix in each class, but, before we prove the theorem, let us review a procedure familiar from college algebra.

EXAMPLE 1

Use a change of variables to reduce the quadratic form $x^2 - 2xy + 4yz - 2y^2 + 4z^2$ to a sum or difference of squares.

Solution

If we "complete the squares" by the usual means of adding and subtracting the necessary terms, the above polynomial form can be expressed as $(x - y)^2 - 4y^2 + (y + 2z)^2$. A change of variables to x', y', z', where $x' = x - y$, $y' = 2y$, $z' = y + 2z$, now transforms the original form into $x'^2 - y'^2 + z'^2$ or, if desired, $x'^2 + z'^2 - y'^2$.

We note that the (symmetric) matrix of the original form in Example 1 is

$$A = \begin{bmatrix} 1 & -1 & 0 \\ -1 & -2 & 2 \\ 0 & 2 & 4 \end{bmatrix}$$

whereas the matrix of the final form is

$$B = \begin{bmatrix} 1 & 0 & 0 \\ 0 & -1 & 0 \\ 0 & 0 & 1 \end{bmatrix}$$

and we observe that B is a diagonal matrix. This result can be given a very simple matrix formulation. If $X = (x,y,z)$ and $Y = (x',y',z')$ are considered general coordinate vectors, respectively, before and after the change of basis of the space, then $X = PY$ or $Y = P^{-1}X$ (with X and Y written as column matrices), where

$$P^{-1} = \begin{bmatrix} 1 & -1 & 0 \\ 0 & 2 & 0 \\ 0 & 1 & 2 \end{bmatrix}$$

A simple computation shows that

$$P = \begin{bmatrix} 1 & \frac{1}{2} & 0 \\ 0 & \frac{1}{2} & 0 \\ 0 & -\frac{1}{4} & \frac{1}{2} \end{bmatrix}$$

the matrix P being nonsingular. But now, if we express the given form as

$$X'AX = \begin{bmatrix} x & y & z \end{bmatrix} \begin{bmatrix} 1 & -1 & 0 \\ -1 & -2 & 2 \\ 0 & 2 & 4 \end{bmatrix} \begin{bmatrix} x \\ y \\ z \end{bmatrix}$$

254 BILINEAR AND QUADRATIC FORMS

a replacement of X by PY transforms $X'AX$ into the equivalent form

$$(PY)'A(PY) = Y'(P'AP)Y$$

$$= [x' \ y' \ z'] \begin{bmatrix} 1 & 0 & 0 \\ \frac{1}{2} & \frac{1}{2} & -\frac{1}{4} \\ 0 & 0 & \frac{1}{2} \end{bmatrix} \begin{bmatrix} 1 & -1 & 0 \\ -1 & -2 & 2 \\ 0 & 2 & 4 \end{bmatrix} \begin{bmatrix} 1 & \frac{1}{2} & 0 \\ 0 & \frac{1}{2} & 0 \\ 0 & -\frac{1}{4} & \frac{1}{2} \end{bmatrix} \begin{bmatrix} x' \\ y' \\ z' \end{bmatrix}$$

$$= [x' \ y' \ z'] \begin{bmatrix} 1 & 0 & 0 \\ \frac{1}{2} & \frac{1}{2} & -\frac{1}{4} \\ 0 & 0 & \frac{1}{2} \end{bmatrix} \begin{bmatrix} 1 & 0 & 0 \\ -1 & -2 & 1 \\ 0 & 0 & 2 \end{bmatrix} \begin{bmatrix} x' \\ y' \\ z' \end{bmatrix}$$

$$= [x' \ y' \ z'] \begin{bmatrix} 1 & 0 & 0 \\ 0 & -1 & 0 \\ 0 & 0 & 1 \end{bmatrix} \begin{bmatrix} x' \\ y' \\ z' \end{bmatrix}$$

The form, expressed in the new variables, is now $x'^2 - y'^2 + z'^2$, in agreement with the college algebra procedure.

The following theorem shows that there is nothing special about the result in Example 1 and that *any* quadratic form is equivalent to a form in which only "square" terms appear.

Theorem 3.1 *Any symmetric matrix is congruent to a diagonal matrix. That is, for any symmetric matrix A, there exists a nonsingular matrix P such that $P'AP$ is in diagonal form.*

Proof
There is no loss in generality if we assume that A is the matrix of the quadratic form $Q(X)$, relative to the $\{E_i\}$ basis, where $X = (x_1, x_2, \ldots, x_n) \in V_n$ and n is the order of A. It was seen in Sec. 5.2 that there exists a bilinear form $B(X,Y)$ which may be associated with a given quadratic form $Q(X)$ as its polar. If $Q(X)$ is the zero form [that is, $Q(X) = 0$, for any $X \in V_n$] it is clear that the matrix A of the form is the zero matrix and so is in diagonal form. On the other hand, if $Q(X)$ is not the zero form, neither is the associated polar bilinear form $B(X,Y)$ the zero form; and this means that there exists a nonzero vector $X_1 \in V_n$ such that $B(X_1, X_1) = Q(X_1) \neq 0$. We now use X_1 as the first member of a new basis for V_n, and select the other members so that the matrix of the form is diagonal.

Our method will be a constructive one, based on the principle of mathematical induction. The set of vectors $Y \in V_n$, such that $B(X_1, Y) = 0$, is the orthogonal complement $W_1^\perp$ of $W_1 = \mathfrak{L}(X_1)$, and $W_1^\perp$, by Theorem 4.1, has dimension $n - 1$ (Prob. 7). If we now take any basis $\{X_2, X_3, \ldots, X_n\}$ of $W_1^\perp$, it is clear that these vectors, along with X_1, will make up a basis of V_n (Prob. 8). Thus, any $X \in V_n$ can be expressed in the form $y_1 X_1 + y_2 X_2 +$

$\cdots + y_n X_n$, so that $(y_1, y_2, \ldots, y_n)$ is the coordinate vector of X, relative to this new basis. But then, recalling how the matrix of a form is related to the form, we see that $B(X, X) = Q(X) = \sum_{i,j=1}^{n} y_i B(X_i, X_j) y_j$, in the new reference system, and we know that $B(X_1, X_j) = 0$, for $j > 1$. Hence, in this system, the matrix of $Q(X)$ is

$$\begin{bmatrix} B(X_1, X_1) & 0 & \cdots & 0 \\ 0 & B(X_2, X_2) & \cdots & B(X_2, X_n) \\ 0 & B(X_3, X_2) & \cdots & B(X_3, X_n) \\ \cdots & \cdots & \cdots & \cdots \\ 0 & B(X_n, X_2) & & B(X_n, X_n) \end{bmatrix}$$

If we let $B(X_1, X_1) = d_1$, this matrix may be abbreviated to $\begin{bmatrix} d_1 & 0 \\ 0 & A_1 \end{bmatrix}$, where A_1 is a square matrix of order $n - 1$, and the two 0s are appropriate row and column zero matrices. If P_1 is the matrix that expresses the transition from the original $\{E_i\}$ to the new basis of V_n, it follows from our earlier theory that

$$P_1' A P_1 = \begin{bmatrix} d_1 & 0 \\ 0 & A_1 \end{bmatrix}$$

The first step of the reduction process is now completed, and the rest of the proof proceeds by induction on n. If $n = 1$, the matrix A is trivially diagonal, and there is nothing to prove. By the usual induction hypothesis, we now assume the existence of a diagonal matrix in the congruence class of any given square matrix of order $n - 1$. In particular, this implies that a nonsingular matrix S_1 exists, such that $S_1' A_1 S_1$ is diag $\{d_2, \ldots, d_n\}$, with $d_2, \ldots, d_n \in \mathbf{R}$. But now, if we denote $\begin{bmatrix} 1 & 0 \\ 0 & S_1' \end{bmatrix}$ by S, the 0s being appropriate row or column zero matrices, it follows easily (Prob. 9) that $S'(P_1' A P_1(S = \text{diag } \{d_1, d_2, \ldots, d_n\} = (P_1 S)' A (P_1 S) = P' A P$, where $P = P_1 S$. We note that P is nonsingular (Prob. 10), and the induction argument is complete. ∎

Corollary 1   Any quadratic form, defined on an n-dimensional vector space, is equivalent to $d_1 x_1^2 + d_2 x_2^2 + \cdots + d_n x_n^2$, with $d_1, d_2, \ldots, d_n \in \mathbf{R}$.

Inasmuch as the reduction of symmetric bilinear forms is identical to that of quadratic forms, we also have the following parallel result.

Corollary 2 *Any symmetric bilinear form, defined on an n-dimensional vector space, is equivalent to $d_1 x_1 y_1 + d_2 x_2 y_2 + \cdots + d_n x_n y_n$, with $d_1, d_2, \ldots, d_n \in \mathbf{R}$.*

It should be noted that, although we have established *existence*, we have not established uniqueness of a diagonal matrix in each equivalence class of congruent matrices. In Sec. 5.4 we shall exhibit a diagonal matrix that is unique—and so canonical—in its class.

There is an obvious application of quadratic forms to geometry, in which the general vector $X = (x, y, z) \in V_3$ is identified with a point of 3-space. If $Q(X) = X'AX$ is a quadratic form on V_3, the equation $X'AX = 1$ will designate the equation of a central conic in some reference system. For example, if

$$A = \begin{bmatrix} 1 & 0 & 2 \\ 0 & -1 & 0 \\ 2 & 0 & 2 \end{bmatrix}$$

$$X'AX = \begin{bmatrix} x & y & z \end{bmatrix} \begin{bmatrix} 1 & 0 & 2 \\ 0 & -1 & 0 \\ 2 & 0 & 2 \end{bmatrix} \begin{bmatrix} x \\ y \\ z \end{bmatrix} = x^2 - y^2 + 4xz + 2z^2 = 1$$

is an equation of such a conic. The result of Theorem 3.1, as applied to the geometry of 3-space, is that, with a proper choice of reference system or axes, the equation of any central conic has the form $f(x, y, z) = 1$, where $f(x, y, z)$ is a sum or difference of squares with constant coefficients.

If we recall from Sec. 3.2 the effect on any matrix of left and right multiplications by elementary matrices, it is clear how to transform a matrix under the rules of congruence: Apply the *same* sequence of elementary column as elementary row operations. We know, by Theorem 3.1, that any symmetric matrix A is congruent to a diagonal matrix $P'AP$, and Theorem 2.1 (Chap. 3) tells us that P is a product of elementary matrices. However, although this procedure is somewhat like that of Sec. 3.2, it is nonetheless awkward, and it is usually easier to apply the method used in the proof of Theorem 3.1. In Sec. 5.6, we shall give a method that is even simpler, based on the so-called *eigenvalues* of A. We shall now make clear the procedure in the theorem with an example.

EXAMPLE 2

Find a diagonal matrix that is congruent to

$$A = \begin{bmatrix} 0 & 4 & 4 \\ 4 & 5 & -1 \\ 4 & -1 & -3 \end{bmatrix}$$

SEC. 5.3 DIAGONAL QUADRATIC FORMS UNDER CONGRUENCE

Solution
We may regard A as the matrix of a bilinear form which, for convenience, is defined on the coordinate space V_3. Thus, with $X = (x_1, x_2, x_3)$ and $Y = (y_1, y_2, y_3)$ in V_3,

$$B(X, Y) = X'AY = \begin{bmatrix} x_1 & x_2 & x_3 \end{bmatrix} \begin{bmatrix} 0 & 4 & 4 \\ 4 & 5 & -1 \\ 4 & -1 & -3 \end{bmatrix} \begin{bmatrix} y_1 \\ y_2 \\ y_3 \end{bmatrix}$$

The problem is to discover an appropriate matrix of transition whose columns are the new basis vectors of V_3. This can be accomplished one step at a time, using the method outlined in the theorem. Since $A \neq 0$, it is easy to find a nonzero vector X_1, by trial and error, such that $B(X_1, X_1) \neq 0$. For instance, if we let $X_1 = (0,1,0)$, we find that $B(X_1, X_1) = 5 \neq 0$. This is the first member of the new basis, and we complete the basis for V_3 by selecting linearly independent vectors from the subspace of vectors orthogonal to X_1. The vectors of this subspace are solutions (s_1, s_2, s_3) of

$$\begin{bmatrix} 0 & 1 & 0 \end{bmatrix} \begin{bmatrix} 0 & 4 & 4 \\ 4 & 5 & -1 \\ 4 & -1 & -3 \end{bmatrix} \begin{bmatrix} s_1 \\ s_2 \\ s_3 \end{bmatrix} = 4s_1 + 5s_2 - s_3 = 0$$

and there are many solutions of this equation. It is easy to find two that are linearly independent: If we let $s_2 = 0$, then $4s_1 = s_3$, so that $X_2 = (1,0,4)$ is one solution; if we let $s_1 = 0$, then $5s_2 = s_3$, and so $X_3 = (0,1,5)$ is another solution, linearly independent of X_2. The desired basis, *to complete the first stage of the inductive process in the theorem*, is then $\{X_1, X_2, X_3\}$, and the associated transition matrix is

$$P_1 = \begin{bmatrix} 0 & 1 & 0 \\ 1 & 0 & 1 \\ 0 & 4 & 5 \end{bmatrix}$$

It is a matter of simple matrix multiplication to verify that

$$P_1'AP_1 = \begin{bmatrix} 5 & 0 & 0 \\ 0 & -16 & -40 \\ 0 & -40 & -80 \end{bmatrix}$$

It is now convenient to regard the submatrix $A_1 = \begin{bmatrix} -16 & -40 \\ -40 & -80 \end{bmatrix}$, in the lower right-hand corner of the above matrix, as the matrix of a new bilinear form on V_2, say $B_1(X',Y')$, with $X' = (x_1',x_2')$ and $Y' = (y_1',y_2')$. In this case, $B_1(X',Y') = \begin{bmatrix} x_1' & x_2' \end{bmatrix} \begin{bmatrix} -16 & -40 \\ -40 & -80 \end{bmatrix} \begin{bmatrix} y_1' \\ y_2' \end{bmatrix}$ and, again, it is easy to find a non-zero vector $X_1' \in V_2$, say $X_1' = (1,0)$, such that $B_1(X_1',X_1') \neq 0$. The subspace of V_2, consisting of the vectors orthogonal to X_1', is one-dimensional and may be determined by solving the equation

$$\begin{bmatrix} 1 & 0 \end{bmatrix} \begin{bmatrix} -16 & -40 \\ -40 & -80 \end{bmatrix} \begin{bmatrix} l_1 \\ l_2 \end{bmatrix} = -16l_1 - 40l_2 = 0$$

for l_1 and l_2. There are many solutions of this equation, and we pick one of them: $l_1 = 5$ and $l_2 = -2$. This second basis vector of V_2 is then $(5,-2)$; and the matrix, which corresponds to S in the proof of the theorem, is then

$$\begin{bmatrix} 1 & 0 & 0 \\ 0 & 1 & 5 \\ 0 & 0 & -2 \end{bmatrix}$$

It may now be readily checked that

$$\begin{bmatrix} 1 & 0 & 0 \\ 0 & 1 & 0 \\ 0 & 5 & -2 \end{bmatrix} \begin{bmatrix} 5 & 0 & 0 \\ 0 & -16 & -40 \\ 0 & -40 & -80 \end{bmatrix} \begin{bmatrix} 1 & 0 & 0 \\ 0 & 1 & 5 \\ 0 & 0 & -2 \end{bmatrix} = \begin{bmatrix} 5 & 0 & 0 \\ 0 & -16 & 0 \\ 0 & 0 & 80 \end{bmatrix}$$

and the final matrix is seen to be in diagonal form. The matrix of transition P, from the initial to the final basis of V_3, is $P_1 S$, and a simple computation shows that

$$P = \begin{bmatrix} 0 & 1 & 5 \\ 1 & 0 & -2 \\ 0 & 4 & 10 \end{bmatrix}$$

and $P'AP = \text{diag}\{5,-16,80\}$. In this example, the process of diagonalization was accomplished in two steps, but, if the original matrix were of larger order, more would likely be needed.

The problem worked out in Example 2 is a matrix-theoretic problem, but we found it convenient to use the context of a bilinear form defined on a vector space. If one's primary interest is in forms, the implication of the result in the example is that the bilinear symmetric or quadratic form on V_3 with matrix A is equivalent to $5x_1 y_1 - 16x_2 y_2 + 80x_3 y_3$ or $5x_1^2 - 16x_2^2 + 80x_3^2$, respectively.

Problems 5.3

1. Use a college algebra change-of-variables procedure to reduce each of the following quadratic forms to a sum or difference of squares with real coefficients: (a) $2x^2 - 4xy + 6yz + 3y^2 - 4z^2$; (b) $x^2 + 6xz + 4yz + 2y^2$.

2. Apply the directions in Prob. 1 to each of the following forms: (a) $x_1^2 - x_1x_2 + 2x_2^2 + 3x_3^2 + x_3x_4$; (b) $4x_1x_2 + 2x_1x_3 - 4x_2x_3 - 2x_3^2$.

3. Use $x_1 = 2y_1 + y_2$ and $x_2 = y_1 - y_2$ to change $2x_1^2 - 4x_1x_2 + 3x_2^2$ into a new quadratic form, and use a matrix computation to check your result. If (x_1,x_2) and (y_1,y_2) are regarded as corresponding points of V_2, relative to two bases, describe the new basis in terms of the old.

4. Explain how the *same* quadratic form can represent different quadratic functions on a vector space V. Illustrate your answer with an example where $V = V_2$.

5. If A in Example 1 is the matrix of a quadratic function on a vector space, relative to the basis $\{\alpha_1,\alpha_2,\alpha_3\}$, describe the basis with respect to which the matrix of the function is B.

6. In Example 2, verify that $B_1(X_1',X_1') = -16$.

Note Problems 7 to 10 refer to the proof of Theorem 3.1.

7. Explain why the dimension of $W_1^\perp$ is $n - 1$.
8. Explain why the collection of vectors, consisting of X_1 and any basis of $W_1^\perp$, constitutes a basis of V.
9. Verify that $S'(P_1'AP_1)S = \text{diag}\{d_1,d_2, \ldots, d_n\}$.
10. Why must P be a nonsingular matrix?

11. Let
$$A = \begin{bmatrix} 2 & 0 & 1 \\ 1 & 2 & 0 \\ 0 & -1 & 1 \end{bmatrix}$$

Then, with S_{ij}, $D_i(r)$, and $P_{ij}(r)$ elementary matrices, as defined in Sec. 3.2, determine the matrix $P'AP$, where $P = S_{12}D_2(3)P_{23}(-1)$.

Note Use the method suggested by Prob. 11 to reduce the matrices in Probs. 12 to 14 to congruent diagonal form.

12. $\begin{bmatrix} 1 & -2 & 3 \\ 2 & 2 & 1 \\ 3 & 2 & 1 \end{bmatrix}$

260   BILINEAR AND QUADRATIC FORMS

13. $\begin{bmatrix} 2 & -1 & 1 \\ 1 & 0 & 2 \\ 1 & 1 & 1 \end{bmatrix}$

14. $\begin{bmatrix} -1 & 1 & 2 \\ 1 & -1 & -2 \\ 2 & -2 & -3 \end{bmatrix}$

Note   Use the method of Example 2 to reduce the matrices, referred to in Probs. 15 to 17 to congruent diagonal form. You should not expect your diagonal matrices to be the same. (Why?)

15. Problem 12.
16. Problem 13.
17. Problem 14.
18. Let A be a matrix that is congruent to diag $\{d_1, d_2, \ldots, d_n\}$, with $d_i \geq 0$, $i = 1, 2, \ldots, n$. Comment on the nature of A and of any quadratic form that it represents.
19. Prove that the quadratic form $ax^2 + 2bxy + cy^2$ is positive if and only if $a > 0$ and $b^2 - ac < 0$.
20. Prove that $f(\alpha, \beta) = x_1 y_2 - x_2 y_1$ defines a bilinear skew-symmetric function on a two-dimensional space V, where (x_1, y_1) and (x_2, y_2) are respective coordinates of $\alpha, \beta \in V$. Moreover, prove that any bilinear skew-symmetric function on V is a scalar multiple of f.
21. If $Q(X) = x^2 + y^2$, with $X = (x, y) \in V_2$, does $Q(X) = 1$ necessarily designate the equation of a circle? Explain your answer.
22. The equation of an ellipse, relative to the $\{E_i\}$ basis of V_2, is given as $5x^2 + 4xy + 8y^2 = 9$. Find the equation of the ellipse, if it is referred to the basis $\{(1, -1), (1, 1/2)\}$.
23. The equation of a conic in V_2 is given as $x^2 - 2xy + 4y^2 = 7$, relative to the $\{E_i\}$ basis. Find a new basis of the space such that the equation of the conic in the new reference system contains no cross-product terms.
24. Apply the directions in Prob. 23 to the equation of a conic in V_3, given as $x^2 - 2xy + y^2 + z^2 = 4$.
25. The equation of a hyperbola is given as $2x^2 - 24xy - 5y^2 + x + 3y + 9 = 0$. Use *in part* the matrix methods of this section to reduce the equation to $14x'^2 - 11y'^2 - 3x' - y' - 9 = 0$, relative to a new reference system. [*Caution:* The center of the conic is not at the origin.]

5.4 Invariants of a Symmetric Matrix Under Congruence

In this section we complete the discussion begun in Sec. 5.3: to find the invariants and the unique canonical form of a symmetric matrix under congruence. In view of the very close relationship between a symmetric bilinear or a quadratic function (or form) and its representing symmetric matrix, it is clear that the invariants of such a matrix will also have important interpretations for the function or form that it represents.

We have already seen that any symmetric matrix A is congruent to a diagonal matrix diag $\{d_1, d_2, \ldots, d_n\}$, with n the order of A. That is, there exists a nonsingular matrix P such that $P'AP =$ diag $\{d_1, d_2, \ldots, d_n\}$. But we can do better than this! We first permute the rows and columns of the diagonal matrix, if necessary, so that the positive diagonal elements (if any) occupy the first positions, followed by the negative diagonal elements (if any); any remaining diagonal positions are occupied by 0s. Since these permutations can be accomplished by a sequence of elementary row operations (of the first kind) *followed by the same sequence of elementary column operations*, the final matrix is congruent to the original diagonal matrix. Since congruent matrices are certainly equivalent, it follows from the lemma in Sec. 5.1 that congruent matrices also have the same rank. A simple argument, based on the dimension of the row or column space of a diagonal matrix, shows immediately (Prob. 10) that the rank of such a matrix is the number of nonzero elements in its diagonal. In view of the foregoing remarks, it then follows that a symmetric matrix of rank r is congruent to a diagonal matrix $D =$ diag $\{d_1, d_2, \ldots, d_p, d_{p+1}, \ldots, d_r, 0, 0, \ldots, 0\}$, where $d_i > 0$, $i = 1, 2, \ldots, p$, if $p > 0$; $d_i < 0$, $i = p+1, p+2, \ldots, r$, if $r - p > 0$; the number of diagonal 0s is $n - r$, where n is the order of A. Now let $P =$ diag $\{1/\sqrt{d_1}, 1/\sqrt{d_2}, \ldots, 1/\sqrt{d_p}, 1/\sqrt{-d_{p+1}}, \ldots, 1/\sqrt{-d_r}, 1, 1, \ldots, 1\}$, where the number of final 1s is $n - r$. It is immediate that P is nonsingular, and a simple computation shows that $P'DP =$ diag $\{1, 1, \ldots, 1, -1, \ldots, -1, 0, 0, \ldots, 0\} = D_0$, where the number of 1s is p, the number of -1s is $r - p$, and the number of 0s is $n - r$. It should be understood that all three kinds of numbers do not necessarily occur in this final diagonal matrix, but not all the diagonal entries are 0 unless $r = 0$. We are now able to sharpen the results stated in the corollaries following Theorem 3.1, and we formulate these improved results as a theorem.

Theorem 4.1 *A quadratic (symmetric bilinear) form on an n-dimensional vector space is the zero form with rank $r = 0$, or $r > 0$ and it is equivalent to $x_1^2 + x_2^2 + \cdots + x_p^2 - x_{p+1}^2 - \cdots - x_r^2$ $(x_1 y_1 + x_2 y_2 + \cdots + x_p y_p - x_{p+1} y_{p+1} - \cdots - x_r y_r)$ if $1 \leq p \leq r \leq n$ and otherwise to $-x_1^2 - x_2^2 - \cdots - x_r^2$ $(-x_1 y_1 - x_2 y_2 - \cdots - x_r y_r)$.*

For historical reasons, the number $2p - r$ or $p - (r - p)$ is called the *signature* of the matrix A, or of a quadratic or symmetric bilinear form that is represented by A. The signature may then be recognized as the excess of the 1s over the -1s in the diagonal matrix congruent to A, as described in Theorem 4.1. We have not yet shown that this is unique, for a given A, but we now do so. The *signature*, *rank*, and *order* constitute a complete set of invariants under congruence for a given symmetric matrix A; that is, these three numbers characterize the congruence equivalence class to which A belongs. It is then an immediate consequence that the diagonal matrix D_0, as described above, which is congruent to A, is the *canonical* matrix of this equivalence class.

It follows from the definition of congruence that congruent matrices have the same order n, and we have already remarked above that they also have the same rank r. Thus, the only nontrivial matter to be established is the invariance of the signature of a matrix under congruence, which is to say that the number p is the same for any diagonal matrix in the same form as D_0 and congruent to A. To verify this, let us assume the existence of two diagonal matrices D_1 and D_2, both congruent to a symmetric matrix A and in the same form as D_0, but such that D_1 has p and D_2 has q elements 1 along the main diagonal, with $p \neq q$. With no loss in generality, we may assume $p > q$ and so $p \geq 1$. Since D_1 and D_2 are in the same congruence class, there exists a nonsingular matrix P such that $D_1 = P'D_2P$. The subset of vectors $(x_1, x_2, \ldots, x_p, 0, 0, \ldots, 0) = X$ constitutes a subspace W_p of V_n of dimension p, and it is clear that $X'D_1X = x_1^2 + x_2^2 + \cdots + x_p^2$ is a positive (definite) quadratic form on W_p (Prob. 11). Since P is a nonsingular matrix, the dimension of the subspace of vectors PX in V_n, with $X \in W_p$, is also p, and we designate this space as W'_p. But every nonzero vector PX in W'_p has the property that $(PX)'D_2(PX) > 0$ (Prob. 12), and every $Y = (0, 0, \ldots, 0, y_{q+1}, y_{q+2}, \ldots, y_n)$ has the property that $Y'D_2Y = -y_{q+1}^2 - y_{q+2}^2 - \cdots - y_r^2 \leq 0$. The vectors Y make up a subspace of V_n of dimension $n - q$, and we designate this subspace W_{n-q}. Since W'_p and W_{n-q} have only the vector 0 in common, $\dim W'_p + \dim W_{n-q} = p + (n - q) = n + (p - q) > n$. However, we know from Theorem 9.2 (Chap. 1) that $\dim W'_p + \dim W_{n-q} = \dim (W'_p \cap W_{n-q}) + \dim (W'_p + W_{n-q})$, while $\dim (W'_p \cap W_{n-q}) = 0$ and $\dim W'_p + \dim W_{n-q} \leq n$ (Prob. 12). It follows that $p + (n - q) \leq 0 + n = n$, and from this we must conclude that $n + (p - q) \leq n$. This is in contradiction with our earlier inequality, and so our assumption that $p \neq q$ is untenable. Hence $p = q$, and, since r is known to be invariant under congruence, so is the signature of the matrix A. We collect the results just established into the form of a second theorem.

Theorem 4.2 *The order, rank, and signature comprise a complete set of invariants under congruence for a symmetric matrix. Moreover, any matrix of order n, rank r, and signature $2p - r$ is congruent to the unique (and hence canonical) matrix* diag $\{1, 1, \ldots, 1, -1, -1, \ldots, -1, 0, 0, \ldots, 0\}$, *in which the number of 1s is p, the number of -1s is $r - p$, and the number of 0s is $n - r$.*

SEC. 5.4 INVARIANTS OF A SYMMETRIC MATRIX UNDER CONGRUENCE

The following corollary is essentially a restatement of the first part of the theorem.

Corollary   *Two symmetric matrices of order n are congruent if and only if they have the same rank and signature.*

Although the canonical matrix, congruent to a symmetric matrix A, is unique, we have implied that the congruence class of A may contain many distinct diagonal matrices. However, the uniqueness of the canonical matrix implies that the number of positive diagonal elements and the number of negative diagonal elements are invariant. When this fact is interpreted in the context of quadratic forms, the result is what is called the *law of inertia* for such forms (Prob. 13). We also note, in passing, that we have obtained anew the result in the corollary to Theorem 7.2 (Chap. 4), since the canonical matrix that is congruent to a positive matrix is the identity matrix of the same order (Prob. 14). A symmetrix matrix, whose congruent canonical matrix contains only -1s (and so is the negative of an identity matrix), may be said to be a *negative* matrix. It is clear (Prob. 15) that the values of a quadratic form $Q(X)$, whose representing matrix is negative, are necessarily negative numbers, except when $X = 0$.

EXAMPLE

Determine the canonical diagonal matrix that is congruent to A in Example 2 of Sec. 5.3.

Solution
We saw in Example 2 of Sec. 5.3 that A is congruent to the matrix diag $\{5, -16, 80\}$. Moreover, since the second and third rows and the second and third columns may be interchanged (a congruence transformation) to obtain $D = \text{diag } \{5, 80, -16\}$, this matrix is also in the congruence class of A. If we now let $P = \text{diag } \{1/\sqrt{5}, 1/\sqrt{80}, 1/4\}$, it is easy to verify that $P'DP = \text{diag } \{1, 1, -1\}$. This is the desired canonical matrix in the congruence class of A, a matrix which could have been written immediately on inspection of D.

We close this section with a final remark of geometric significance. With the proper choice of coordinate axes, it is possible to express the equation of any central conic in 3-space in one of the following ways: $x^2 + y^2 + z^2 = 1$; $x^2 + y^2 - z^2 = 1$; $x^2 - y^2 - z^2 = 1$. The extension of this result to n-dimensional geometry is immediate.

Problems 5.4

1. If $D = \text{diag } \{2,1,3\}$, find the matrix P such that (a) $P'DP = \text{diag } \{1,2,3\}$; (b) $P'DP = \text{diag } \{3,2,1\}$.

2. If $D = \text{diag } \{1,-2,3,-3\}$, find the matrix P such that (a) $P'DP = \text{diag } \{1,3,-2,-3\}$; (b) $P'AP = \{-2,-3,1,3\}$.
3. Show that the following matrices are congruent:

$$\begin{bmatrix} 1 & 0 & 1 \\ 0 & 2 & -1 \\ 1 & -1 & 0 \end{bmatrix} \quad \begin{bmatrix} 1 & 1 & 2 \\ 1 & 4 & -1 \\ 2 & -1 & 3 \end{bmatrix}$$

4. Determine the canonical matrix congruent to (a) diag $\{2,-3,-4,5\}$; (b) diag $\{-3,2,9,-4,0,0\}$.
5. If $A = \text{diag } \{3,-4,-9\}$ is the matrix of a quadratic form on a vector space V, determine the basis of V, relative to which the equivalent form is canonical.
6. Determine the canonical matrix congruent to

(a) $\begin{bmatrix} 1 & 1 & 1 \\ 0 & 1 & 0 \\ 1 & 1 & 1 \end{bmatrix}$ (b) $\begin{bmatrix} 3 & 2 & 4 \\ 2 & 0 & 2 \\ 4 & 2 & 3 \end{bmatrix}$

7. Determine the canonical form equivalent to (a) $x_1^2 - 2x_1x_2 + 3x_2^2 - 4x_2x_3 + x_3^2$; (b) $2x^2 + 3xy + y^2 + 2yz + z^2$.
8. Explain why a symmetric matrix is positive if and only if its canonical congruent matrix is an identity matrix.

9. Explain why the order, rank, and signature of a symmetric matrix must constitute a *complete* set of invariants under congruence.
10. Explain why the rank of a diagonal matrix is the number of nonzero elements in its diagonal.
11. Why is $x_1^2 + x_2^2 + \cdots + x_p^2$ $(p > 0)$ a positive quadratic form?
12. In the discussion leading up to the statement of Theorem 4.2, explain why (a) $(PX)'D_2(PX) > 0$; (b) $\dim W_p' + \dim W_{n-q} \leq n$.
13. Give a statement of the law of inertia for quadratic forms.
14. Use the results in this section to prove that any symmetric matrix A may be expressed as $A = P'P$, for some non-singular matrix P.
15. If the matrix of a quadratic form $Q(X)$ is negative, explain why $Q(X) < 0$, except for $X = 0$.

16. Determine the number of congruence classes in the set of all 3×3 symmetric matrices.

17. Prove that there are $(n+1)(n+2)/2$ quadratic forms in n variables, no two of which are equivalent to each other.
18. Prove that a quadratic form on a space of dimension n (or an $n \times n$ matrix) is positive if and only if both its rank and signature are n.
19. Use the methods of this section to verify that the form $5x^2 - 4xy + 3y^2 - 2yz + z^2$ is positive (definite).
20. If V is the vector space of all $n \times n$ matrices, show (see Prob. 20 of Sec. 2.10) that f is a bilinear function on V where
    (a) $f(A,B) = \text{tr}\,(A'CB)$, for a fixed $C \in V$
    (b) $f(A,B) = n\,\text{tr}\,(AB) - [\text{tr}(A)][\text{tr}(B)]$
    In the case of (b), assume that $\text{tr}(AB) = \text{tr}(BA)$, and use the simplest basis of V to describe the matrix of the quadratic form $f(A,A)$.
21. Prove that the plane conic, whose equation is given as $ax^2 + 2bxy + cy^2 = 1$, is an ellipse or hyperbola according as the rank and signature of the form are 2 and 2 or 2 and 0, respectively.
22. Let
$$A = \begin{bmatrix} 1 & 2 & 0 \\ 2 & 0 & -1 \\ 0 & -1 & 0 \end{bmatrix}$$
be the matrix of a bilinear form. Prove that the form is nondegenerate (see problems in Sec. 5.1), and determine its signature and the canonical form to which it is equivalent.
23. If A is a real symmetric matrix such that $A^2 = 0$, prove that $A = 0$.

5.5 Eigenvalues and Eigenvectors

In Chap. 4 we defined the abstract concepts of *length* and *angle* as they apply to vectors of any euclidean space V. If T is a linear operator on V, and α is an arbitrary vector in V, it is generally the case that $T\alpha$ and α are different in both length and direction. It can happen, however, that T affects the lengths but not the directions—apart from a possible reversal in sense—of some vectors. For such a vector, $T\alpha = \lambda\alpha$, for a real number λ, and this leads to the following definition.

Definition Let T be a linear operator on a vector space V. Then, if $T\alpha = \lambda\alpha$, for a nonzero vector α and a real number λ, the number λ is called an eigenvalue and α is called an eigenvector (belonging to λ) of T.

BILINEAR AND QUADRATIC FORMS

There are several other names in common use for the concepts just defined, the prefix *eigen* being, of course, of German origin. For example, the values and associated vectors are often called *characteristic*, *latent*, or *principal*, the preference for each of these words depending for the most part on the geographic location or origin of its user. However, it is believed that the German terminology, which we have adopted, is the one most widely used by both mathematicians and applied scientists throughout the world.

EXAMPLE 1

If T is a linear operator on a vector space V, such that $T\alpha = 3\alpha$, for every $\alpha \in V$, it is immediate that every vector of the space is an eigenvector and is associated with the eigenvalue 3. There are no other eigenvalues for this operator.

EXAMPLE 2

On V_3, we can define a linear operator T so that $T(x_1,x_2,x_3) = (2x_1,2x_2,x_3)$. In this case, we observe that any vector $(0,0,x_3)$ is unchanged by T and so is an eigenvector belonging to the eigenvalue 1. On the other hand, the vectors $(x_1,x_2,0)$ are doubled in length by T, and this implies that they are eigenvectors that belong to the eigenvalue 2.

It should be observed from Example 2 that many different eigenvectors can belong to the same eigenvalue. In other words, *by no means* is there a one-to-one correspondence between eigenvalues and eigenvectors—contrary to what one might surmise from the similar names.

EXAMPLE 3

If we define a linear operator T on V_2, so that $T(x_1,x_2) = (-x_2,x_1)$, we see that the length of (x_1,x_2) is unchanged by T, but each vector is rotated through an angle of $\pi/2$ rad (Prob. 12). Hence this operator T can have no *real* eigenvalues nor any associated real eigenvectors.

EXAMPLE 4

Let us define T on V_3 so that $T(x_1,x_2,x_3) = (3x_1 + 2x_2 + 4x_3, 2x_1 + 2x_3, 4x_1 + 2x_2 + 3x_3)$. Then it is not difficult to check (Prob. 13) that $T(a,0,-a) = (-a,0,a)$ and $T(2a,a,2a) = (16a,8a,16a) = 8(2a,a,2a)$, for any real number a. Hence, by inspection, we have discovered that all vectors of the form $(a,0,-a)$ and $(2a,a,2a)$ are eigenvectors of T, and these belong, respectively, to the eigenvalues -1 and 8.

The eigenvectors and eigenvalues, as illustrated in the preceding examples, were discovered by inspection of the simple operators involved, but we have given no clue as to how such a discovery might be made for an operator with a relatively complicated description. We shall give our attention to this important matter later in the section but first we present an important theorem. The proof of the theorem is elementary, and its validity can be checked in the examples just given.

Theorem 5.1   *Let* T *be a linear operator on a vector space V. Then, if the vector 0 is included with the set of eigenvectors that belong to a given eigenvalue λ of* T, *this augmented set comprises a subspace of V, called the subspace of eigenvectors belonging to λ. No one vector of V can belong to distinct eigenvalues.*

Proof
Let S_λ be the given set of vectors α, including 0, such that $T\alpha = \lambda\alpha$. Then, with $\alpha, \beta \in S_\lambda$ and a, b any real numbers, $T(a\alpha + b\beta) = a(T\alpha) + b(T\beta) = a(\lambda\alpha) + b(\lambda\beta) = \lambda(a\alpha + b\beta)$, so that $a\alpha + b\beta \in S_\lambda$. It follows from this closure property that S_λ may be considered a subspace of V. To prove the final assertion in the theorem, let us suppose that α is a vector in both S_{λ_1} and S_{λ_2}, where λ_1 and λ_2 are distinct eigenvalues of T. Then $T\alpha = \lambda_1\alpha = \lambda_2\alpha$, and so $(\lambda_1 - \lambda_2)\alpha = 0$. But this implies that $\alpha = 0$, whence $S_{\lambda_1} \cap S_{\lambda_2} = \emptyset$, as asserted. ∎

A brief comment about both the real number 0, as an eigenvalue, and the vector 0, as an eigenvector, may be in order. Since $T0 = \lambda 0 = 0$, for *any* real number λ, it might seem reasonable to regard any real number *as* an eigenvalue, with the vector 0 one of the eigenvectors that belong to it. Since 0 would then be an eigenvector of any operator T, *it seems desirable to exclude 0 from the set of eigenvectors* of an operator, and this was done in our definition. However, *we do not exclude the number 0 as a possible eigenvalue.* In fact, the null space of any linear operator T contains all eigenvectors that belong to the eigenvalue 0, and this null space (provided T is singular) will contain nonzero vectors.

Before proceeding further, let us take another look at the examples at the beginning of this section. The conclusions and assertions found in Examples 1 and 2 are quite clear, but those in Examples 3 and 4 were somewhat more tenuous. In Example 3, we are seeking a real number λ, such that $T(x_1, x_2) = (-x_2, x_1) = (\lambda x_1, \lambda x_2)$. But then $-\lambda x_1 - x_2 = 0$, $x_1 - \lambda x_2 = 0$, and this system of equations has a nonzero solution for x_1 and x_2 if and only if $\lambda^2 + 1 = 0$. Since the latter equation is impossible for a *real* number λ, no real eigenvalue can exist for T, which is the assertion made in the example. A similar analysis can be made for Example 4. In this case, we are trying to find a real number λ and triples (x_1,x_2,x_3) such that $T(x_1,x_2,x_3) = (\lambda x_1, \lambda x_2, \lambda x_3) = (3x_1 + 2x_2 + 4x_3,\ 2x_1 + 2x_3,\ 4x_1 + 2x_2 + 3x_3)$. But then $[(3 - \lambda)x_1 + 2x_2 + 4x_3,\ 2x_1 - \lambda x_2 + 2x_3,\ 4x_1 + 2x_2 + (3 - \lambda)x_3] = (0,0,0)$, so that the following system of equations must be satisfied:

$$(3 - \lambda)x_1 + 2x_2 + 4x_3 = 0$$
$$2x_1 - \lambda x_2 + 2x_3 = 0$$
$$4x_1 + 2x_2 + (3 - \lambda)x_3 = 0$$

This system has a nontrivial solution for x_1, x_2, x_3 if and only if

$$\begin{vmatrix} 3 - \lambda & 2 & 4 \\ 2 & -\lambda & 2 \\ 4 & 2 & 3 - \lambda \end{vmatrix} = 0$$

and this determinantal equation can be easily reduced to $(\lambda + 1)^2(\lambda - 8) = 0$. Hence the only possible eigenvalues are -1 and 8, and we can complete the problem by considering these two in succession. If $\lambda = -1$, the conditions to be satisfied reduce to the single condition $2x_1 + x_2 + 2x_3 = 0$, and we note that the vectors $(a, 0, -a)$, for any real a, are among the associated eigenvectors (Prob. 14). If $\lambda = 8$, the conditions reduce to $x_1 = x_3 = 2x_2$, and so the associated eigenvectors have the form $(2a, a, 2a)$. These results are in agreement with those obtained less formally in Example 4.

We now generalize the procedure just described in connection with the discussion of Example 4 and consider any linear operator T on an arbitrary vector space V of dimension n. Moreover, let us suppose that, relative to some assumed basis of V, $A = [a_{ij}]$ is the matrix of T. If α is an eigenvector of T, belonging to an eigenvalue λ, then $T\alpha = \lambda\alpha$; or, in matrix notation, $AX = \lambda X$, where X is the coordinate vector of α. But then $(A - \lambda I)X = 0$, and this matrix equation is equivalent to a system of n equations in the n unknown coordinates of α. This system has a nontrivial solution if and only if

$$|A - \lambda I| = f(\lambda) = \begin{vmatrix} a_{11} - \lambda & a_{12} & \cdots & a_{1n} \\ a_{21} & a_{22} - \lambda & \cdots & a_{2n} \\ \cdots & \cdots & \cdots & \cdots \\ a_{n1} & a_{n2} & \cdots & a_{nn} - \lambda \end{vmatrix} = 0$$

In other words, the existence of real eigenvalues and eigenvectors of T is dependent on the existence of real solutions of this equation. (Of course, complex eigenvalues *always* exist, but we shall postpone use for such numbers until Chap. 6.) An expansion of this determinant, according to any of the rules given in Chaps. 1 or 3, gives the *characteristic polynomial* $f(\lambda)$ of A. A minor difficulty arises at this point! If we think of λ as a real number, albeit an unknown one, the expansion of the above determinant is in keeping with earlier work, since all elements involved can be considered to be real numbers. On the other hand, if we think of λ as a real variable or symbol, the coefficients of the equations must be considered elements of the commutative ring of real polynomials in λ, and we *assume* that it still makes sense to talk about the "determinant" of these coefficients and that its vanishing is a necessary and sufficient condition for the equations to have a nonzero solution. It is not a difficult matter to prove that this criterion *is* valid for equations with coefficients in *any field* and, moreover, that the polynomials in λ can be embedded in the field of all "rational functions" or quotients of such polynomials. However, we omit the details of these arguments. In the sequel, we shall usually replace λ by x in the characteristic polynomial of A and the eigenvalues of T will then be the solutions of the characteristic equation $f(x) = |A - xI| = 0$. For matrices of large order, the actual computation of the characteristic polynomial may be quite tedious, and the reader is referred to the books by Gantmachter (pp. 87–89; pp. 202–214) and Pipes (pp. 61–63) listed at the end of the chapter.

Although the characteristic polynomial is defined for any square matrix, our central interest is in linear operators rather than matrices as such, and it is a priori conceivable that the characteristic polynomials of matrices that represent even the same linear operator may differ. Since similar matrices represent the same linear operator, relative to different bases, the following theorem gives a negative answer to this considered possibility.

Theorem 5.2 Let A and $B = P^{-1}AP$ be similar matrices, both representing a linear operator T on a vector space V. Then

(1) A and B have the same characteristic polynomial.

(2) If X is the coordinate vector of an eigenvector of T, in the reference frame associated with A, then $P^{-1}X$ is also the coordinate vector of an eigenvector of T, in the reference frame associated with B; and these eigenvectors belong to the same eigenvalue of T.

Proof

(1) We first use Theorem 4.1 (Chap. 3) in order to obtain the following sequence of equalities: $|B - xI| = |P^{-1}AP - xI| = |P^{-1}AP - P^{-1}(xI)P| = |P^{-1}(A - xI)P| = |P^{-1}| |A - xI| |P|$. Since $|P^{-1}|$ and $|P|$ are real numbers and $|A - xI|$ is a real polynomial in x, it follows that $|B - xI| = |P^{-1}| |P| |A - xI| = |P^{-1}P| |A - xI| = |A - xI|$, as asserted.

(2) Let $\alpha \in V$ be an eigenvector belonging to the eigenvalue λ, so that

$T\alpha = \lambda\alpha$. If we "translate" this into an equality involving A and the coordinate vector X of α, we obtain $AX = \lambda X$. But then $P^{-1}APP^{-1}X = \lambda P^{-1}X$, so that $B(P^{-1}X) = \lambda(P^{-1}X)$, completing the proof of (2). ∎

Since the characteristic polynomials of similar matrices are the same, it is meaningful to speak of the *characteristic polynomial* of a linear operator: It is the characteristic polynomial of *any* matrix that represents the operator, and its solutions may be regarded as the eigenvalues of *either* the operator or the matrix. Theorem 5.2 then makes the concept of an eigenvalue of a matrix compatible with the definition given near the beginning of this section for a linear operator. Although the *eigenvalues* of a linear operator T on V are quite independent of the basis used for V, it must be understood that the eigenvectors of T have different representations according to the basis, as detailed in (2) of the theorem.

It may have seemed that this section is somewhat out of place in the context of the present chapter. However, we shall see in the following section how the notions of eigenvector and eigenvalue can be used to advance our study of bilinear and quadratic forms. In Chap. 6, we shall use them again as key concepts in our final investigations of linear operators.

Problems 5.5

1. If T is the linear operator defined on a vector space V such that $T\alpha = 2\alpha$, for each $\alpha \in V$, explain why 2 is the only eigenvalue of T. What are the eigenvectors belonging to 2?
2. Review the reason why we prefer not to consider the zero vector an eigenvector.
3. Let $T(x_1,x_2,x_3,x_4) = (x_1, 3x_2, 3x_3, x_4)$ define a linear operator T on V_4. Determine two eigenvalues of T, without any use of its characteristic polynomial, and also find a subspace of eigenvectors belonging to each of these eigenvalues.
4. Explain why the characteristic polynomial, as developed in this section, is in fact a *real polynomial* in λ.
5. Find the characteristic polynomial of

    (a) $\begin{bmatrix} 2 & 1 & 1 \\ 0 & 1 & 2 \\ -1 & 0 & 1 \end{bmatrix}$ (b) $\begin{bmatrix} 1 & 2 & 0 \\ 1 & -1 & 1 \\ 0 & -2 & 1 \end{bmatrix}$

6. Find the characteristic polynomial and eigenvalues of

    (a) $\begin{bmatrix} 2 & 3 \\ 3 & 2 \end{bmatrix}$ (b) $\begin{bmatrix} 1 & 2 \\ -3 & 4 \end{bmatrix}$ (c) $\begin{bmatrix} 3 & 2 & 0 \\ 2 & 0 & 0 \\ 1 & 0 & 2 \end{bmatrix}$

7. Use the definitions (not the characteristic polynomials) to find the eigenvalues and eigenvectors of the operator T defined on V_2 by (a) $T(x_1,x_2) = (x_1,x_2)$; (b) $T(x_1,x_2) = (3x_1 + 2x_2, x_2)$.
8. Apply the directions in Prob. 7 to the operator T defined on V_3 by the following:
   (a) $T(x_1,x_2,x_3) = (x_1 - x_2, x_2 + x_1, x_2 - x_3)$
   (b) $T(x_1,x_2,x_3) = (x_1 - 2x_3, 2x_1 + x_2, x_3)$
9. If a linear operator T on a vector space V is represented by the matrix A, relative to the basis $\{\alpha_1,\alpha_2,\alpha_3\}$, determine the real eigenvalues of T when

   (a) $A = \begin{bmatrix} 3 & 2 & 4 \\ 2 & 0 & 2 \\ 4 & 2 & 3 \end{bmatrix}$ (b) $A = \begin{bmatrix} 1 & 2 & 0 \\ 2 & 1 & -1 \\ 0 & -1 & 1 \end{bmatrix}$

10. If $B = P^{-1}AP$, and (2,1,3) is a coordinate eigenvector of the linear transformation represented by A, use Theorem 5.2 to find the corresponding coordinate eigenvector after the basis has been changed by the transition matrix P, where

$$P = \begin{bmatrix} 1 & 2 & 1 \\ -1 & 1 & 0 \\ 2 & 0 & 1 \end{bmatrix}$$

11. Review the argument, used at the end of the proof of Theorem 5.1, that $k\alpha = 0$, for a real number $k \neq 0$, implies that $\alpha = 0$.
12. Prove the assertion made in the first sentence of Example 3.
13. Do the necessary checking, as suggested in Example 4.
14. Verify that $(a,0,-a)$, for any $a \in R$, is a solution of the equation $2x_1 + x_2 + 2x_3 = 0$.

15. If $A = [a_{ij}]$ is a triangular matrix (that is, $a_{ij} = 0$ for $i < j$, or $a_{ij} = 0$, for $j < i$), prove that the eigenvalues are the numbers along the main diagonal.
16. Prove that a square matrix and its transpose have the same eigenvalues. What information does this give concerning the eigenvectors of a linear operator T and its transpose T'? What about the adjoint T* of T?
17. Give an alternative proof of (1) of Theorem 5.2, based on the fact that A and B can represent the same linear operator on a vector space, and recalling that T must be independent of any coordinate system. [Hint: Argue that $|A - \lambda I| = k|B - \lambda I|$, for a real number k, and then let $\lambda = 0$.]

272 BILINEAR AND QUADRATIC FORMS

18. If X is an eigenvector (regarded as a column matrix) belonging to the eigenvalue λ of a matrix A, prove that $a_0 + a_1\lambda + a_2\lambda^2 + \cdots + a_r\lambda^r$ is an eigenvalue of the matrix $a_0 I + a_1 A + a_2 A^2 + \cdots + a_r A^r$. [*Hint:* Show first that $A^2 X = \lambda^2 X$.]

19. Was it established in Prob. 18 that *all* eigenvalues of the matrix polynomial have the form designated in the problem?

20. Prove that the real eigenvalues of an orthogonal matrix are ± 1.

21. If A is a nonsingular matrix, prove that the eigenvalues of A^{-1} are the reciprocals of the eigenvalues of A.

22. If $A^n = 0$, for a matrix A, prove that the only eigenvalue of A is 0. (Compare with Prob. 15 of Sec. 3.4.)

23. A Markov matrix $A = [a_{ij}]$ has the property that $0 \leq a_{ij} \leq 1$, and $\sum_{j=1}^{n} a_{ij} = 1$, for $i = 1, 2, \ldots, n$. Prove that $\lambda = 1$ is an eigenvalue of any Markov matrix.

24. If α_1 and α_2 are eigenvectors, belonging to distinct eigenvalues of a linear operator, prove that α_1 and α_2 are linearly independent.

25. Two differential equations of frequent occurrence in applied mathematics are $d^2y/dx^2 + k^2y = 0$ and $d^2y/dx^2 - k^2y = 0$, for a real number k. If D is the usual differentiation operator on the space (not of finite dimension) of infinitely differentiable functions, the solution of these equations become "eigenvalue problems" when they are written as $D^2y = -k^2y$ and $D^2y = k^2y$. Use familiar methods to solve each of the equations, and identify the eigenvectors (*eigenfunctions*) of the operator D^2, in both cases.

26. Let V be the (infinite-dimensional) space of continuous functions f on **R**, with T a linear operator on V defined by $(Tf)(x) = \int_0^x f(x)\, dx$. Prove that T has no eigenvalues.

27. If A and B are nonsingular $n \times n$ matrices, prove that AB and BA have the same eigenvalues. Does the same conclusion hold if *either* A or B is nonsingular? Consider the same proposition with $A = \begin{bmatrix} 0 & 0 \\ 1 & 0 \end{bmatrix}$ and $B = \begin{bmatrix} 0 & 0 \\ 0 & 1 \end{bmatrix}$.

5.6 Orthogonal Reduction of Quadratic Forms

It has already been pointed out that *congruence* and *similarity* are two of the three most important equivalence relations that occur in matrix theory.

SEC. 5.6 ORTHOGONAL REDUCTION OF QUADRATIC FORMS

Congruent matrices may be used to represent the same quadratic or symmetric bilinear form, whereas *similar* matrices may be used to represent the same linear transformation or operator on a vector space, relative to various bases. Although it may seem that forms and operators have very little in common, the concepts of *congruence* and *similarity* sometimes coincide in a matrix-theoretic context. To be precise, the matrices $P'AP$ and $P^{-1}AP$ coincide whenever $P' = P^{-1}$, and this is true by definition if and only if P is an orthogonal matrix. In such an instance, the matrices A and $P^{-1}AP$ (or $P'AP$) are sometimes said to be *orthogonally similar*. It is the principal result in this section to prove that *any symmetric matrix is orthogonally similar to a diagonal matrix*, though we do not claim this to be the canonical matrix of Sec. 5.5. Before stating this theorem, however, it is useful to have available two results, one of which we shall state without proof at this time. We have seen that the eigenvalues of a matrix are the zeros of its characteristic equation, and it is clear that some or all zeros of a polynomial equation may be nonreal complex numbers. If no real eigenvalues exist, neither can there exist any real eigenvectors, but the following fact is of essential interest in the present context: *All eigenvalues of a real symmetric matrix are real numbers.* The simplest proofs of this result involve the use of vectors over the field of complex numbers; and so, in line with our plan announced earlier, we shall postpone our proof of it until Chap. 6 (see Theorem 3.1 of that chapter). The other preliminary result, which we desire at this time, is an easy consequence of the following lemma.

Lemma *If T is a self-adjoint linear operator on a euclidean space V, eigenvectors that belong to distinct eigenvalues of T are orthogonal.*

Proof
Let α_1, α_2 be eigenvectors of T that belong to λ_1 and λ_2, respectively, where $\lambda_1 \neq \lambda_2$. Then $T\alpha_1 = \lambda_1\alpha_1$ and $T\alpha_2 = \lambda_2\alpha_2$, and, from the self-adjoint property of T, $(T\alpha_1, \alpha_2) = (\alpha_1, T^*\alpha_2) = (\alpha_1, T\alpha_2)$. Hence $\lambda_1(\alpha_1, \alpha_2) = \lambda_2(\alpha_1, \alpha_2)$ or, equivalently, $(\lambda_1 - \lambda_2)(\alpha_1, \alpha_2) = 0$. Since $\lambda_1 \neq \lambda_2$, we conclude that $(\alpha_1, \alpha_2) = 0$, as asserted. ∎

Since any symmetric matrix of order n can be regarded as the matrix (relative to an orthonormal basis) of a self-adjoint linear operator on the euclidean space V_n, the lemma implies the following result which we shall use immediately: *Any two elements of V_n, that are associated as eigenvectors with distinct eigenvalues of a symmetric matrix of order n, are orthogonal.*

We now come to the central result in this section, known as the *principal-axis theorem*. The name of the theorem arises from its geometric significance, and, although we shall state it in matrix form, the relevance of the result to geometry and quadratic forms will be noted in the corollaries that follow it.

Theorem 6.1 (Principal-axis Theorem)

If A is a symmetric matrix, there exists an orthogonal matrix P such that $P'AP = P^{-1}AP = D$, where D is a diagonal matrix with the eigenvalues of A along the diagonal.

Proof
There is no loss in generality in assuming that A represents a linear operator T on the euclidean space V_n, relative to the $\{E_i\}$ basis. Let $\lambda_1, \lambda_2, \ldots, \lambda_n$ be all solutions of the characteristic equation of T, so that these real numbers (A is symmetric) comprise the eigenvalues of T, *but possibly with some repetitions*. If X is an eigenvector belonging to λ_1, the vector $X_1 = X/|X|$ is a unit eigenvector belonging to the same eigenvalue as X. We now expand X_1 to an orthonormal basis of V_n. Since this basis is orthonormal, it follows that the matrix of T is $P_1'AP_1 = P_1^{-1}AP_1$, where P_1 is an orthogonal matrix (see Theorem 5.1 and Prob. 15 of Chap. 4). If we denote this new representing matrix by B, it is clear that B is real and symmetric (Prob. 7), and the structure of P_1 requires that $B = \begin{bmatrix} \lambda_1 & 0 \\ 0 & B_1 \end{bmatrix}$, where B_1 is a real symmetric matrix of order $n - 1$. (Note that the first column of P_1 is the eigenvector X_1.) It is now important to recall that similar matrices have the same eigenvalues, so that the eigenvalues of B are the same as those of A. In particular, this implies that λ_2 is an eigenvalue of B_1 (quite independent of whether $\lambda_1 = \lambda_2$ or $\lambda_1 \neq \lambda_2$), and a repetition of the preceding argument yields an orthogonal matrix Q_1 such that $Q_1'B_1Q_1 = Q_1^{-1}B_1Q_1 = \begin{bmatrix} \lambda_2 & 0 \\ 0 & B_2 \end{bmatrix}$, where B_2 is a real symmetric matrix of order $n - 2$. Since Q_1 is orthogonal, so is $\begin{bmatrix} 1 & 0 \\ 0 & Q_1 \end{bmatrix} = P_2$, and the product $P_2P_1 = R$ is also an orthogonal matrix of order n. A simple check reveals that

$$R'AR = R^{-1}AR = \begin{bmatrix} \lambda_1 & 0 & 0 \\ 0 & \lambda_2 & 0 \\ 0 & 0 & B_2 \end{bmatrix}$$

and we have completed two stages of the reduction process. It is clear that, after at most n stages, we shall have established the existence of an orthogonal matrix P of order n, such that $P'AP = P^{-1}AP = \text{diag}\{\lambda_1, \lambda_2, \ldots, \lambda_n\}$. This completes the proof of the theorem. ∎

Although the method outlined in the above proof *could* be used to construct the orthogonal matrix P, this is not recommended, for the *existence* and *not the construction* of P is the important contribution of the theorem. Since the ith column of P is necessarily an eigenvector of T, belonging to the eigenvalue λ_i, $i = 1, 2, \ldots, n$, the problem of constructing P is one of dis-

SEC. 5.6 ORTHOGONAL REDUCTION OF QUADRATIC FORMS 275

covering n *mutually orthogonal unit* eigenvectors, with at least one belonging to each of the distinct eigenvalues of T. The lemma asserts that eigenvectors belonging to *distinct* eigenvalues are always orthogonal in a euclidean space and the existence of the matrix P in the theorem may be used as an indirect proof that the subspace of V_n, consisting of zero and all eigenvectors belonging to a given eigenvalue (see Theorem 5.1) of multiplicity r, has dimension r. We are then assured of the *existence* of enough eigenvectors for the construction of P, and the problem is simply to find them. We shall illustrate this method of discovery with several examples, after we state some immediate corollaries of the theorem.

If we define two quadratic or symmetric bilinear forms as *orthogonally equivalent* if their matrices are orthogonally similar, the following corollary is evident.

Corollary 1 *Any quadratic (symmetric bilinear) form on a vector space of dimension n is orthogonally equivalent to $\lambda_1 x_1^2 + \lambda_2 x_2^2 + \cdots + \lambda_n x_n^2$ ($\lambda_1 x_1 y_1 + \lambda_2 x_2 y_2 + \cdots + \lambda_n x_n y_n$), where $\lambda_1, \lambda_2, \ldots, \lambda_n$ are the eigenvalues of the matrix associated with the form.*

From the point of view of geometry, the important property of an *orthogonal* linear operator on a vector space is that inner products are preserved. The next corollary follows from the fact that such an operator maps orthonormal bases onto orthonormal bases.

Corollary 2 *There exists a cartesian coordinate system, relative to which the equation of any central conic in 3-space (2-space) has the form $ax^2 + by^2 + cz^2 = 1$ ($ax^2 + by^2 = 1$), for real numbers a, b, c. The principal axes of the conic then lie along the coordinate axes, this result giving the theorem its name.*

We are now able to fulfill a promise made in Sec. 5.3, that in the sequel an even simpler method would be forthcoming for determining a diagonal matrix that is congruent to a given symmetric matrix. As a result of Theorem 6.1, once the characteristic equation of a matrix is solved, a congruent diagonal matrix is known. In many cases, the solution of this equation may be quite laborious. Inasmuch as *any* diagonal matrix can be used to obtain the canonical matrix described in Theorem 4.2, the following criterion may (Prob. 8) be stated as the final corollary.

Corollary 3 *A symmetric matrix or associated quadratic form is positive or negative according as to whether all of its eigenvalues are positive or negative.*

We close this section with three examples that will illustrate the procedure for constructing an orthogonal matrix P, to be associated with a given matrix A, such that $P'AP$ is in the diagonal form of Theorem 6.1. As was noted earlier, manifestly the problem is to find n mutually orthogonal unit eigenvectors, at least one of which belongs to each of the distinct eigenvalues of A.

EXAMPLE 1

If $\lambda_1, \lambda_2, \lambda_3$ are the eigenvalues of the matrix

$$A = \begin{bmatrix} -1 & -3 & 0 \\ -3 & -1 & 0 \\ 0 & 0 & 1 \end{bmatrix}$$

construct an orthogonal matrix P such that $P'AP = P^{-1}AP = \text{diag}\{\lambda_1,\lambda_2,\lambda_3\}$.

Solution

It is easy to find that the characteristic polynomial of A is $-(x-1)(x-2)(x+4)$, so that the eigenvalues are 1, 2, -4. We now regard A as the matrix of a linear operator on the euclidean space V_3, use the definition of an eigenvector to find one belonging to each of the eigenvalues, and then normalize them to unit length. If we let $X = (x_1,x_2,x_3)$, and express it as a column matrix, the problem is to solve $AX = \lambda X$ [or, equivalently, $(A - \lambda I)X = 0$], for $\lambda = 1, 2, -4$. In matrix form, these equations become the following:

$$\begin{bmatrix} -2 & -3 & 0 \\ -3 & -2 & 0 \\ 0 & 0 & 0 \end{bmatrix} \begin{bmatrix} x_1 \\ x_2 \\ x_3 \end{bmatrix} = 0 \qquad \begin{bmatrix} -3 & -3 & 0 \\ -3 & -3 & 0 \\ 0 & 0 & -1 \end{bmatrix} \begin{bmatrix} x_1 \\ x_2 \\ x_3 \end{bmatrix} = 0$$

$$\begin{bmatrix} 3 & -3 & 0 \\ -3 & 3 & 0 \\ 0 & 0 & 5 \end{bmatrix} \begin{bmatrix} x_1 \\ x_2 \\ x_3 \end{bmatrix} = 0$$

or, in the form of equations in x_1, x_2, and x_3,

$$\begin{aligned} -2x_1 - 3x_2 &= 0 & -3x_1 - 3x_2 &= 0 & 3x_1 - 3x_2 &= 0 \\ -3x_1 - 2x_2 &= 0 & -x_3 &= 0 & 5x_3 &= 0 \end{aligned}$$

From these sets of equations, it is easy to find respective solutions as $(0,0,1)$, $(1,-1,0)$, and $(1,1,0)$ which, normalized, become $(0,0,1)$, $(1/\sqrt{2}, -1/\sqrt{2}, 0)$, and $(1/\sqrt{2}, 1/\sqrt{2}, 0)$. On using these vectors for the columns of the matrix P, we find that

$$P = \begin{bmatrix} 0 & 1/\sqrt{2} & 1/\sqrt{2} \\ 0 & -1/\sqrt{2} & 1/\sqrt{2} \\ 1 & 0 & 0 \end{bmatrix}$$

It is now a matter of simple matrix computation to check that $P'AP = $ diag $\{1,2,-4\}$, observing that P is (as it *must* be by construction) an orthogonal matrix.

EXAMPLE 2

If $\lambda_1, \lambda_2, \lambda_3$ are the eigenvalues of the matrix

$$A = \begin{bmatrix} 1 & 1 & 1 \\ 1 & 2 & 0 \\ 1 & 0 & 2 \end{bmatrix}$$

construct an orthogonal matrix P such that $P'AP = P^{-1}AP = $ diag $\{\lambda_1,\lambda_2,\lambda_3\}$.

Solution
As in Example 1, we assume that A is the matrix of a linear operator on the euclidean space V_3, and it is not difficult to find that the eigenvalues are 0, 2, 3. If we use each of these eigenvalues separately for λ and solve the equation $(A - \lambda I)X = 0$, we find (Prob. 10) that associated solutions are $(2,-1,-1)$, $(0,1,-1)$, and $(1,1,1)$. It may be noted, as a partial check, that these vectors are orthogonal in the euclidean space V_3, and if we normalize them to unit vectors and use them for the columns of P, our construction leads to

$$P = \begin{bmatrix} 2/\sqrt{6} & 0 & 1/\sqrt{3} \\ -1/\sqrt{6} & 1/\sqrt{2} & 1/\sqrt{3} \\ -1/\sqrt{6} & -1/\sqrt{2} & 1/\sqrt{3} \end{bmatrix}$$

It may now be verified that $P'AP = P^{-1}AP = $ diag $\{0,2,3\}$, as desired.

In both of the two preceding examples, the eigenvalues of the matrix were distinct. It was observed, however, in the discussion following the proof of Theorem 6.1, that no difficulty arises in the case of eigenvalues of multiplicities $r > 1$, because the subspace of associated eigenvectors has dimension r. This situation arises in our final example.

EXAMPLE 3

If $\lambda_1, \lambda_2, \lambda_3$ are the eigenvalues of the matrix

$$A = \begin{bmatrix} 3 & 0 & 0 \\ 0 & 4 & \sqrt{3} \\ 0 & \sqrt{3} & 6 \end{bmatrix}$$

construct an orthogonal matrix P such that $P'AP = P^{-1}AP = \text{diag }\{\lambda_1,\lambda_2,\lambda_3\}$.

Solution

We assume that A is the matrix of a linear operator on the euclidean space V_3, and, after discovering that the solutions of the characteristic equation of A are 3, 3, 7, we proceed as in Examples 1 and 2. For $\lambda = 3$, we must find *two* orthogonal solutions of

$$(A - 3I)X = 0 = \begin{bmatrix} 0 & 0 & 0 \\ 0 & 1 & \sqrt{3} \\ 0 & \sqrt{3} & 3 \end{bmatrix} \begin{bmatrix} x_1 \\ x_2 \\ x_3 \end{bmatrix}$$

Since the only condition arising from this equation is $\sqrt{3}\,x_2 + 3x_3 = 0$, the general solution of the matrix equation is $X = (a, -\sqrt{3}\,b, b)$, with a, b arbitrary real numbers (Prob. 11). An elementary computation (Prob. 12) shows that $X_1 = (1,0,0)$ and $X_2 = (0,\sqrt{3},-1)$ are orthogonal solutions, and we note that we have discovered two orthogonal vectors belonging to the same eigenvalue 3. If we solve $(A - 7I)X = 0$, we find quite easily that $X_3 = (0,\sqrt{3},3)$ is an eigenvector belonging to the eigenvalue 7. We now normalize the three orthogonal vectors X_1, X_2, X_3 and use them for the columns of P, and see that

$$P = \begin{bmatrix} 1 & 0 & 0 \\ 0 & \sqrt{3}/2 & \frac{1}{2} \\ 0 & -\frac{1}{2} & \sqrt{3}/2 \end{bmatrix}$$

A matrix check now shows that $P'AP = P^{-1}AP = \text{diag }\{3,3,7\}$.

Problems 5.6

1. Find an orthogonal matrix P such that $P'AP$ is in diagonal form, where (a) $A = \begin{bmatrix} 3 & -1 \\ -1 & 3 \end{bmatrix}$; (b) $A = \begin{bmatrix} -2 & 2\sqrt{3} \\ 2\sqrt{3} & 2 \end{bmatrix}$; (c) $A = \begin{bmatrix} 1 & 4\sqrt{5} \\ 4\sqrt{5} & -1 \end{bmatrix}$. Check your results.

2. Apply the directions given in Prob. 1, where

$$A = \begin{bmatrix} 1 & 2 & 0 \\ 2 & 3 & -2 \\ 0 & -2 & 1 \end{bmatrix}$$

3. Apply the directions given in Prob. 1, where

$$A = \begin{bmatrix} 5 & 3\sqrt{3} & 0 \\ 3\sqrt{3} & -1 & 0 \\ 0 & 0 & 4 \end{bmatrix}$$

4. Determine a diagonal quadratic form that is orthogonally equivalent to: $-2x_1^2 - 4\sqrt{3}\, x_1 x_2 + 2x_2^2 + 4x_3^2$.
5. Review the reason why any eigenvalue of a linear operator on a vector space V has at least one associated eigenvector in V.
6. Is there a *unique* diagonal matrix to which a given symmetric matrix is congruent? Comment on this.

7. In the proof of Theorem 6.1, explain why B is a real symmetric matrix.
8. Explain why Corollary 3 follows from Theorem 6.1.
9. Check the computation suggested at the end of each of the three examples given in this section.
10. Solve the system $(A - \lambda I)X = 0$, for $\lambda = 0, 2, 3$, in Example 2 and so check the solutions given.
11. Solve the system $(A - 3I)X = 0$ in Example 3 and so check the solution given.
12. In the subspace of the euclidean space V_3, consisting of vectors of the form $(a, -\sqrt{3}\, b, b)$, with $a, b \in \mathbf{R}$, use a systematic method to determine two orthogonal vectors.

13. Find an orthogonal matrix P such that $P'AP = \text{diag}\,\{1,1,6,6\}$, where $\{1,1,6,6\}$ is the complete set of solutions of the characteristic equation of

$$A = \begin{bmatrix} 5 & 2 & 0 & 0 \\ 2 & 2 & 0 & 0 \\ 0 & 0 & 5 & -2 \\ 0 & 0 & -2 & 2 \end{bmatrix}$$

14. Give a complete inductive proof of Theorem 6.1.
15. Find a change of variables that will reduce the following form to a sum of squares with real coefficients:

$$4x^2 - 2y^2 + z^2 - 2xy + 4xz - 5yz$$

16. Determine a system of cartesian axes that lie along the principal axes of the conic whose equation, relative to a given cartesian system, is $4x^2 + 4xy + y^2 = 1$.
17. Apply the directions in Prob. 16 to the conic whose equation is

$$x^2 + y^2 - z^2 - 4xy + 2xz + 4yz = 1$$

18. Review Descartes' *rule of signs* in an elementary algebra book, and prove that a real symmetric matrix is positive if and only if the coefficients of its characteristic equations alternate in signs.
19. If A and B are permuting matrices, prove that there exists an orthogonal matrix P such that $P'AP$ and $P'BP$ are *both* in diagonal form.
20. If A is a real symmetric matrix of odd order, prove that 1 or -1 is an eigenvalue of A, according as $|A| = 1$ or $|A| = -1$. What is the interpretation of this result for quadratic forms?
21. If an eigenvalue λ_1 of a linear operator on V_n has multiplicity r, give a proof (independent of Theorem 6.1) that the subspace of eigenvectors belonging to λ_1 has dimension r. [*Hint:* We know that an orthogonal matrix P exists such that $P^{-1}AP = \text{diag } \{\lambda_1, \lambda_1, \ldots, \lambda_1, \ldots\}$, with r occurrences of λ_1 in the diagonal form. Now consider the subspace $\mathcal{L}\{P^{-1}E_1, P^{-1}E_2, \ldots, P^{-1}E_r\}$.]
22. Prove that all eigenvalues of a symmetric matrix are equal if and only if the matrix is a scalar multiple of the identity.
23. Prove that the set of real symmetric matrices of order n comprises a vector space, and determine its dimension.
24. If A is a positive symmetric matrix, prove that, for any integer $k > 0$, there exists a symmetric matrix B such that $A = B^k$.
25. In the Taylor expansion of a function f on $\mathbf{R} \times \mathbf{R}$ at the point (a,b), the term involving the second derivatives (apart from the factor $\frac{1}{2}$) is $h^2 f_{xx}(a,b) + 2hk f_{xy}(a,b) + k^2 f_{yy}(a,b)$, which is seen to be a quadratic form in h and k. Assume that the rank of the form is 2, and prove that $f(a,b)$ is a relative minimum or maximum of f according as the signature of the form is 2 or -2.

Selected Readings

Cullen, C. G.: A Method of Calculating the Characteristic Equation of a Matrix, *Amer. Math. Monthly*, **67**(9): 889–890 (November, 1960).

Gantmacher, F. R.: "The Theory of Matrices," vol. 1, Chelsea Publishing Company, New York, 1959.

Gleason, Andrew M.: The Definition of a Quadratic Form, *Amer. Math. Monthly*, **73**(10): 1049–1056 (December, 1966).

Hersh, Reuben: A Well-known Eigenvalue Problem, *Amer. Math. Monthly*, **72**(2): 190 (February, 1965).

Pipes, Louis A.: "Matrix Methods for Engineering," Prentice-Hall, Inc., Englewood Cliffs, N.J., 1963.

Samuels, S. M.: A Simplified Proof of a Sufficient Condition for a Positive Definite Quadratic Form, *Amer. Math. Monthly*, **73**(3): 297–298 (March, 1966).

6

SIMILARITY AND NORMAL OPERATORS

6.1 The Cayley-Hamilton Theorem

In this final chapter of the book, it is our overall objective to examine the relation of *similarity*, in an attempt to discover some of the simplest—and hence most useful—matrix representations of a linear operator on a vector space. We begin this study with the proof of one of the most famous theorems in all matrix theory. This theorem is known as either the *Hamilton-Cayley* or *Cayley-Hamilton theorem*, and it states loosely that "any square matrix satisfies its characteristic equation." That is, if $f(x) = a_n x^n + a_{n-1} x^{n-1} + \cdots + a_1 x + a_0 = 0$ is the characteristic equation, as defined in Chap. 5, of an $n \times n$ matrix A, then $f(A) = a_n A^n + a_{n-1} A^{n-1} + \cdots + a_1 A + a_0 I = 0$. We note that, in order to make "sense" of the polynomial $f(A)$, it is necessary to regard the scalar a_0 as the scalar matrix $a_0 I$. As a matter of fact, although the other scalar products are well defined, it is appropriate to consider *all* the scalars $a_n, a_{n-1}, \ldots, a_1, a_0$ as scalar matrices $a_n I, a_{n-1} I, \ldots, a_1 I, a_0 I$, so that the equation "satisfied" by A is really one with scalar matrices as coefficients. The loose identification of the two equations should cause no difficulty, however, and this will always be understood when the variable of a real equation is replaced by a matrix.

As a prelude to the proof of the theorem, it should be recalled that the set

SEC. 6.1 THE CAYLEY-HAMILTON THEOREM   283

of all $n \times n$ matrices forms a vector space of dimension n^2. This implies, for any matrix A of order n, that $I = A^0, A, A^2, \ldots, A^{n^2}$ constitute a linearly dependent set, and, by Theorem 7.1 (Chap. 1), some element of this ordered set is expressible as a linear combination of those which precede it. If A^m is the *first* element for which this is true, there must exist scalars $b_{m-1}, \ldots, b_1, b_0$ such that $A^m + b_{m-1}A^{m-1} + \cdots + b_1 A + b_0 I = 0$, but A can satisfy no such polynomial equation of lower degree. This monic polynomial $m(x) = x^m + b_{m-1}x^{m-1} + \cdots + b_1 x + b_0$, of smallest degree and satisfied by A, is called the *minimum* polynomial of A. We leave it to the reader to verify that the minimum polynomial of any square matrix is unique (Prob. 12). We have then shown that it is an elementary consequence of the dimension properties of a vector space that there is associated with each square matrix A a unique monic polynomial $m(x)$ of minimal degree such that $m(A) = 0$. There does not appear to be any a priori reason why A should also satisfy its characteristic equation, but this is the statement of the following theorem.

Theorem 1.1 (Cayley-Hamilton)
> *If $f(x)$ is the characteristic polynomial of a square matrix A, then $f(A) = 0$.*

Proof
The proof will be based on simple matrix considerations. Let $B = A - xI$, and express the characteristic polynomial of A as $f(x) = |A - xI| = b_0 + b_1 x + b_2 x^2 + \cdots + b_{n-1} x^{n-1} + b_n x^n$, for real numbers $b_n = (-1)^n$, $b_{n-1}, \ldots, b_2, b_1, b_0$. We recall that the classical adjoint of B, designated adj B, is a matrix whose entries are polynomials in x of degrees not in excess of $n - 1$; and so adj $B = B_0 + xB_1 + x^2 B_2 + \cdots + x^{n-1} B_{n-1}$, where $B_0, B_1, B_2, \ldots, B_{n-1}$ are matrices, defined by the equation and with real entries. The basic property of adj B is that $B(\text{adj } B) = |B|I$, and so $(A - xI)(\text{adj } B) = |B|I = f(x)I = A(\text{adj } B) - x(\text{adj } B)$. From the expression for $f(x)$, we now obtain $b_0 I + b_1 xI + b_2 x^2 I + \cdots + b_{n-1} x^{n-1} I + b_n x^n I = AB_0 + xAB_1 + x^2 AB_2 + \cdots + x^{n-1} AB_{n-1} - xB_0 - x^2 B_1 - x^3 B_2 - \cdots - x^{n-1} B_{n-2} - x^n B_{n-1}$, which may be regarded as a polynomial identity in x with matrix coefficients. Since coefficients of like powers of x must be equal, the following equalities result:

$$\begin{aligned} b_0 I &= AB_0 \\ b_1 I &= AB_1 - B_0 \\ b_2 I &= AB_2 - B_1 \\ &\cdots\cdots\cdots \\ b_{n-1} I &= AB_{n-1} - B_{n-2} \\ b_n I &= -B_{n-1} \end{aligned}$$

If we now perform a left multiplication on both members of these equalities by $I, A, A^2, \ldots, A^{n-1}, A^n$, respectively, the result is

$$b_0 I = A B_0$$
$$b_1 A = A^2 B_1 - A B_0$$
$$b_2 A^2 = A^3 B_2 - A^2 B_1$$
$$\cdots\cdots\cdots\cdots\cdots$$
$$b_{n-1} A^{n-1} = A^n B_{n-1} - A^{n-1} B_{n-2}$$
$$b_n A^n = -A^n B_{n-1}$$

On adding all members on the right we obtain O, and the sum of the left members is $b_0 I + b_1 A + b_2 A^2 + \cdots + b_{n-1} A^{n-1} + b_n A^n$; and we recognize the latter sum as $f(A)$. Hence $f(A) = O$, as was asserted in the theorem. ∎

There are many ways to prove the Cayley-Hamilton theorem, and one of the easiest is suggested as Prob. 23 in Sec. 6.7. However, we prefer to have the result available at this time and so have given the above direct—but not particularly elegant—proof. In view of Theorem 5.2 (Chap. 5) and our familiar matrix-operator isomorphism results, we may also refer to $f(x)$ as the *characteristic polynomial* of any linear operator T that A represents, and the following is an immediate consequence of the theorem.

Corollary 1   *If $f(x)$ is the characteristic polynomial of a linear operator T, then $f(T) = 0$.*

It is to be understood that, although coefficients are regarded as scalar *matrices* in the case of $f(A) = O$, the equality $f(T) = 0$ "makes sense" only if we regard the coefficients as scalar multiples of the identity *operator*.

We saw above that any $n \times n$ matrix satisfies its minimum equation $m(x) = 0$, the degree of the polynomial being not greater than n^2. As a result of Corollary 2, which we now state, we are able to obtain much better information concerning the degree of $m(x)$.

Corollary 2   *The minimum polynomial of a square matrix A of order n is a divisor of its characteristic polynomial and so has a degree not greater than n.*

Proof

If $f(x)$ and $m(x)$ are the characteristic and minimum polynomials of A, respectively, there exist polynomials $q(x)$ and $r(x)$ such that $f(x) = m(x) q(x) + r(x)$, where the degree of $r(x)$ is less than that of $m(x)$. However, since $f(A) = m(A) = O$, we must have $r(A) = O$, and the minimal nature of $m(x)$ requires that $r(x)$ be the zero polynomial. Hence $f(x) = m(x) q(x)$, so that $m(x)$ divides $f(x)$ as asserted. ∎

SEC. 6.1 THE CAYLEY-HAMILTON THEOREM

There is, of course, a parallel result for linear operators, the minimum polynomial of an operator being identified with the minimum polynomial of any matrix representing it. (See Prob. 12.)

Corollary 3 *The minimum polynomial of a linear operator T on a vector space of dimension n is a divisor of its characteristic polynomial and so has degree not greater than n.*

EXAMPLE 1

Let us see if we can determine the minimum polynomial of the matrix

$$A = \begin{bmatrix} 5 & -6 & -6 \\ -1 & 4 & 2 \\ 3 & -6 & -4 \end{bmatrix}$$

It is not difficult to determine that the characteristic polynomial $f(x)$ of A is $-(x-1)(x-2)^2$, and, in view of Corollary 2, the minimum polynomial $m(x)$ of A must be either $-f(x) = (x-1)(x-2)^2$, $(x-1)(x-2)$, $(x-2)^2$, $x-1$ or $x-2$. It is clear that $A - I \neq O$ and $A - 2I \neq O$, and an elementary check verifies that

$$(A - 2I)^2 = \begin{bmatrix} 3 & -6 & -6 \\ -1 & 2 & 2 \\ 3 & -6 & -6 \end{bmatrix} \begin{bmatrix} 3 & -6 & -6 \\ -1 & 2 & 2 \\ 3 & -6 & -6 \end{bmatrix} \neq O$$

However,

$$(A - I)(A - 2I) = \begin{bmatrix} 4 & -6 & -6 \\ -1 & 3 & 2 \\ 3 & -6 & -5 \end{bmatrix} \begin{bmatrix} 3 & -6 & -6 \\ -1 & 2 & 2 \\ 3 & -6 & -6 \end{bmatrix}$$

$$= \begin{bmatrix} 0 & 0 & 0 \\ 0 & 0 & 0 \\ 0 & 0 & 0 \end{bmatrix} = O$$

so that $m(x) = (x-1)(x-2)$.

The fact that $m(x)$ divides $f(x)$ is useful information, and Example 1 illustrates how this can be used to determine $m(x)$ in a somewhat laborious way. The following lemma is a prelude to a much more practical result.

Lemma Let $f(x)$ and $m(x)$ be, respectively, the characteristic and minimum polynomials of a square matrix A. Then, if $B = A - xI$, and $g(x)$ is the gcd of all the entries of adj B, we have $m(x) = f(x)/g(x)$.

Proof
Since $B(\text{adj } B) = f(x)I$, it follows from the definition of $g(x)$ that $g(x)$ divides $f(x)$, and we let $f(x)/g(x) = h(x)$, for some polynomial $h(x)$. Since adj $B = g(x)C$, where C is a matrix whose entries have no factor common to all except for scalars, we have $B[g(x)C] = f(x)I$, or, equivalently, $g(x)BC = f(x)I$. Thus $BC = h(x)I$. (See Prob. 13.) If we now repeat the argument given in the proof of the Cayley-Hamilton theorem, with C replacing adj B, it is easy to verify (Prob. 14) that $h(A) = 0$. Hence $m(x)$ divides $h(x)$.

On the other hand, if we consider the polynomial $m(x) - m(y)$, for indeterminate symbols x and y, it is clear (Prob. 15) that it is divisible by $y - x$. That is, $m(x) - m(y) = (y - x)q(x,y)$ for some polynomial $q(x,y)$ in x and y. Since this is an identity in x and y, the equality remains valid if the indeterminates are replaced by square matrices of equal orders: In particular, we replace x by xI and y by A, the order of I being the same as that of A. The equation now becomes $m(xI) - m(A) = m(x)I = (A - xI)q(xI,A) = Bq(xI,A)$, so that $m(x)I = Bq(xI,A)$. If we now multiply both members of the latter equation on the left by adj B, the result is $m(x)(\text{adj } B) = (\text{adj } B)Bq(xI,A) = f(x)q(xI,A)$. But adj $B = g(x)C$, and so $m(x)g(x)C = h(x)g(x)q(xI,A)$, whence $m(x)C = h(x)q(xI,A)$. Since $h(x)$ divides each entry of $m(x)C$, and the only common divisors of all entries of C are scalars, it follows that $h(x)$ divides $m(x)$. Since we showed earlier that $m(x)$ divides $h(x)$, we conclude that $h(x) = m(x)$, and the proof of the lemma is complete. ∎

Although the lemma gives a precise description of the minimum polynomial of a matrix, the computation involved in the process leaves much to be desired! The following theorem, which is an easy consequence of the proof of the lemma, is usually more useful and often completely definitive.

Theorem 1.2 Every irreducible factor of the characteristic polynomial of an $n \times n$ matrix A (linear operator T on an n-dimensional vector space) is a factor of its minimum polynomial.

Proof
We give the proof for the matrix case. If we use the notation in the proof of the lemma, we may capitalize on the equality $m(x)I = Bq(xI,A)$. But then $\det [m(x)I] = [m(x)]^n = [\det B][\det q(xI,A)] = f(x)[\det q(xI,A)]$. It follows from this equation that every irreducible factor of $f(x)$ must be a divisor of $[m(x)]^n$ and so also a divisor of $m(x)$. This completes the proof of the theorem. ∎

Corollary *If the characteristic polynomial of a square matrix, or linear operator on a finite-dimensional vector space, is the product of distinct irreducible factors, it differs from the associated minimum polynomial at most in sign.*

EXAMPLE 2

If we take a new look at Example 1, in the light of Theorem 1.2, we see that the possibilities for $m(x)$ are reduced to two: either $m(x) = -f(x)$ or $m(x) = (x-1)(x-2)$. It may then be ascertained, after a *single* check, that the latter is the correct minimum polynomial.

We shall see later that the minimum polynomial will play an important role in our subsequent developments, and the Cayley-Hamilton theorem has initiated this line of investigation. In Probs. 19 and 20, we indicate some other simple applications of this important theorem.

Problems 6.1

1. If the following are the characteristic polynomials of certain matrices or linear operators on finite-dimensional spaces, list the possible candidates for the minimum polynomials:
   (a) $-(x-1)^2(x+2)$; (b) $(x+2)^2(x^2+x+1)^2$; (c) $-(x-1)(x^2+1)$; (d) $-(x-2)^3(x^2-x-1)$.
2. Apply the directions in Prob. 1 to the following polynomials:
   (a) $-(x+2)(x+1)^2$; (b) $(x-1)(x+1)(x+2)^2$.
3. Use matrix computation to verify the Cayley-Hamilton theorem for each of the following matrices:

   (a) $\begin{bmatrix} 1 & -1 \\ 2 & 1 \end{bmatrix}$ (b) $\begin{bmatrix} 2 & -1 & 1 \\ 0 & 1 & 0 \\ 1 & -2 & 1 \end{bmatrix}$

4. Let $g(x) = x - 2$ be the gcd of the entries of adj B, where $B = A - xI$ for a square matrix A and identity matrix I of the same order. If the characteristic polynomial of A is $x^4 - 2x^2 - 3x - 2$, determine its minimum polynomial.
5. Use the lemma in this section to determine the minimum polynomial of the matrix A, where

   (a) $A = \begin{bmatrix} 2 & 1 \\ 2 & 3 \end{bmatrix}$ (b) $A = \begin{bmatrix} 1 & -4 & 2 \\ -4 & 1 & -2 \\ 2 & -2 & -2 \end{bmatrix}$

6. (a) If the characteristic polynomial of a matrix is $-x^3 + 2x^2 + x - 2$, what is its minimum polynomial?
   (b) If the minimum polynomial of a 3×3 matrix is $x^3 + 2x^2 - 4x - 8$, what is its characteristic polynomial?

7. (a) Prove that the characteristic and minimum polynomials of the matrix

$$\begin{bmatrix} 0 & 0 & 2 \\ 1 & 0 & -1 \\ 0 & 1 & 1 \end{bmatrix}$$

differ only in sign.
   (b) Replace the third column of the matrix in (a) by (a,b,c), and prove that the same conclusion holds.

8. Explain why the minimum polynomial of a scalar matrix aI must have degree 1.

9. Why is the dimension of the space of all $n \times n$ matrices exactly n^2?

10. Explain what is meant by a certain polynomial $m(x)$ being the minimum polynomial of a linear operator T on a vector space.

11. Use the method of Theorem 1.1 to give a proof of the Cayley-Hamilton theorem for 2×2 matrices by direct calculations.

12. Prove that the minimum polynomial of a square matrix (or linear operator on a finite-dimensional vector space) is unique.

13. In the proof of the lemma, explain why $g(x)$ is not the zero polynomial.

14. Give the details of the argument, as suggested in the lemma, which is similar to that in the proof of the Cayley-Hamilton theorem.

15. If m is an *arbitrary* polynomial, explain why $m(x) - m(y)$ is divisible by $y - x$ for any indeterminates x and y.

16. Determine all 2×2 matrices that are solutions of the equation $x^2 + I = 0$, regarding 0 and I, respectively, as the zero and identity matrix of order 2.

17. Prove that there are no 3×3 matrices (with real entries) that are solutions of the equation $x^2 + I = 0$. (See Prob. 16.)

18. Prove that any (matrix) solution of the equation $x^2 + x + I = 0$ is nonsingular.

19. Let $f(x) = a_0 + a_1 x + \cdots + a_n x^n$ be the characteristic polynomial of a square matrix A. Then use the Cayley-

Hamilton theorem to show that any positive integral power of A can be expressed as a linear combination (with real coefficients) of $I, A, A^2, \ldots, A^{n-1}$.

20. If A is a nonsingular matrix, use Prob. 19 to show that

$$A^{-1} = -\frac{a_1}{a_0}I - \frac{a_2}{a_0}A - \cdots - \frac{a_n}{a_0}A^{n-1}$$

and extend the result in Prob. 19 to include all negative integral powers of A.

21. Use the method indicated in Prob. 20 to determine A^{-1}, where

$$A = \begin{bmatrix} 1 & 1 & -1 \\ 0 & -1 & 2 \\ 1 & 0 & 0 \end{bmatrix}$$

22. If $m(x)$ is the minimum polynomial of a square matrix A, and $g(x)$ is any real polynomial, prove that $g(A)$ is nonsingular if and only if $g(x)$ and $m(x)$ are relatively prime.

23. If $A = [a_{ij}]$ is an $n \times n$ matrix, show that $A = B + C$, where $C = cI$ is a scalar matrix, for some $c \in \mathbf{R}$, and $B = [b_{ij}]$ is an $n \times n$ matrix whose trace is 0. (See Prob. 20 of Sec. 2.10.)

24. Let

$$A = \begin{bmatrix} 1 & 1 & 0 \\ -1 & 1 & 1 \\ 0 & 1 & -1 \end{bmatrix}$$

and assume that $-x^3 + ax^2 + bx + c$ is its characteristic polynomial. Now use the Cayley-Hamilton theorem to determine a, b, and c and therefore the characteristic equation of A. Check by the usual method.

6.2 Similarity and Diagonal Matrices

With this section, we begin a serious discussion of the principal theme of the chapter: to discover a basis of a vector space such that a given linear operator on the space has as simple a matrix representation as possible. Throughout the book, we have tried to emphasize the parallelism that exists between matrix theory and the theory of linear transformations of vector spaces of finite dimension, and the matrix-theoretic equivalent of our present objective is to find the matrix in simplest form that is *similar* to a given square matrix.

The scalar matrices aI, with $a \in \mathbf{R}$, are undoubtedly the simplest matrices, but each such matrix is the only member of its equivalence class under similarity (Prob. 7), and so a nonscalar matrix is never similar to a scalar matrix. With scalar matrices effectively ruled out of our present study, the next candidate for a matrix in simple form is a diagonal matrix diag $\{a_1, a_2, \ldots, a_n\}$, with $a_1, a_2, \ldots, a_n \in \mathbf{R}$. It turns out that the number of equivalence classes under similarity, that may be represented by diagonal matrices, is quite large, and to these classes we direct our immediate attention. Inasmuch as similar matrices may be used to represent the same linear operator on a vector space, we shall study simultaneously the circumstances under which a linear operator can be represented by a diagonal matrix. An operator of this sort is said to be *diagonalizable*.

Our study of eigenvectors in Chap. 5 has already thrown considerable light on the question of whether a linear operator T on a vector space V is diagonalizable. In fact, if T is represented by a matrix A, the existence of a nonsingular matrix P, such that $P^{-1}AP$ is diagonal, depends upon the existence of a basis of eigenvectors for V. In view of its basic importance in the present context, we restate this familiar fact as our first theorem, assuming that the reader will recall its proof (Prob. 8).

Theorem 2.1 *A linear operator on a vector space of finite dimension is diagonalizable if and only if there exists a basis of eigenvectors for the space.*

Since there can be no eigenvectors independent of those in a basis of eigenvectors, the following corollary is immediate.

Corollary *If a linear operator T is represented by a diagonal matrix D, the elements along the main diagonal of D are the eigenvalues of* T.

As we have already noted, the question of whether a linear operator is diagonalizable depends on whether a basis of eigenvectors exists. If we define a *symmetric operator* as a linear operator that may be represented by a symmetric matrix, we know from Chap. 5 that every symmetric operator is diagonalizable. It should also be recalled that a very important property of a symmetric matrix—which played a key role in our obtaining this result—is that all its eigenvalues are real. However, if a linear operator T on a space V is represented by a matrix A, the fact that the eigenvalues of T are all real is not *by itself* sufficient to guarantee the existence of an eigenvector basis for V. For example, if $A = \begin{bmatrix} 1 & 1 \\ 0 & 1 \end{bmatrix}$ and T operates on V_2, the only eigenvalue of T is the real number 1, and it is easy to see that the associated eigenvectors

$(b,0)$, with $b \in \mathbf{R}$, do not span V_2. In this example, the single eigenvalue occurs with a multiplicity of 2, but, if all eigenvalues of an operator are distinct, the following theorem shows that a basis of eigenvectors does exist for the space on which the operator is defined.

Theorem 2.2 *Any linear operator* T *on a vector space of dimension n is diagonalizable if it has n distinct eigenvalues.*

Proof
Since we know that each eigenvalue has an eigenvector belonging to it, the only point to establish is that a set of n eigenvectors, in one-to-one correspondence with n distinct eigenvalues to which they belong, is linearly independent. The proof of this was suggested earlier (see Prob. 24 of Sec. 5.5), but we give the argument here. We shall use an indirect method of proof and tentatively assume that the eigenvectors $\alpha_1, \alpha_2, \ldots, \alpha_n$, belonging respectively to the distinct eigenvalues $\lambda_1, \lambda_2, \ldots, \lambda_n$, are linearly dependent; i.e., for some k, $1 \leq k < n$, and certain real numbers $c_1, c_2, \ldots, c_k$, we assume that the vectors $\alpha_1, \alpha_2, \ldots, \alpha_k$ are linearly independent but $\alpha_{k+1} = c_1\alpha_1 + c_2\alpha_2 + \cdots + c_k\alpha_k$. It follows that $T\alpha_{k+1} = c_1(T\alpha_1) + c_2(T\alpha_2) + \cdots + c_k(T\alpha_k)$, and so $\lambda_{k+1}\alpha_{k+1} = \lambda_1 c_1 \alpha_1 + \lambda_2 c_2 \alpha_2 + \cdots + \lambda_k c_k \alpha_k$. However, it is also true, as a result of direct multiplication of the expression for α_{k+1} by λ_{k+1}, that $\lambda_{k+1}\alpha_{k+1} = \lambda_{k+1} c_1 \alpha_1 + \lambda_{k+1} c_2 \alpha_2 + \cdots + \lambda_{k+1} c_k \alpha_k$, so that $(\lambda_{k+1} - \lambda_1)c_1\alpha_1 + (\lambda_{k+1} - \lambda_2)c_2\alpha_2 + \cdots + (\lambda_{k+1} - \lambda_k)c_k\alpha_k = 0$. In view of the linear independence of $\alpha_1, \alpha_2, \ldots, \alpha_k$, we must conclude that $c_1 = c_2 = \cdots = c_k = 0$. But then $\alpha_{k+1} = 0$, contrary to our assumption that α_{k+1} is an eigenvector. Hence the n eigenvectors $\alpha_1, \alpha_2, \ldots, \alpha_n$ must be linearly independent and so form a basis for the space. The operator T is then diagonalizable by Theorem 2.1. ∎

Corollary *A matrix* A *of order* n *is similar to* diag $\{\lambda_1, \lambda_2, \ldots, \lambda_n\}$ *if* $\lambda_1, \lambda_2, \ldots, \lambda_n$ *are the eigenvalues of* A, *all distinct.*

Proof
If we regard A as the matrix that represents a linear operator T on a vector space V, relative to some basis, the corollary is an immediate consequence of the theorem. Our only comment is that the columns of the matrix P, such that $P^{-1}AP = $ diag $\{\lambda_1, \lambda_2, \ldots, \lambda_n\}$, are the coordinates of n eigenvectors, one of which is associated with each of the n eigenvalues of A. ∎

EXAMPLE 1

If λ_1, λ_2 are the eigenvalues of the matrix $A = \begin{bmatrix} 1 & 1 \\ 3 & -1 \end{bmatrix}$, determine a nonsingular matrix P such that $P^{-1}AP = $ diag $\{\lambda_1, \lambda_2\}$.

Solution
The characteristic polynomial of A is easily found to be $(x - 2)(x + 2)$, so that its eigenvalues are ± 2. If we now think of A as a representation of a linear operator T on V_2, a simple computation shows that the eigenvectors belonging, respectively, to 2 and -2 are (a,a) and $(a,-3a)$ for any $a \in \mathbf{R}$. A suitable eigenvector basis of V_2 is then $\{(1,1),(1,-3)\}$, and so $P = \begin{bmatrix} 1 & 1 \\ 1 & -3 \end{bmatrix}$ is the corresponding matrix of transition. We find that $P^{-1} = \begin{bmatrix} \frac{3}{4} & \frac{1}{4} \\ \frac{1}{4} & -\frac{1}{4} \end{bmatrix}$, and a simple check shows that

$$P^{-1}AP = \begin{bmatrix} \frac{3}{4} & \frac{1}{4} \\ \frac{1}{4} & -\frac{1}{4} \end{bmatrix} \begin{bmatrix} 1 & 1 \\ 3 & -1 \end{bmatrix} \begin{bmatrix} 1 & 1 \\ 1 & -3 \end{bmatrix} = \begin{bmatrix} \frac{3}{4} & \frac{1}{4} \\ \frac{1}{4} & -\frac{1}{4} \end{bmatrix} \begin{bmatrix} 2 & -2 \\ 2 & 6 \end{bmatrix} = \begin{bmatrix} 2 & 0 \\ 0 & -2 \end{bmatrix}$$

as desired.

EXAMPLE 2

If we regard

$$A = \begin{bmatrix} -1 & 2 & 2 \\ 2 & 2 & 2 \\ -3 & -6 & -6 \end{bmatrix}$$

as the matrix of a linear operator T on V_3, it is not difficult to discover that $(1,0,-1)$, $(0,1,-1)$, and $(2,-1,0)$ are three linearly independent eigenvectors of T in the space. The matrix of transition whose columns are these eigenvectors, is then

$$P = \begin{bmatrix} 1 & 0 & 2 \\ 0 & 1 & -1 \\ -1 & -1 & 0 \end{bmatrix}$$

with

$$P^{-1} = \begin{bmatrix} -1 & -2 & -2 \\ 1 & 2 & 1 \\ 1 & 1 & 1 \end{bmatrix}$$

An elementary matrix calculation shows that $P^{-1}AP = $ diag $\{-3,0,-2\}$, a diagonal matrix that lists the eigenvalues of T along its main diagonal.

In both the examples given, it may be noticed that no eigenvalue has a multiplicity greater than 1, a circumstance that we would normally expect in a random matrix. However, it is still possible for a matrix of order n to be similar to a diagonal matrix, even though it does not have n distinct eigenvalues. The following theorem provides what is probably the most general criterion for deciding whether a given linear operator is diagonalizable, although its application is not always the most practical way to make this decision.

Theorem 2.3 *A necessary and sufficient condition that a linear operator on a finite-dimensional vector space is diagonalizable is that its minimum polynomial factors into distinct linear factors.*

Proof
Let us first suppose that the linear operator T is diagonalizable, so that it can be represented by a diagonal matrix D, relative to some basis. The characteristic polynomial $f(x)$ of D is then a product of linear factors and, by Theorem 1.2, each of these distinct factors must be a factor of the minimum polynomial $m(x)$ of T. However, it is a consequence of the lemma in Sec. 6.1 (it can also be seen by direct substitution) that the minimum polynomial of D cannot have any repeated factors (Prob. 9), and so $m(x)$ is a product of distinct linear factors.

Conversely, let us suppose that the minimum polynomial $m(x)$ of T has the form $m(x) = (x - \lambda_1)(x - \lambda_2) \cdots (x - \lambda_t)$, where $\lambda_1, \lambda_2, \ldots, \lambda_t$ are the distinct eigenvalues of T. If V is the space on which T operates, we define K_i to be the *kernel* of the operator $T - \lambda_i I$; that is, K_i is the subspace of vectors $\alpha \in V$, such that $(T - \lambda_i I)\alpha = 0$, and so its nonzero members are the eigenvectors of T that belong to λ_i. Since eigenvectors that belong to distinct eigenvalues are linearly independent (see the proof of Theorem 2.2), the sum $K_1 + K_2 + \cdots + K_t$ is direct (Prob. 10). Now $K_1 \oplus K_2 \oplus \cdots \oplus K_t \subset V$, so that $\sum_{i=1}^{t} r_i \leq n$, where $n = \dim V$ and $r_i = \dim K_i$, $i = 1, 2, \ldots, t$. By Theorem 2.2 (Chap. 2), $\dim (T - \lambda_1 I) V = n - r_1$, and a further application of the same theorem implies (Prob. 11) that $\dim (T - \lambda_2 I)[(T - \lambda_1 I)] V \geq (n - r_1) - r_2 = n - (r_1 + r_2)$. If we continue in a similar way with $T - \lambda_3 I, \ldots, T - \lambda_t I$, we find ultimately that $0 = \dim [m(T) V] \geq n - (r_1 + r_2 + \cdots + r_t)$, so that $r_1 + r_2 + \cdots + r_t = n$. Hence $V = K_1 \oplus K_2 \oplus \cdots \oplus K_t$, every vector in V is a linear combination of eigenvectors, and so a basis of eigenvectors of T exists for V. It then follows from Theorem 2.1 that T is diagonalizable. ∎

Corollary *A necessary and sufficient condition that a square matrix A is similar to a diagonal matrix is that its minimum polynomial factors into distinct linear factors.*

Proof
Let A represent a linear operator T on a vector space of appropriate dimension, and apply the theorem. ∎

Just prior to the statement of Theorem 2.3, we indicated that, although the theorem has great theoretical interest, the practicality of its application leaves something to be desired. Even if we determine the minimum polynomial of a matrix and are able to conclude that a basis of eigenvectors exists, the work we have done is of no help in the actual determination of such a basis. If we are able to solve the characteristic equation, it is usually easier to take a direct approach and make an actual determination of an eigenvector basis—provided one exists. However, the matter of solving the characteristic equation may also present some serious practical difficulties. In Probs. 22 to 25, we discuss some of the available (but admittedly meager) aids in solving polynomial equations of degree greater than 2. It should be understood that no formula involving radicals exists for the solution of such general equations of degree greater than 4, and the formulas for solving cubic and quartic equations are far from simple! In general, one must be satisfied with approximations to the eigenvalues of a matrix, usually obtained by some iterative process—if possible on a high-speed computer. For such processes, the reader should consult some of the many modern texts on numerical analysis.

We stated that the principal objective of this chapter was to examine the problem of simple matrix representations of linear operators. We are able to condense our results obtained so far into the following parallel forms:

(1) A linear operator is diagonalizable if it is symmetric or either its characteristic or minimum polynomial factors into distinct linear factors.

(2) A square matrix is similar to a diagonal matrix if it is symmetric or either its characteristic or minimum polynomial factors into distinct linear factors.

We shall continue this investigation in Sec. 6.4, in a somewhat more specialized context which will lead to the spectral theorem. But, before doing this, we introduce complex vector spaces.

Problems 6.2

1. Explain why the matrices $\begin{bmatrix} 1 & 0 \\ 0 & 1 \end{bmatrix}$ and $\begin{bmatrix} 1 & 1 \\ 0 & 1 \end{bmatrix}$, although having the same minimum polynomial, could not be similar.
2. If the eigenvalues of an $n \times n$ matrix A are $\lambda_1, \lambda_2, \ldots, \lambda_n$, and A is similar to a diagonal matrix D, explain why the diagonal elements of D, in some order, must be $\lambda_1, \lambda_2, \ldots, \lambda_n$.

3. Use a 2 × 2 matrix to illustrate how two linearly independent eigenvectors can belong to the same eigenvalue.
4. Decide whether a linear operator is diagonalizable, given that its minimum polynomial is (a) $x^2 - 5x - 6$; (b) $x^3 + x + 1$; (c) $x^3 - x^2 - 4x + 4$; (d) $x^4 - 2x^3 + x^2 + x - 1$.
5. Determine a matrix P such that $P^{-1}AP$ is diagonal, given that

   (a) $A = \begin{bmatrix} 2 & 1 \\ 2 & 1 \end{bmatrix}$ (b) $\begin{bmatrix} 1 & 0 & 0 \\ 1 & 0 & -2 \\ -1 & 1 & -3 \end{bmatrix}$

6. Prove that the following matrix is not similar to a diagonal matrix:

   (a) $\begin{bmatrix} 1 & -1 \\ 2 & 1 \end{bmatrix}$ (b) $\begin{bmatrix} 1 & 1 & -1 \\ -1 & 3 & -1 \\ -1 & 2 & 0 \end{bmatrix}$ (c) $\begin{bmatrix} 1 & 0 \\ a & 1 \end{bmatrix}$

   for any a ($\neq 0$) $\in$ **R**.

7. Why is any scalar matrix aI the *only* member of its equivalence class under similarity?
8. Review the proof of Theorem 2.1.
9. Use the lemma in Sec. 6.1 to see why the minimum polynomial of a diagonal matrix cannot have repeated factors. Explain why this also follows by direct substitution, illustrating with diag $\{a,a,b\}$.
10. In the proof of Theorem 2.3, give the details of the argument which proves that the sum $K_1 + K_2 + \cdots + K_t$ is direct.
11. In the proof of Theorem 2.3, explain why dim $(T - \lambda_2 I)[(T - \lambda_1 I)]V \geq (n - r_1) - r_2$.
12. Explain why a set of n linearly independent eigenvectors of a linear operator on an n-dimensional space must contain *at least one* belonging to each of the eigenvalues.

13. If P is a nonsingular matrix, prove that PA is similar to AP, for any matrix A of the same order as P.
14. Find a condition that is both necessary and sufficient for the partitioned matrix $\begin{bmatrix} A & O \\ O & B \end{bmatrix}$ to be similar to a diagonal matrix, where the zero matrices have orders consistent with those of the square matrices A and B.
15. Prove that similar matrices have the same trace (see Prob. 20 of Sec. 2.10).

16. If A is similar to a diagonal matrix, prove that A' has the same property.

17. If a matrix A is similar to a diagonal matrix, and $g(x)$ is any real polynomial, prove that $g(A)$ is also similar to a diagonal matrix.

18. If A is a triangular matrix (that is, $A = [a_{ij}]$, where either $a_{ij} = 0$ if $i < j$ or $a_{ij} = 0$ if $i > j$), prove that A is similar to a diagonal matrix if the diagonal entries of A are distinct.

19. If T is a diagonalizable operator on a finite-dimensional vector space V, prove that T must map each subspace, that is a direct summand of V, into itself. That is, if $V = W_1 \oplus W_2 \oplus \cdots \oplus W_k$, for subspaces W_i, $i = 1, 2, \ldots, k$, then $TW_i \subset W_i$.

20. If T is a diagonalizable operator on a vector space V of dimension n (see Prob. 19), with $c_1, c_2, \ldots, c_n$ the distinct eigenvalues of T, establish the existence of operators (*projections*) $E_1, E_2, \ldots, E_n$ on V such that (a) $T = c_1 E_1 + c_2 E_2 + \cdots + c_n E_n$; (b) $E_1 + E_2 + \cdots + E_n$ is the identity operator on V; (c) $E_i E_j = 0$, $i \neq j$.

21. With reference to Prob. 20, prove that the range of E_i is the subspace of eigenvectors of T, belonging to c_i, $i = 1, 2, \ldots, n$.

Note The following problems are concerned with the mechanics of solving polynomial equations, and, in particular, the characteristic equations of matrices or linear operators. For the purposes of these problems, the solutions should not be assumed to be necessarily real numbers.

22. Let $f(x) = (x - c)^r g(x)$ be a real polynomial, where $r \geq 1$ and $x - c$ is prime to $g(x)$. Then prove that the derivative $f'(x)$ has c as a zero of multiplicity $r - 1$, and hence a simple zero of $f(x)$ is not a zero of $f'(x)$. What is the significance of this insofar as characteristic polynomials are concerned?

23. If $d(x)$ is the gcd of a polynomial $f(x)$ and its derivative $f'(x)$, use the result in Prob. 22 to prove that $f(x)/d(x)$ has only simple zeros—the distinct zeros of $f(x)$. Illustrate this with the use of the polynomial $f(x) = x^3 - x^2 - x + 1$.

24. If $f(x) = a_0 + a_1 x + a_2 x^2 + \cdots + a_n x^n$, with $a_0, a_1, a_2, \ldots, a_n \in \mathbf{Z}$, has a rational zero a/b, with $a, b \in \mathbf{Z}$, prove that a must divide a_0 and b must divide a_n. Use this result (a) to find all rational zeros of $6x^3 + 7x^2 - x - 2$; (b) to show that $x^4 - 2x^3 + x^2 - x - 1$ has no rational zeros.

25. Show that a polynomial $f(x) = a_0 + a_1 x + \cdots + a_n x^n$, with $a_0, a_1, \ldots, a_n$ integers, has no integral zero if both $f(0)$ and $f(1)$ are odd integers.

6.3 Complex Vector Spaces

Up to this point in the text, by a *vector space* we have always meant a *real* vector space, the understanding being that the scalars are *real* numbers. In fact we gave a formal definition only of this type of space, in Sec. 1.1, with mere passing mention of the existence of vector spaces over other fields. Our reason for choosing the field of real numbers **R** as the source of scalars was that it is the *most familiar* field that possesses *nearly all the useful properties*. In one respect, however, it is somewhat unsatisfactory: It is not algebraically closed, and so there exist equations with real coefficients not all of whose solutions are real numbers. This defect is overcome by using the field **C** of complex numbers, and it is the purpose of this section to discuss the slight changes in the foregoing theory, which arise from the use of the complex number field. Although we *could* discuss our complete theory of vector spaces over a very general algebraically closed field, there would be very little advantage in such a general program (for we would lose the familiar aspect of the scalars) and no significant new results would be obtained, provided we assume that the characteristic of the general field is 0. Although the results would be different for vector spaces over a field of prime characteristic, it would take us too far off our intended course to go into this more complicated setting. (The reader may check in any book on abstract algebra for the meaning of characteristic, as applied to a ring or field.)

We shall assume at least a working familiarity with the field **C** of complex numbers. Any complex number has the representation $a + bi$, for real numbers a and b where $i^2 = -1$; any two complex numbers can be added, subtracted, multiplied, or divided provided—in the case of division—that the divisor is not 0. It is often convenient to refer to a as the *real part*, and occasionally to b as the *imaginary part*, of the number $a + bi$, but only historical significance should be attached to the word *imaginary*. The complex numbers of the form $a + 0i$ are not to be distinguished from the real numbers a, and so we consider **R** to be a *subfield* of **C**, or **C** to be an *extension field* of **R**.

Essentially everything in the first three chapters of the text carry over to vector spaces over **C** (that is, *complex vector spaces*), if we merely regard the scalars as *complex* rather than *real* numbers. There is no useful complex analog to the geometric space of line vectors, discussed briefly in Chap. 1, but the only other changes are notational. The important space of *complex n*-tuples is defined quite like the real coordinate space V_n, except for the presence of complex numbers as coordinates. Although there has been no danger of ambiguity in our previous use of the symbol V_n, we must now distinguish between the real and complex *coordinate* spaces: They will be symbolized as $V_n(\mathbf{R})$ and $V_n(\mathbf{C})$, respectively. Even the *same* $\{E_i\}$ basis is available for both spaces. A *linear transformation* on a complex vector space V must now be defined as a mapping T of V, such that $T(c_1\alpha + c_2\beta) = c_1T(\alpha) + c_2T(\beta)$, for any $c_1, c_2 \in \mathbf{C}$, and any vectors $\alpha, \beta \in V$. But again, this differs from the familiar definition of a linear transformation of a real vector space only in that the scalars now are understood to be complex numbers. It would be tedious to discuss again *the notions that remain the same* for complex as for real scalars, such as the whole theory of systems of linear equations, determinants, isomorphism of vector spaces, operations on matrices with complex entries, matrix inversion, change of basis, coordinates, etc. We shall merely note the few places where there are significant changes, and these occur in a simple way in the material that begins with Chap. 4.

All the results, which are peculiar to our use of complex numbers as scalars, are due to the fact that every polynomial equation with either real or nonreal complex coefficients can be completely solved in $\mathbf{C}$. It is a fact, familiar from college algebra, that complex solutions of a *real* polynomial equation always occur in *conjugate* pairs; that is, if $c = a + bi$ is a solution, so is its *conjugate* $a - bi$, this conjugate of c being usually designated $\bar{c}$. The familiar properties of conjugation may be listed as follows for arbitrary $c, d \in \mathbf{C}$: $\bar{\bar{c}} = c$; $\overline{c + d} = \bar{c} + \bar{d}$; $\overline{cd} = \bar{c}\bar{d}$; c is real if and only if $\bar{c} = c$; $c\bar{c} = a^2 + b^2$, where $c = a + bi$; the real part of c is $(c + \bar{c})/2$, and its imaginary part is $(c - \bar{c})/2i$. By the *conjugate* of a matrix $A = [a_{ij}]$, we mean $\bar{A} = [\bar{a}_{ij}]$, as is natural (Prob. 10). The reader will also be familiar with the notion of *absolute value* of a complex number c, denoted by $|c|$. If $c = a + bi$, we define $|c| = \sqrt{a^2 + b^2}$, and we list its well-known properties: $|c| > 0$ if $c \neq 0$, $|0| = 0$; $|c|^2 = c\bar{c}$; $|cd| = |c|\,|d|$; $|c + d| \leq |c| + |d|$.

The first important difference in complex spaces arises with our definition of an *inner product* on such a space. Even here, the change is not great, but, for the sake of clarity, we give the complete new definition.

Definition   An inner product on a complex vector space V is a complex-valued function $(\ ,\ )$ on $V \times V$ such that, for any $\alpha, \beta, \gamma \in V$ and any $c, d \in \mathbf{C}$, the following conditions are satisfied:
(1) $(\alpha, \beta) = \overline{(\beta, \alpha)}$
(2) $(c\alpha + d\beta, \gamma) = c(\alpha, \gamma) + d(\beta, \gamma)$
(3) $(\alpha, c\beta + d\gamma) = \bar{c}(\alpha, \beta) + \bar{d}(\alpha, \gamma)$
(4) $(\alpha, \alpha) > 0$, if $\alpha \neq 0$

It should be observed that it is the first and third of these properties (see Prob. 11) that distinguish a complex inner product from the familiar real case. The two agree, of course, for a vector space over the real field $\mathbf{R} \subset \mathbf{C}$. We note that $(\alpha, \alpha) > 0$, for any vector $\alpha \neq 0$, and so (as in the real case) we denote the real nonnegative number $(\alpha,\alpha)^{\frac{1}{2}}$ by $|\alpha|$ and call it the *norm* of α. With the notion of an inner product now introduced into a complex vector space, it is possible to use the same *language* as for real spaces and refer to the *angle between* two vectors, *orthogonal* vectors, the *parallelogram law*, the *triangle inequality*, and other concepts that owe their origins to the existence of an inner product. The *polarization identity*, which connects inner products with the norms defined by them, varies slightly for the real and complex cases: In the real case, $(\alpha,\beta) = |\alpha + \beta|^2/4 - |\alpha - \beta|^2/4$; in the complex case, $(\alpha,\beta) = |\alpha + \beta|^2/4 - |\alpha - \beta|^2/4 + |\alpha + i\beta|^2 i/4 - |\alpha - i\beta|^2 i/4$. In Prob. 12, we suggest that the reader check the more complicated complex case of the identity. Our earlier discussions pertaining to orthogonal and orthonormal bases, including the Gram-Schmidt construction, carry over to complex vector spaces.

The fourth condition in the definition of a complex inner product, requires slight changes in two particular inner products that have played a very important role in our previous developments. We define the *standard* inner product of two vectors X, Y in $V_n(\mathbf{C})$ as follows: If $X = (x_1, x_2, \ldots, x_n)$ and $Y = (y_1, y_2, \ldots, y_n)$, then $(X,Y) = x_1 \bar{y}_1 + x_2 \bar{y}_2 + \cdots + x_n \bar{y}_n$. The other case occurs when we define an inner product in a vector space of continuous functions or polynomials of bounded degrees. If f and g are two such vectors, instead of $\int_0^1 f(x)g(x)\,dx$, which we have regularly used for (f,g), it is now appropriate to use $\int_0^1 f(x)\overline{g(x)}\,dx$. It should be clear that we are using $\overline{g(x)}$ to indicate the function or polynomial which is the complex conjugate of that denoted by $g(x)$. Notice, however, that the variable x, as used here, is still real. If an inner product has been defined on a complex vector space, the space is usually called *unitary*. In other words, a *unitary* space is the complex analog of a *euclidean* space.

A slight change occurs in connection with an *adjoint* operator on a complex space. We recall that the definition of the adjoint T*, associated with a linear operator T on a space V, requires that $(T\alpha,\beta) = (\alpha, T^*\beta)$, for any $\alpha, \beta \in V$. Moreover, if V is finite-dimensional and $A = [a_{ij}]$ and $B = [b_{ij}]$ are matrix representations of T and T*, respectively, relative to any orthonormal basis $\{\alpha_i\}$, then $a_{ij} = (T\alpha_i, \alpha_j)$ and $b_{ij} = (T^*\alpha_i, \alpha_j)$. But $(T\alpha_i, \alpha_j) = (\alpha_i, T^*\alpha_j) = \overline{(T^*\alpha_j, \alpha_i)}$, so that $b_{ij} = \bar{a}_{ji}$, and B is called the *conjugate transpose* or *matrix adjoint* of A. We denote this conjugate transpose of A by A^*, and note that $A^* = A'$ in the real case. We have shown that *if A is the matrix of* T, *relative to an orthonormal basis, the matrix of* T*, *relative to the same basis, is A^*.* If T = T*, the self-adjoint operator T on a complex space is said to be *hermitian*, and $A = A^*$ defines A as a *hermitian matrix*. It is clear that the diagonal entries of a hermitian matrix are all real. The complex analog of an *orthogonal* operator on a euclidean space is a *unitary* operator on a unitary space; that is, T is a *unitary* operator on a unitary space if inner products are preserved by T, or, equivalently, if T* exists and TT* = T*T = I. Analogously, a square matrix A is called *unitary* if its entries are complex and $AA^* = A^*A = I$, so that $A^* = A^{-1}$. In the presence of an orthonormal basis for the underlying space, the relationship existing between unitary operators and unitary matrices is quite like that known to exist between orthogonal operators and orthogonal matrices. The following is a partial listing of concepts that have analogous meanings in real and complex vector spaces:

$$\text{Euclidean space} \longleftrightarrow \text{unitary space}$$
$$\textit{Symmetric operator (matrix)} \longleftrightarrow \textit{hermitian operator (matrix)}$$
$$\textit{Orthogonal operator (matrix)} \longleftrightarrow \textit{unitary operator (matrix)}$$

Finally, let us take a brief look at the three basic equivalence relations that have played a central role in our treatment of real vector spaces and matrices.

(1) *Equivalence.* *Equivalent* matrices may be used to represent the same bilinear function; and *real* matrices A and B are *equivalent* if there exist real nonsingular matrices P and Q such that $B = PAQ$. The only invariants of a class of equivalent matrices are the *order n* and *rank r* of the matrices in the class, and the canonical matrix of such a class is $\begin{bmatrix} I_r & 0 \\ 0 & 0 \end{bmatrix}$, where the 0s are zero matrices of order $n - r$. There is no change, in either the invariants or the canonical matrix, when the scalars are taken from the complex field **C**.

(2) *Congruence.* We recall that *congruent* matrices arose in our discussions from an attempt to represent the same quadratic or bilinear function or form on a space V by matrices, relative to different bases. Two real matrices A and B were then defined to be *congruent* if $B = P'AP$, for some nonsingular matrix P. The invariants of a class of congruent real symmetric matrices (the only ones of interest) are the *order n*, the *rank r*, and the *signature s*, and the canonical matrix of such a class is diag $\{1, 1, \ldots, 1, -1, -1, \ldots, -1, 0, 0, \ldots, 0\}$, where there are r nonzero diagonal elements, with s the excess number of 1's over -1's. If we allow our scalars to be complex numbers, everything remains the same, except for one simplification: The existence of the scalar i, as a square root of -1, allows the elimination of the -1's in the canonical matrix of a class. Thus, the *order n* and *rank r* are the only invariants of a class of complex congruent symmetric matrices, and the canonical matrix in the class is $\begin{bmatrix} I_r & 0 \\ 0 & 0 \end{bmatrix}$, which we note is the same as the canonical matrix of order n and rank r under *equivalence*. In the case of complex vector spaces, however, it is possible to use the standard inner product as a model and define a *conjugate-bilinear* (and an associated *hermitian*) function, the general value of which is expressed as a

conjugate-bilinear (and a *hermitian*) *form*, represented by a hermitian matrix. A study of these functions and forms leads to the notion of two matrices being *conjunctive:* Two complex matrices A and B are said to be *conjunctive* if there exists a nonsingular matrix P such that $B = P^*AP$. (See Prob. 13.)

(3) *Similarity.* With a consideration of *similarity* and *similar* matrices, our survey of complex vector spaces has brought us to the analogous problem with which we are also concerned in the context of real spaces: What is the simplest matrix representation of a linear operator on a complex space of finite dimension, and, if possible, what are the invariants and the canonical matrix in the similarity class of any given square matrix? The results already obtained for diagonalizable operators remain valid for complex scalars, but we shall see that certain simplifications will result in the sequel because of the factorization of the characteristic and minimum polynomial of any complex square matrix into linear factors.

We close this section with a proof (in more general form) of a result that we have been using, and whose proof was promised in Sec. 5.6. In the next section, we conclude our discussion of diagonalizable operators, now regarded as operators on a complex vector space.

Theorem 3.1   *Every eigenvalue of a hermitian operator (matrix) is real.*

Proof

Let λ be an eigenvalue of the hermitian operator T, so that $T\alpha = \lambda\alpha$, for some eigenvector α belonging to λ. The definition of an adjoint operator, the properties of a complex inner product, and the fact that T is hermitian now imply that $(T\alpha,\alpha) = (\lambda\alpha,\alpha) = \lambda(\alpha,\alpha) = (\alpha,T^*\alpha) = (\alpha,T\alpha) = (\alpha,\lambda\alpha) = \bar{\lambda}(\alpha,\alpha)$. Since $(\alpha,\alpha) \neq 0$, we conclude that $\bar{\lambda} = \lambda$, so that λ is real. Since any hermitian matrix can be considered to represent a hermitian operator on a coordinate space, *relative to an orthonormal basis*, the corresponding matrix theorem follows. ∎

Corollary   *The eigenvalues of a real symmetric matrix are real.*

(See Prob. 14, for an alternative suggested proof of the corollary.)

Problems 6.3

1. Find (X,Y), if X and Y are elements of $V_3(\mathbf{C})$, defined by (a) $X = (1, -i, 1+i)$, $Y = (2i, 2, -i)$; (b) $X = (2,i,1)$, $Y = (1, 1-i, i)$; (c) $X = (1-i, 1+i, 2)$, $Y = (1+i, 1-i, -2)$.

2. Determine $\int_0^1 f(x)\overline{g(x)}\,dx$ if (a) $f(x) = 1 + ix - x^2$, $g(x) = (1+i)x + 2x^2$; (b) $f(x) = x + (1+i)x^2$, $g(x) = 1 + x - ix^2$.

3. Write A', $\bar{A}$, and A^*, where

(a) $A = \begin{bmatrix} 2 & 1-i \\ i & 3i \end{bmatrix}$
(b) $A = \begin{bmatrix} i & 1 & -i \\ 2i & 2 & 1 \\ 2 & -i & 1+i \end{bmatrix}$

4. Use any method to determine A^{-1}, where

(a) $A = \begin{bmatrix} 1 & -i \\ i & -2 \end{bmatrix}$
(b) $A = \begin{bmatrix} 1 & 0 & i \\ -i & -2 & 0 \\ i & 2 & 1 \end{bmatrix}$

5. Show that any complex matrix can be expressed uniquely in the form $A + iB$, where A and B are hermitian matrices.

6. Verify that the eigenvalues of the following hermitian matrices are real:

(a) $\begin{bmatrix} 2 & 1-i \\ 1+i & 1 \end{bmatrix}$
(b) $\begin{bmatrix} 1 & i & 2 \\ -i & 0 & 1+i \\ 2 & 1-i & 1 \end{bmatrix}$

7. Verify that (a) the matrix $\begin{bmatrix} (1+i)/\sqrt{3} & i/\sqrt{3} \\ i/\sqrt{3} & (1-i)/\sqrt{3} \end{bmatrix}$ is unitary; (b) the matrix diag $\{c_1, c_2, \ldots, c_n\}$ is unitary if and only if $|c_i| = 1$, $i = 1, 2, \ldots, n$.

8. A matrix A is said to be *involutary* if $A^2 = I$. Verify that the following *Pauli spin* matrices (which occur in quantum mechanics) are involutary, hermitian, and unitary:

$$\sigma_x = \begin{bmatrix} 0 & 1 \\ 1 & 0 \end{bmatrix} \quad \sigma_y = \begin{bmatrix} 0 & -i \\ i & 0 \end{bmatrix} \quad \sigma_z = \begin{bmatrix} 1 & 0 \\ 0 & -1 \end{bmatrix}$$

9. If we define $AB - BA$ to be the *commutator* (Why?) of two matrices A and B of the same order, show that the commutators of σ_x and σ_y, σ_y and σ_z, σ_z and σ_x are $2i\sigma_z$, $2i\sigma_x$, and $2i\sigma_y$, respectively.

10. Why is it "natural" to define the conjugate of $A = [a_{ij}]$ to be $\bar{A} = [\bar{a}_{ij}]$?

11. Show that the third property listed in the definition of a complex inner product follows from the first two.

12. Verify the complex polarization identity, that relates an inner product with its associated norm.

13. A *conjugate-bilinear* function is a function f, defined on $V \times V$ for a complex vector space V, subject to the same conditions as an inner-product function. The function h, defined on a finite-dimensional complex space V so that $h(\alpha) = f(\alpha,\alpha)$, is an associated *hermitian function* with matrix $H = H^*$. The general value of h, expressed in terms of coordinates of α, relative to a basis of V, is called a *hermitian form*. Now prove the following:
    (a) If A is a complex matrix, both AA^* and A^*A are hermitian.
    (b) If H is a hermitian matrix, then $\bar{X}'HX$ is a real number, for any complex coordinate vector X of the appropriate length.
    (c) If H is a hermitian matrix, there exists a unitary matrix P such that $P^*HP = P^{-1}HP$ is a diagonal matrix that displays the eigenvalues of H along its diagonal.
    (d) Conjunctivity is an equivalence relation in the set of all $n \times n$ square complex matrices.
    (e) Any hermitian matrix of order n is conjunctive to one of the canonical matrices in the classes of real congruent matrices of order n.
    (f) A hermitian form in n variables is positive (definite) if and only if both its rank and signature are n, where all definitions are analogous to those in the real symmetric case.

14. Use the following guide to give another proof of the corollary to Theorem 3.1: If $c = a + bi$ is an eigenvalue of A, and $p(c) = 0$ is an eigenvalue of $B = p(A) = A^2 - 2aA + (a^2 + b^2)I = (A - aI)^2 + b^2I$ (see Prob. 18 of Sec. 5.5), there exists $X \, (\neq 0)$ in the appropriate coordinate space, such that $BX = 0$, and so $(BX,0) = 0$. Express (BX,X) as $(BX)'X$, replace B by the equivalent expression involving A, and then use $A' = A$ to conclude—after some reduction—that $b = 0$ and c is real.

15. If A and B are square matrices such that the indicated operations are defined, establish the following properties pertaining to adjoint matrices: (a) $(A + B)^* = A^* + B^*$; (b) $(cA)^* = \bar{c}A^*$, for any $c \in \mathbf{C}$; (c) $(AB)^* = B^*A^*$; (d) $(A^*)^* = A$.

16. If A and B are permutable unitary matrices, show that
$$\frac{1}{\sqrt{2}}\begin{bmatrix} A & -B \\ B^* & A^* \end{bmatrix}$$ is also a unitary matrix.

17. Prove that a square matrix of order n is unitary if and only if its columns are mutually orthogonal unit vectors of $V_n(\mathbf{C})$. Then conclude that these vectors comprise an orthonormal basis for the space.

18. If $A = \begin{bmatrix} 1 & 2 \\ -1 & -1 \end{bmatrix}$ is the matrix of a linear operator T, show that T is diagonalizable as an operator on $V_2(\mathbf{C})$ but not as an operator on $V_2(\mathbf{R})$.

19. Find the rank and nullity of an operator represented by each of the following matrices. If the matrix is nonsingular, find its inverse.

    (a) $\begin{bmatrix} 0 & 0 & i \\ 1 & i & i \\ 1 & 1 & i \end{bmatrix}$ (b) $\begin{bmatrix} 1 & -i & i \\ 0 & 0 & i \\ i & 1 & i \end{bmatrix}$ (c) $\begin{bmatrix} 0 & i & -i \\ -i & 0 & i \\ i & -i & 0 \end{bmatrix}$

20. Use elementary matrices to determine nonsingular matrices P and Q, such that PAQ is the equivalent canonical matrix, for each of the matrices A in Prob. 19.

21. (a) Find a basis for $V_3(\mathbf{C})$ which includes (i) $\{(1,1,1),(1,i,1)\}$; (ii) $\{(1,0,1),(i,1,1)\}$.
    (b) Use the Gram-Schmidt procedure and the bases in (a) to determine the associated orthonormal bases.

22. If the linear operator T is defined on $V_3(\mathbf{C})$ by $T(x_1,x_2,x_3) = (2x_1 + x_2, ix_2, x_2 - ix_3)$, find the matrix of T relative to each of the following bases: (a) $\{E_1,E_2,E_3\}$; (b) $\{(1,i,1),(i,0,-1),(i,1,i)\}$.

23. If $X = (x_1,x_2,x_3)$, determine the associated coordinate vector, relative to the basis in Prob. 22b, where (a) $X = (1,i,1)$; (b) $X = (i,1,-1)$.

24. Prove that $\det A^* = \overline{\det A}$.

25. Use the Cayley-Hamilton theorem to determine A^5, where
$$A = \begin{bmatrix} 1 - 2i & 1 - i \\ i & 0 \end{bmatrix}.$$

26. Prove that a linear operator T on a finite-dimensional unitary space V is self-adjoint if and only if $(T\alpha,\alpha)$ is real, for each $\alpha \in V$. [Hint: Use the polarization identity.]

27. If T is a self-adjoint operator on a unitary space V, prove that (a) $|\alpha + i(T\alpha)| = |\alpha - i(T\alpha)|$, for any $\alpha \in V$; (b) $I \pm iT$ is invertible; (c) $(I - iT)(I + iT)^{-1}$ is unitary, if V is finite-dimensional.

28. (a) Prove that the mapping U: $(x,y,z,t) \to \begin{bmatrix} t+x & y+zi \\ y-zi & t-x \end{bmatrix}$
    is an isomorphism of $V_4(\mathbf{R})$ into the real space of 2×2 hermitian matrices.
    (b) A *Lorentz transformation* is a linear operator on $V_4(\mathbf{R})$ that preserves $t^2 - x^2 - y^2 - z^2$, for any $(x,y,z,t) \in V_4(\mathbf{R})$. With U defined in (a), prove that $t^2 - x^2 - y^2 - z^2 = \det U(x,y,z,t)$.

6.4 The Spectral Theorem

We now take a final look at the problem under study in the first two sections of this chapter but in the context of complex vector spaces: If T is a linear operator on a complex vector space V of finite dimension, under what conditions is T diagonalizable, or, more precisely, what are necessary and sufficient conditions for the existence of an orthonormal basis of V that consists of eigenvectors of T? In the case of real vector spaces, a very satisfactory answer is provided by Theorem 2.3, and it is not difficult to check that this theorem is valid for vector spaces over **C**. However, there is another way in which to formulate the answer, and it is the purpose of this section to investigate this different approach, which will lead to the very famous result known as the *spectral theorem*. There are many ways to prove this theorem, each of which involves its own peculiar blend of algebra and geometry, and there is even an extension of the theorem to certain spaces of infinite dimension known as *Hilbert spaces*. One of the most elegant methods of proof for the finite-dimensional case involves a study of V as a "module" over a ring of polynomials in T. However, this would entail a development of too much background theory for our present study of complex vector spaces and would, in fact, prepare us to solve the associated problem in a much more general environment. It is not our intent to be concerned with the various ramifications of proof that exist for the theorem nor with its extension to spaces of infinite dimension. Rather, it will be our aim to obtain the theorem for complex spaces with a minimum of effort and by making maximal use of the foregoing developments.

One of the key results which we shall need here is the complex analog of Theorem 2.3, as mentioned above. We shall not repeat the proof of this theorem for the complex case but *shall assume its validity* and leave its verification to the reader (Prob. 13). It may be well, however, to recall some of the matters that should be checked: Each irreducible (i.e., linear) factor of the minimum polynomial of T on a complex space V divides its characteristic polynomial; each eigenvalue of T has at least one eigenvector in V that belongs to it; eigenvectors that belong to distinct eigenvalues of T are linearly independent; rank of T + nullity of T = dim V.

If $\{\alpha_i\}$ is an orthonormal basis of V such that $T\alpha_i = c_i\alpha_i$, for complex numbers c_i, $i = 1, 2, \ldots, n$, it is clear that the matrix of T, relative to this basis, is diag $\{c_1, c_2, \ldots, c_n\}$. Since the matrix of T*, relative to the same basis, is the adjoint of the matrix of T, it can be seen that this matrix of T* is diag $\{\bar{c}_1, \bar{c}_2, \ldots, \bar{c}_n\}$. If V is a *real* (euclidean) space, $\bar{c}_i = c_i$, $i = 1, 2, \ldots, n$, and so T* = T. In other words, if T is a linear operator on a euclidean space of finite dimension for which there exists an orthonormal basis of eigenvectors, then T *must be self-adjoint*. This conclusion does not follow for a unitary space V, but, even in this case, the fact that any two diagonal matrices of the same order commute permits the assertion that TT* = T*T. It may now be somewhat surprising that this condition of the commutativity of T with T* is not only necessary but also *sufficient* for the existence in V of an orthonormal basis of eigenvectors. We formalize this important property with a definition.

Definition *If* T *is a linear operator on a finite-dimensional unitary space, we say that* T *is normal if it commutes with its adjoint; that is,* TT* = T*T.

The following *matrix* definition should be quite natural, in view of the fact that the representation of T by a matrix A, *relative to an orthonormal basis*, implies that A^* is the corresponding matrix representation of T*.

Definition *A square matrix* A *is normal if* $AA^* = A^*A$.

It is clear that any self-adjoint (T* = T) or unitary (TT* = I) operator T is normal, as well as any scalar multiple of any normal operator. However, it is not true in general that sums and products of normal operators are normal. Before deriving some of the most useful properties of normal operators, we call attention to the fact that Theorem 6.4 (Chap. 4) remains true for unitary spaces, if (*b*) of the theorem is altered to read: $(cT)^* = \bar{c}T^*$, for $c \in \mathbf{C}$ (Prob. 14). In particular, if $T_1 = T_2 = T$ in (*c*), a simple induction (Prob. 15) on k shows that $(T^*)^k = (T^k)^*$, for any positive integer k.

Theorem 4.1 *The following are properties of any normal operator* T *on a unitary space* V, *with* α, β *arbitrary vectors in* V.
  (1) $|T\alpha| = |T^*\alpha|$.
  (2) *If* $T\alpha = c\alpha$, *then* $T^*\alpha = \bar{c}\alpha$.
  (3) *If* $T\alpha = c_1\alpha$ *and* $T\beta = c_2\beta$ ($c_1 \neq c_2$) *then* $(\alpha,\beta) = 0$.
  (4) *If* $(T^m)\alpha = 0$, *for* $\alpha \in V$ *and some integer* $m \geq 1$, *then* $T\alpha = 0$.
  (5) $g(T)$ *is normal for any complex polynomial* $g(x)$.

Proof
  (1) Since $(T\alpha, T\alpha) = (\alpha, T^*T\alpha) = (\alpha, TT^*\alpha) = (T^*\alpha, T^*\alpha)$, for any $\alpha \in V$, we conclude that $|T\alpha| = |T^*\alpha|$.
  (2) We assume that $(T - cI)\alpha = 0$. But $(T - cI)^* = T^* - \bar{c}I$, so that the normality of T implies that $(T - cI)(T - cI)^* = (T - cI)(T^* - \bar{c}I) = (T^* - \bar{c}I)(T - cI) = (T - cI)^*(T - cI)$, whence T − cI is a normal operator. But now, by Property (1), $0 = |(T - cI)\alpha| = |(T^* - \bar{c}I)\alpha|$, for any $\alpha \in V$, so that $(T^* - \bar{c}I)\alpha = 0$ and $T^*\alpha = \bar{c}\alpha$.

(3) It is a consequence of Property (2) that $c_1(\alpha,\beta) = (c_1\alpha,\beta) = (T\alpha,\beta) = (\alpha,T^*\beta) = (\alpha,\bar{c}_2\beta) = c_2(\alpha,\beta)$. Since $c_1 \neq c_2$, we conclude that $(\alpha,\beta) = 0$.

(4) We may suppose that m is the smallest positive integer such that $(T^m)\alpha = 0$, and let us make the tentative assumption that $m > 1$, so that $(T^{m-1})\alpha = \beta \neq 0$. But $T\beta = (T^m)\alpha = 0$, and Property (1) implies that $|T^*\beta| = |T\beta| = 0$. Hence $T^*\beta = 0$, and $0 = (T^*\beta, T^{m-2}\alpha) = (\beta, T(T^{m-2})\alpha) = (\beta,\beta)$, which implies that $\beta = 0 = (T^{m-1})\alpha$. This contradicts the minimal nature of m, and we conclude that $m = 1$ and hence $T\alpha = 0$.

(5) Let $g(x) = c_0 + c_1 x + \cdots + c_t x^t$, for complex numbers $c_1, c_2, \ldots, c_t$. Then, by the generalization of Theorem 6.4 (Chap. 4), as referred to above, $[g(T)]^* = (c_0 I + c_1 T + \cdots + c_t T^t)^* = \bar{c}_0 I + \bar{c}_1 T^* + \cdots + \bar{c}_t (T^*)^t$. But since T commutes with T^*, it now follows immediately that $g(T)$ commutes with $[g(T)]^*$ and so $g(T)$ is normal. ∎

The result we are seeking follows easily after we establish an elementary but crucial lemma.

Lemma *If $m(x)$ is the minimum polynomial of a normal operator on a finite-dimensional unitary space V, $m(x)$ may be decomposed into a product of distinct linear factors.*

Proof
Since any complex polynomial can be decomposed into a product of linear factors, we have merely to verify that the multiplicity of each factor of $m(x)$ is 1. Let us write $m(x) = p^s(x)Q(x)$, where $p(x)$ is any linear factor of $m(x)$ and $Q(x)$ is relatively prime to $p(x)$. We know that $p^s(T)Q(T) = m(T) = 0$ on V, so that $[p^s(T)Q(T)]\alpha = 0$, for any $\alpha \in V$. The operator $p(T)$ is normal, by either Property (5) or the *proof* of Property (2). Hence, if we use Property (4) as applied to $p(T)$ and the vector $[Q(T)]\alpha$, we see from the preceding equality that $[p(T)][Q(T)]\alpha = [p(T)Q(T)]\alpha = 0$. Since α is an arbitrary vector of V, it follows that $p(T)Q(T) = 0$, and the minimal nature of $m(x)$ requires that $s = 1$. ∎

We are now in a position to state the spectral theorem, which is the ultimate characterization of diagonalizable operators.

Theorem 4.2 (Spectral Theorem) *Let T be a linear operator on a unitary space V of dimension n, with eigenvalues $c_1, c_2, \ldots, c_n$ (possibly not all distinct). Then there exists an orthonormal basis (of eigenvectors) of V, relative to which T is represented by diag $\{c_1, c_2, \ldots, c_n\}$, if and only if T is normal.*

Proof
The proof is essentially already completed. If an orthonormal basis of eigenvectors exists, we noted near the beginning of this section that T is normal.

On the other hand, if T is normal, it follows from the lemma and the complex version of Theorem 2.3 that T is diagonalizable; and any diagonal matrix representation of T must exhibit along its diagonal all the eigenvalues of T with their proper multiplicities. As in the proof of Theorem 2.3, we see that $V = V_1 \oplus V_2 \oplus \cdots \oplus V_k$, where V_i is the null space of the operator $T - c_i I$, $i = 1, 2, \ldots, k$, with $c_1, c_2, \ldots, c_k$ the *distinct* eigenvalues of T. It is a consequence of Property (3) in Theorem 4.1 that V_i and V_j are orthogonal, for $i \neq j$. Thus, if we select an orthonormal basis for V_i, $i = 1, 2, \ldots, k$, the collection of all these basis vectors constitutes a basis for V with the desired property. ∎

Since an $n \times n$ complex matrix is unitary if and only if both its rows and columns are (Prob. 16) orthonormal sets of vectors in $V_n(\mathbf{C})$, the following corollary is immediate.

Corollary 1   *For any normal matrix A, there exists a unitary matrix C such that C^*AC is a diagonal matrix, the diagonal entries of which are the eigenvalues of A occurring with their correct multiplicities.*

We know that the eigenvalues of a hermitian matrix are real, and so, for this special kind of normal matrix, Corollary 1 takes the following form.

Corollary 2   *For any hermitian matrix A, there exists a unitary matrix C such that C^*AC is a real diagonal matrix, the diagonal entries of which are the eigenvalues of A occurring with their correct multiplicities.*

The spectral theorem is often stated as the resolution of a normal operator into a sum of orthogonal projections. In general, a *projection* is an *idempotent* linear operator (i.e., it is equal to its square); an *orthogonal* projection has the additional property that its null space is the orthogonal complement of its range. There is a very natural way to define orthogonal projections on a vector space V, if V has a direct sum decomposition, as occurred in the proof of the spectral theorem. If $V = V_1 \oplus V_2 \oplus \cdots \oplus V_k$, for orthogonal subspaces $V_1, V_2, \ldots, V_k$ of V, any $\alpha \in V$ can be expressed in the form $\alpha = \alpha_1 + \alpha_2 + \cdots + \alpha_k$, with α_i some *unique* vector in V_i ($i = 1, 2, \ldots, k$) dependent on α. If we now define a mapping E_i of V, so that $E_i \alpha = \alpha_i$, it is clear that E_i is linear (Prob. 17), $i = 1, 2, \ldots, k$, and that these k mappings are orthogonal projections. In fact, it is immediate that the range of E_i is V_i, and its null space is $V_1 \oplus \cdots \oplus V_{i-1} \oplus V_{i+1} \oplus \cdots \oplus V_k$. The most important part of Theorem 4.2 can now be stated in the following form, but we leave the details of the proof to the reader as Prob. 18. (See also Prob. 29.)

Theorem 4.3 (Spectral Theorem)
Let T *be a normal operator on a unitary space of finite dimension, with* $c_1, c_2, \ldots, c_k$ *the distinct eigenvalues of* T. *Then, if* E_j *is the orthogonal projection of* T *on the subspace of eigenvectors belonging to* c_j ($j = 1, 2, \ldots, k$), *the following hold:* (1) $T = c_1 E_1 + c_2 E_2 + \cdots + c_k E_k$; (2) $E_1 + E_2 + \cdots + E_k = 1$; (3) $E_i E_j = 0$, $i \neq j$.

The set of distinct eigenvalues of a linear operator is called its *spectrum*, and the decomposition expressed in (1) above, subject to (2) and (3), is called the *spectral resolution* of the normal operator T. It is an interesting application of Lagrange polynomials that the projections in this spectral resolution are completely determined by T and its spectrum, and so the spectral resolution of T is unique. The Lagrange polynomials, associated with the numbers $c_1, c_2, \ldots, c_k$, are $p_j(x)$, where

$$p_j(x) = \frac{(x - c_1) \cdots (x - c_{j-1})(x - c_{j+1}) \cdots (x - c_k)}{(c_j - c_1) \cdots (c_j - c_{j-1})(c_j - c_{j+1}) \cdots (c_j - c_k)}$$
$$j = 1, 2, \ldots, k$$

These polynomials have many interesting properties (see Prob. 21), but the one of present interest is that $p_j(c_i) = \delta_{ij}$. If $T = c_1 E_1 + c_2 E_2 + \cdots + c_k E_k$ is the spectral resolution of T, as described in the theorem, it is a simple application (Prob. 19) of the properties listed for $E_1, E_2, \ldots, E_k$ that $g(T) = \sum_{j=1}^{k} g(c_j) E_j$, for any polynomial $g(x)$. We now use the polynomial $p_j(x)$ and see that $p_j(T) = \sum_{i=1}^{k} p_j(c_i) E_i = \sum_{i=1}^{k} \delta_{ij} E_i = E_j$, as asserted above. This shows not only that E_j is determined by T and its spectrum but that E_j is a well-defined *polynomial* in the operator T: *in order to obtain the projections that occur in the spectral resolution of* T, *we have only formally to replace* x *by* T *in each of the Lagrange polynomials associated with the spectrum of* T.

There is another sense in which we may speak of the "uniqueness" of the spectral resolution of T. If $c_1, c_2, \ldots, c_k$ are distinct complex numbers, and $E_1, E_2, \ldots, E_k$ are nonzero linear operators on V such that (1), (2), and (3) of Theorem 4.3 are satisfied, then the complex numbers are the eigenvalues of T, and E_i is the orthogonal projection of T on the subspace of eigenvectors that belong to c_i, $i = 1, 2, \ldots, k$. It should be noted that here we are not *assuming* that $E_1, E_2, \ldots, E_k$ are orthogonal projections but merely linear operators. The proof of this "uniqueness" result is left to the reader as Prob. 30.

Problems 6.4

1. Let T be a normal operator on a unitary space of dimension 4. Give the spectral resolution of T and list the diagonal matrices, one of which will represent T relative to some orthonormal basis, if (a) $\{1, 2, -1\}$ is the spectrum of T; (b) the minimum polynomial of T is $(x^2 + 1)(x - 2)$.

2. Give two arguments, one without any computational use of matrices, to prove that if A is a normal matrix so is $P^{-1}AP$ for any unitary matrix P.

3. If $A = \begin{bmatrix} 1 & 2 \\ 2 & 1 \end{bmatrix}$, determine A^{25} without laborious matrix computation. [Hint: Use similarity and the eigenvalues of A.]

4. Decide if the operator T on the indicated complex coordinate space is normal: (a) $T(x,y) = (x + iy, -ix + y)$; (b) $T(x,y) = (x + (1 + i)y, (1 - i)x + y)$; (c) $T(x,y,z) = (y, z, ix + iy + iz)$; (d) $T(x,y,z) = (x + iy, -ix + y + (1 - i)z, (1 + i)y + z)$.

5. If $T = E_1 + iE_2 - iE_3$ is the spectral resolution of a normal operator on a vector space V, describe the associated decomposition of V.

6. Use the definition to find a 2×2 *normal* matrix (a) all of whose entries are real; (b) not all of whose entries are real.

7. Find and compare the eigenvalues of each normal matrix and its adjoint, as discovered in Prob. 6. [See Property (2) of Theorem 4.1.]

8. If $\{1, i, 1 + i\}$ is the spectrum of a normal operator T on a unitary space V of dimension 4, find the minimum polynomials of both T and T*. Write the spectral resolution of each operator.

9. Discover a normal matrix of order $n > 1$ that is neither hermitian nor the negative of a hermitian matrix.

10. Show that $A = \begin{bmatrix} 1 & i \\ -i & 1 \end{bmatrix}$ and $B = \begin{bmatrix} 0 & i \\ i & 0 \end{bmatrix}$ are normal matrices, but that neither $A + B$ nor AB is normal.

11. If T is any linear operator on a finite-dimensional vector space V, and a, b are complex numbers of equal absolute values, show that $aT + bT^*$ is normal. [Hint: Use the revision of (b) of Theorem 6.4 (Chap. 4), as stated just prior to Theorem 4.1.]

12. Use the result in Prob. 11 to find a 3 × 3 normal matrix, not all the entries of which are real (cf. Prob. 6).

13. Check the details needed to extend the validity of Theorem 2.3 to operators on complex spaces.
14. Prove the modified extension of Theorem 6.4 (Chap. 4) to operators on a unitary space. [*Hint:* Apply the defining property of an adjoint to $((T_1 + T_2)\alpha, \beta)$ and $((T_1 T_2)\alpha, \beta)$.]
15. If T is a linear operator on a finite-dimensional unitary space, use Prob. 14 and induction to prove that $(T^*)^k = (T^k)^*$, for any integer $k > 0$.
16. Prove that an $n \times n$ complex matrix is unitary if and only if its rows and columns are both orthonormal sets of vectors of $V_n(\mathbf{C})$.
17. If $V = V_1 \oplus V_2 \oplus \cdots \oplus V_k$ is a decomposition of a vector space V into a direct sum of subspaces, each $\alpha \in V$ may be uniquely expressed as $\alpha = \alpha_1 + \alpha_2 + \cdots + \alpha_k$, with $\alpha_i \in V_i$, $i = 1, 2, \ldots, k$. Then prove that the mapping E_i, defined so that $E_i \alpha = \alpha_i$, is linear for each i.
18. Show that Theorem 4.3 is a direct consequence of Theorem 4.2.
19. If the normal operator T has the spectral resolution described in Theorem 4.3, prove that $g(T) = \sum_{j=1}^{k} g(c_j) E_i$, for any polynomial $g(x)$. [*Hint:* First see that $T^2 = \sum_{j=1}^{k} c_j^2 E_j$; then extend the argument to T^n and finally to $g(T)$.]

20. Show that any self-adjoint operator T on a finite-dimensional vector space V may be expressed in the form $T = P_1 - P_2$, where P_1 and P_2 are positive operators on V. [*Note:* A *positive* operator P on any inner-product space V has the property that $(P\alpha, \alpha) > 0$, for any $\alpha \ (\neq 0) \in V$.]
21. Prove that the Lagrange polynomials, associated with k distinct complex numbers, form a basis for the space of polynomials of degree not greater than $k - 1$. [*Hint:* For any polynomial $f(x)$ in the space, consider the number of zeros of $f(x) - f(c_1)p_1(x) - f(c_2)p_2(x) - \cdots - f(c_k)p_k(x)$, where $c_1, c_2, \ldots, c_k$ are the given complex numbers.]
22. Each of the following matrices A may be used to represent a normal operator T on a coordinate space, relative to an ortho-

normal basis. In each case, determine the spectral resolution of T, and find the associated decomposition of A as a linear combination of matrices of orthogonal projections.

(a) $A = \begin{bmatrix} 1 & 2 \\ 2 & 1 \end{bmatrix}$ (b) $A = \begin{bmatrix} 1 & i \\ -i & 1 \end{bmatrix}$

(c) $A = \begin{bmatrix} 1 & i & i \\ -i & -1 & 2 \\ -i & 2 & 1 \end{bmatrix}$

23. For each of the following matrices A, determine a unitary matrix C, such that C^*AC is diagonal:

(a) $A = \begin{bmatrix} 1 & 1+i \\ 1-i & 2 \end{bmatrix}$ (b) $A = \begin{bmatrix} i & 1+i \\ -1+i & 2i \end{bmatrix}$

24. Prove that a linear operator T on a unitary space V is normal if and only if $|T\alpha| = |T^*\alpha|$, for every $\alpha \in V$. [See (1) of Theorem 4.1.]

25. If T is a normal operator on a unitary space, prove the following:
    (a) T is hermitian if and only if its spectrum contains only real numbers.
    (b) T is unitary if and only if the absolute value of every number in its spectrum is 1.
    (c) T is a projection if and only if its spectrum is either $\{0\}$, $\{1\}$, or $\{0,1\}$.
    (d) T $= 0$ if and only if its spectrum is $\{0\}$.

26. If T is a normal operator such that $T^k = 0$, for some integer $k > 0$, prove that T $= 0$.

27. If T is a self-adjoint operator such that $T^k = $ I, for some integer $k > 0$, prove that $T^2 = $ I.

28. If A and B are hermitian matrices of the same order, prove that $A + iB$ is normal if and only if $AB = BA$.

29. With $p_j(x)$, $j = 1, 2, \ldots, k$, the Lagrange polynomials associated with the spectrum $c_1, c_2, \ldots, c_k$ of a normal operator T on a unitary space V of finite dimension, define $E_j(T) = p_j(T)$, and obtain an alternative proof of Theorem 4.3 by proving the following:
    (a) If $f(x) = 1$ and then $f(x) = x$ in Prob. 21, the result in Prob. 19 can be used to obtain (1) and (2) of the theorem.
    (b) $E_j \neq 0$, for each j.
    (c) $E_i E_j = 0$, for $i \neq j$.
    (d) Each E_j is a projection on V.
    (e) Each E_j is orthogonal. [*Hint:* Use $E_j = p_j(T)$ to see that E_j is normal; show that any normal projection is self-adjoint; show that any self-adjoint projection is orthogonal.]

(f) The subspace of eigenvectors belonging to c_j is the range of E_j.

30. If $c_1, c_2, \ldots, c_k$ are merely distinct complex numbers, and $E_1, E_2, \ldots, E_k$ are nonzero linear operators subject to (1), (2), (3) of Theorem 4.3, prove that $\{c_1, c_2, \ldots, c_k\}$ is the spectrum of the normal operator T. [*Hint:* With $\alpha \, (\neq 0) \in R_{E_j}$, show that $T\alpha = c_j \alpha$; with c an eigenvalue of T and α in the null space of $T - cI$, show that $E_i[(T - cI)\alpha] = (c_i - c)E_j\alpha = 0$]. The arguments in (c), (d), (e), (f) of Prob. 29 may now be repeated to obtain the decomposition of T asserted in Theorem 4.3. This is the "uniqueness" result, to which reference was made in the paragraph just prior to the beginning of this set of problems.

6.5 Invariant Subspaces and Primary Decomposition

It is an important observation, based on the results in Sec. 6.4, that each similarity equivalence class of *normal* matrices of a given order contains a diagonal matrix, the diagonal entries of which are the eigenvalues of the matrices in the class. Moreover, in this matrix, each eigenvalue occurs as a diagonal entry as many times as its multiplicity in the associated characteristic equation. If we prescribe some order for the occurrence of the diagonal entries (say in increasing natural order) this diagonal matrix becomes uniquely defined and so may be regarded as the *canonical* member of the class. The invariants of any one of these similarity classes are then the *order* of the matrices and the set of *eigenvalues*, along with their *multiplicities*, since these are the same for all matrices of the class. Our basic question on *canonical* matrices and *invariants* under similarity has then been answered for the case of *normal* operators and matrices.

It would be unreasonable to expect that all linear operators might be diagonalizable, for, if so, the significance of a normal operator would evaporate. A fundamental criterion for the normality of an operator was supplied in Theorem 2.3 for the case of real vector spaces, and the validity of the same result was noted in Sec. 6.4 for complex spaces. There are then two circumstances under which an operator will fail to be diagonalizable: in the real case, if not all the zeros of the minimum polynomial are real; in either the real or complex case, if not all the irreducible factors of this polynomial are distinct. For example, if a linear operator T is represented by the matrix

$$A = \begin{bmatrix} 2 & 1 & 0 \\ 0 & 2 & 0 \\ 0 & 0 & -1 \end{bmatrix}$$

it is easy to check that the minimum polynomial of T is $(x - 2)^2(x + 1)$, and, since the multiplicity of the factor $x - 2$ is 2, the operator cannot be diagonalized. It may be noted that *if* T were diagonalizable, the null space of T − 2I would of necessity have dimension 2, whereas it is easy to check (Prob. 11) that the dimension of this space is 1. In other words, it is impossible to represent T by a diagonal matrix because there are not enough linearly independent eigenvectors to form a basis for the space V on which T operates. It may be discovered (Prob. 11), however, that if a basis for the null space of $(T - 2I)^2$ is combined with a basis for the null space of T + I, the result is a basis for V. It is one of the major results in this section to prove that this situation is not special but is illustrative of a very general result.

Even if it is known that an operator is not diagonalizable, it may still be possible to represent it by a matrix that is in some sense nearly diagonal. It is often useful to let diag $\{A_1, A_2\}$ denote the partitioned matrix $\begin{bmatrix} A_1 & 0 \\ 0 & A_2 \end{bmatrix}$, in which A_1 and A_2 are square submatrices, while the notation diag $\{A_1, A_2, \ldots, A_k\}$ is a meaningful description for a more general partitioned matrix. (A matrix partitioned in this way is often called a *direct sum* of the matrices $A_1, A_2, \ldots, A_k$ and is denoted by $A_1 \oplus A_2 \oplus \cdots \oplus A_k$. However, since this is not the usual kind of matrix sum, we shall not use this language or notation.) It is clear that, for a given matrix $A = $ diag $\{A_1, A_2, \ldots, A_k\}$ of order n, the larger $k(\leq n)$, the more nearly A comes to being a diagonal matrix. In the limiting case, when $k = n$, A is in fact diagonal. Our problem then is to represent a given linear operator by a partitioned matrix of near-diagonal type, with a maximal number of "diagonal" submatrices. The solution of any matrix representation problem lies in the possibility of finding an appropriate basis for the underlying vector space, and the following concept is basic in this investigation.

Definition *If* T *is a linear operator on a vector space* V, *a subspace* W *of* V *is said to be invariant under* T (*or* T-*invariant*) *if* $T\alpha \in W$, *for an arbitrary* $\alpha \in W$.

The characteristic property of a T-invariant subspace is that every vector in the subspace is mapped by T onto a vector in the *same* subspace. The range and zero subspaces are both T-invariant for any operator T, but there are often other *nontrivial* T-invariant subspaces.

The importance of T-invariant subspaces in our present study is the result of the effect they produce on matrix representations of T. Suppose a vector space V of dimension n has the decomposition $V = W_1 \oplus W_2$, for nontrivial T-invariant subspaces W_1 and W_2. If we now obtain a basis for V by combining a basis for W_1 with a basis for W_2, the matrix of T *relative to this basis*, will have the form diag $\{A_1, A_2\}$, where the orders of A_1 and A_2 are the respective dimensions of W_1 and W_2. (See Prob. 12.) The extension of this result to a general decomposition of V as a direct sum of T-invariant subspaces is evident: If $V = W_1 \oplus W_2 \oplus \cdots \oplus W_k$, where $W_1, W_2, \ldots, W_k$ are T-invariant subspaces of V, it is possible to represent T by a matrix diag $\{A_1, A_2, \ldots, A_k\}$, where the order of A_i is equal to the dimension of W_i, $i = 1, 2, \ldots, k$. The following theorem, which is valid for vector spaces over any field, provides us with a decomposition of this type.

Theorem 5.1 (*Primary Decomposition Theorem*)
Let $m(x) = p_1^{r_1}(x) p_2^{r_2}(x) \cdots p_k^{r_k}(x)$ be the minimum polynomial of a linear operator T on a finite-dimensional vector space V, where the $p_i(x)$ are distinct irreducible polynomials and the r_i are positive integers. Then, if W_i is the null space of $p_i^{r_i}(T)$, $i = 1, 2, \ldots, k$, the following hold:
  (1) $V = W_1 \oplus W_2 \oplus \cdots \oplus W_k$.
  (2) *W_i is invariant under T.*
  (3) *If T_i is the operator induced on W_i by T, then $p_i^{r_i}(x)$ is the minimum polynomial of T_i.*

Proof
Since the decomposition asserted in (1) implies the existence of a projection of V onto each of the subspaces $W_1, W_2, \ldots, W_k$, it should not be surprising to find that our method of proof has some resemblance to that of the spectral theorem. We first define $f_i(x) = m(x)/p_i^{r_i}(x) = \prod_{j \neq i} p_j^{r_j}(x)$, and it is easily seen that $f_1, f_2, \ldots, f_k$ are relatively prime polynomials. It then follows, from what is loosely referred to as the *euclidean algorithm* for polynomial rings, that there exist polynomials $g_1(x), g_2(x), \ldots, g_k(x)$ such that $f_1(x)g_1(x) + f_2(x)g_2(x) + \cdots + f_k(x)g_k(x) = 1$. It will now be shown that the polynomials $h_i(x) = f_i(x)g_i(x)$, $i = 1, 2, \ldots, k$, provide the desired projections, just as the Lagrange polynomials do in the case of a normal operator.

We define the mappings E_i of V, so that $E_i = h_i(T)$ for $i = 1, 2, \ldots, k$. Since $h_1(x) + h_2(x) + \cdots + h_k(x) = 1$, and the minimum polynomial $m(x)$ divides $f_i(x)f_j(x)$ $(i \neq j)$, the following equalities hold:

$$E_1 + E_2 + \cdots + E_k = I$$
$$E_i E_j = 0 \quad i \neq j$$

It is an easy consequence of these two properties (Prob. 13) that T_i is a projection, and we shall show that its range R_{E_i} is precisely the subspace W_i described in the theorem. If $\alpha \in R_{E_i}$, then $\alpha = E_i \alpha$; and, since $m(x)$

divides $p_i^{r_i}(x)f_i(x)g_i(x)$, $[p_i^{r_i}(\mathsf{T})]\alpha = [p_i^{r_i}(\mathsf{T})\mathsf{E}_i]\alpha = [p_i^{r_i}(\mathsf{T})f_i(\mathsf{T})g_i(\mathsf{T})]\alpha = 0$. Hence $R_{\mathsf{E}_i} \subset W_i$. Conversely, suppose α is in the null space of $p_i^{r_i}(\mathsf{T})$. If $j \neq i$, then $p_i^{r_i}(x)$ divides $f_j(x)g_j(x)$, so that $\mathsf{E}_j\alpha = [f_j(\mathsf{T})g_j(\mathsf{T})]\alpha = 0$, $\alpha \in R_{\mathsf{E}_i}$, and $W_i \subset R_{\mathsf{E}_i}$. Hence $R_{\mathsf{E}_i} = W_i$, $i = 1, 2, \ldots, k$. Since $V = IV = (\mathsf{E}_1 + \mathsf{E}_2 + \cdots + \mathsf{E}_k)V = \mathsf{E}_1 V + \mathsf{E}_2 V + \cdots + \mathsf{E}_k V$, and the properties of $\mathsf{E}_1, \mathsf{E}_2, \ldots, \mathsf{E}_k$ imply (Prob. 13) that this sum is direct, the decomposition in (1) is established.

Since $[p_i^{r_i}(\mathsf{T})]\mathsf{T}\alpha = \mathsf{T}[p_i^{r_i}(\mathsf{T})]\alpha$, it is clear that $\mathsf{T}\alpha \in W_i$ if $\alpha \in W_i$, so that each of the subspaces W_i is T-invariant, as asserted in (2). Finally, the fact that $p_i^{r_i}(\mathsf{T})$ is the zero operator on W_i implies that the minimum polynomial of the operator T_i induced on W_i by T divides $p_i^{r_i}(x)$. On the other hand, if $g(x)$ is any polynomial such that $g(\mathsf{T}_i) = 0$, it follows that $g(\mathsf{T})f_i(\mathsf{T}) = 0$. Hence $m(x)$, which may be expressed in the form $p_i^{r_i}(x)f_i(x)$, must divide $g(x)f_i(x)$, and so $p_i^{r_i}(x)$ divides $g(x)$. The minimum polynomial of T_i must then be $p_i^{r_i}(x)$, and the proof of the theorem is complete. ∎

In case V is a complex space, each of the irreducible polynomials $p_i(x)$ has the form $x - c_i$, for some $c_i \in \mathbf{C}$, and in this case we are able to obtain more quantitative information about the decomposition.

Corollary *If V is a vector space over $\mathbf{C}$, and $m(x) = (x - c_1)^{r_1}(x - c_2)^{r_2} \cdots (x - c_k)^{r_k}$, with $c_1, c_2, \ldots, c_k$ distinct complex numbers, then these numbers make up the spectrum of T and $\dim W_i$ is equal to the multiplicity of c_i ($i = 1, 2, \ldots, k$) in the characteristic polynomial of T.*

Proof
Let A_i be a matrix that represents T_i in (3) of Theorem 5.1 so that T may be represented by diag $\{A_1, A_2, \ldots, A_k\}$. The corollary then follows by noting that the characteristic polynomial of the latter matrix is the product of the characteristic polynomials of the matrices $A_1, A_2, \ldots, A_k$. ∎

After the primary decomposition of a vector space V, relative to a linear operator T, has been effected, it would be now only natural to examine the primary components for their possible decomposition into direct sums of T-invariant subspaces of even smaller dimensions. It is quite possible to do this, but a thorough investigation along these lines is quite tedious and we prefer to omit it. However, we shall include one very useful matrix form for the representation of a linear operator, and this matrix will be associated with an "ultimate" decomposition of the vector space. Since the vector spaces to be studied in this connection are complex, *we shall assume complex scalars for all vector spaces from this point on in the text.*

The concept, which will provide the connecting link between normal or diagonalizable operators and the Jordan canonical form to be studied in Sec. 6.6, is provided in the following matrix-operator definition.

Definition *A square matrix A (linear operator T on a vector space V) is said to be nilpotent if $A^r = O$ ($T^r = 0$) for some positive integer r.*

We continue to use the notation of Theorem 5.1, including the corollary, and assume that V is a finite-dimensional complex vector space. Let us define the operator D_T on V so that $D_T = c_1 E_1 + c_2 E_2 + \cdots + c_k E_k$. It is easy to verify (Prob. 14) that every vector in the range of E_i is in the null space of $D_T - c_i I$, so that the nonzero vectors in this range are eigenvectors of D_T. Since $I = E_1 + E_2 + \cdots + E_k$, these eigenvectors span V, and a linearly independent subset forms a basis. Hence D_T is a diagonalizable operator. We now examine the operator $N_T = T - D_T$, where we may write $T = TE_1 + TE_2 + \cdots + TE_k$, and $D_T = c_1 E_1 + c_2 E_2 + \cdots + c_k E_k$. Then $N_T = (T - c_1 I)E_1 + (T - c_2 I)E_2 + \cdots + (T - c_k I)E_k$, and the properties of the projections E_i allow us (Prob. 14) to write

$$N_T^r = (T - c_1 I)^r E_1 + (T - c_2 I)^r E_2 + \cdots + (T - c_k I)^r E_k$$

for any integer $r > 0$. If $r \geq \max \{r_1, r_2, \ldots, r_k\}$, then $(T - c_i I)^r$ maps the range of E_i onto 0, for each $i = 1, 2, \ldots, k$, and so $N_T^r = 0$. Hence N_T is nilpotent, and we have established the following result.

Theorem 5.2 *Any linear operator T on a finite-dimensional complex vector space can be expressed in the form $T = D_T + N_T$, where D_T is diagonalizable and N_T is nilpotent.*

If T is a normal operator, the nilpotent part N_T of T is, of course, 0. On the other hand, for an operator T that is not normal, $N_T \neq 0$, and, in this case, the nilpotent part may be regarded in some sense as a measure of the divergence of T from normality. In any event, this theorem reduces the study of linear operators on a finite-dimensional complex space to a study of nilpotent operators. After a brief look at these operators in the first part of Sec. 6.6, we shall be able to obtain the desired canonical form and to make our final remarks on the subject of invariants.

EXAMPLE

The primary decomposition theorem remains valid (Prob. 17), insofar as statements (1) and (2) are concerned, even if the space V is not of finite dimension and for *any* monic polynomial $m(x)$ such that $m(T) = 0$ for a linear operator T. A familiar instance of this occurs in the solution of a differential equation of the form $d^n f/dx^n + a_{n-1} d^{n-1} f/dx^{n-1} + \cdots + a_1 df/dx + a_0 f = 0$, for some positive integer n and in the space of infinitely differentiable complex functions. In this case, $m(x) = x^n + a_{n-1}x^{n-1} + \cdots + a_1 x + a_0 = (x - c_1)^{r_1}(x - c_2)^{r_2} \cdots (x - c_k)^{r_k}$, for distinct $c_1, c_2, \ldots, c_k \in \mathbf{C}$ and positive integers $r_1, r_2, \ldots, r_k$. According to the decomposition theorem, as generalized, the solution of the differential equation is a matter of solving $(D - c_i 1)^{r_i} f = 0$, for each $i = 1, 2, \ldots, k$, with D the differentiation operator. If W_i is the null space of $(D - c_i 1)^{r_i}$, the complete solution space is $W = W_1 \oplus W_2 \oplus \cdots \oplus W_k$. Each solution f of the given equation is then expressible uniquely in the form $f = f_1 + f_2 + \cdots + f_k$, where $f_i \in W_i$, $i = 1, 2, \ldots, k$. If it is recalled that the general solution of an equation of the form $(D - c1)^r f = 0$ is $e^{ct}(b_0 + b_1 t + \cdots + b_{r-1} t^{r-1})$, where t is a complex variable, it is easy to obtain the general solution of the original equation. Moreover, it may be observed that the dimension of the solution space is finite and equal to the degree of the polynomial $m(x)$ (Prob. 16).

Problems 6.5

1. If $A_1 = [2]$, $A_2 = \begin{bmatrix} 1 & 1 \\ 2 & -1 \end{bmatrix}$, and $A_3 = \begin{bmatrix} 1 & 0 \\ -1 & 2 \end{bmatrix}$, give the usual matrix form of (a) diag $\{A_1, A_1, A_3\}$; (b) diag $\{A_2, A_2, A_3\}$; (c) diag $\{A_1, A_2, A_3\}$.

2. Each of the following matrices represents a linear operator T on a vector space V, relative to some basis. Determine the bases of the T-invariant subspaces that effect the primary decomposition of V. (a) $\begin{bmatrix} 2 & 0 \\ -1 & 2 \end{bmatrix}$; (b) $\begin{bmatrix} 1 & 0 \\ -1 & 2 \end{bmatrix}$; (c) $\begin{bmatrix} 1 & 3 \\ 2 & 2 \end{bmatrix}$.

3. Apply the directions given in Prob. 2 to each of the following matrices:

    (a) $\begin{bmatrix} 1 & 0 & 1 \\ -1 & 1 & 0 \\ 0 & 0 & 2 \end{bmatrix}$ (b) $\begin{bmatrix} 1 & 1 & -2 \\ 0 & 1 & 0 \\ -2 & 1 & 1 \end{bmatrix}$

4. If P_n is the vector space of all complex polynomials of degree less than n, prove that the differentiation operator D on P_n is nilpotent.

5. Verify that each of the following matrices is nilpotent:

    (a) $\begin{bmatrix} 0 & 1 & -1 \\ 0 & 0 & 1 \\ 0 & 0 & 0 \end{bmatrix}$ (b) $\begin{bmatrix} 0 & 0 & 0 \\ 1 & 0 & 0 \\ 2 & -2 & 0 \end{bmatrix}$

6. Discover all 2 × 2 matrices with complex entries such that $A^2 = O$.
7. Find a nilpotent 3 × 3 matrix A such that (a) $A^3 = O$, but $A^2 \neq O$; (b) $A^2 = O$, but $A \neq O$.
8. Show that $\mathcal{L}\{(1,1,1),(1,2,1)\}$ is a T-invariant subspace of $V_3(\mathbf{R})$, where (a) $T(x,y,z) = (x+y-z, \; x+y, \; x+y-z)$; (b) $T(x,y,z) = (-2x+y+2z, \; -4x+3y+2z, \; -5x+y+5z)$.
9. Let $T = D_T + N_T$, where D_T and N_T are, respectively, the diagonalizable and nilpotent portions of the operator T on $V_3(\mathbf{C})$. If the matrix representations of D_T and N_T, relative to the $\{E_i\}$ basis, are

$$\begin{bmatrix} 1 & 0 & 0 \\ 0 & 2 & 0 \\ 0 & 0 & -1 \end{bmatrix} \quad \text{and} \quad \begin{bmatrix} 0 & 1 & -1 \\ 0 & 0 & 2 \\ 0 & 0 & 0 \end{bmatrix}$$

respectively, determine TX, where (a) $X = (1,2,1)$; (b) $X = (3,-1,0)$; (c) $X = (1,1,-2)$.
10. Use the method indicated in the example to find the complete solution of each of the following differential equations: (a) $(D^2 - 2D + I)f = 0$; (b) $(D^2 - 3D - 2I)f = 0$; (c) $(D - I)(D^2 + I)f = 0$. [Caution: I is the identity operator.]

11. With reference to the discussion in the second paragraph of this section, check that (a) the dimension of the null space of $T - 2I$ is 1; (b) a basis for the null space of $(T - 2I)^2$, when combined with a basis for the null space of $T + I$, makes up a basis for V. [See Caution in Prob. 10.]
12. Let $V = W_1 \oplus W_2$ be a decomposition of a vector space V in terms of subspaces W_1 and W_2, that are invariant under some linear operator T on V. Then explain why, if a basis for V is obtained by combining bases for W_1 and W_2, the associated representation matrix of T has the form diag $\{A_1, A_2\}$, where the orders of A_1 and A_2 are, respectively, the dimensions of W_1 and W_2.
13. With reference to the proof of Theorem 5.1, explain why the two equalities involving $E_1, E_2, \ldots, E_k$ imply that (a) E_i is a projection, for $i = 1, 2, \ldots, k$; (b) the sum $E_1 V + E_2 V + \cdots + E_k V$ is direct.
14. Refer to the paragraph just prior to the statement of Theorem 5.2 and prove the following:
    (a) Every vector in the range of E_i is in the null space of $D_T - c_i I$.
    (b) $N_T^r = (T - c_1 I)^r E_1 + (T - c_2 I)^r E_2 + \cdots + (T - c_k I)^r E_k$.

15. If a linear operator T is expressed in the form $T = D_T + N_T$, where D_T and N_T are, respectively, diagonalizable and nilpotent linear operators, explain why $D_T N_T = N_T D_T$.
16. Check the statement concerning the dimension of the solution space in the example by exhibiting a basis for the space.
17. Check that the proof of statements (1) and (2) in Theorem 5.1 is independent of whether V is of finite dimension or $m(x)$ is the *minimum* polynomial such that $m(T) = 0$.

18. The minimum polynomial of

$$A = \begin{bmatrix} 3 & -1 & -1 & -2 \\ 1 & 1 & -1 & -1 \\ 1 & 0 & 0 & -1 \\ 0 & -1 & 1 & 1 \end{bmatrix}$$

is known to be $(x - 1)^2(x - 2)$. If A represents a linear operator T on $V_4(\mathbf{C})$, relative to the $\{E_i\}$ basis, determine a basis for this space by combining bases for the null spaces of $(T - I)^2$ and $T - 2I$. Finally, use this basis to obtain a nonsingular matrix P such that $P^{-1}AP$ is in the form diag $\{A_1, A_2\}$, for submatrices A_1, A_2. [See Caution in Prob. 10.]

19. Prove that the sum and product of commutative nilpotent operators are nilpotent.
20. If T is a self-adjoint operator on a vector space V, and W is a T-invariant subspace of V, prove that the orthogonal complement of W is also T-invariant.
21. Use nilpotent matrices to illustrate the fact that nonsimilar matrices may have the same characteristic and minimum polynomials.
22. Describe the subspace of V_2 that is invariant under the operator T where (a) T is a rotation about the origin; (b) T is a reflection in the x axis.
23. Prove that a linear operator T is (a) singular if and only if 0 is an eigenvalue of T; (b) nilpotent if and only if 0 is the only eigenvalue of T.
24. If T is nilpotent so that $T^k = 0$, but $T^{k-1} \neq 0$, prove that $I + T$ is invertible and $(I + T)^{-1} = I - T + T^2 - T^3 + \cdots + (-1)^{k-1}T^{k-1}$.
25. A linear operator T on a vector space $V(\mathbf{C})$ is represented by the matrix

$$\begin{bmatrix} 1 & 1 & -1 \\ 1 & 0 & -1 \\ 1 & -1 & 0 \end{bmatrix}$$

relative to some basis. Use the same basis to represent the operators D_T and N_T in the decomposition $T = D_T + N_T$.

26. Let T be a linear operator on a finite-dimensional vector space over **C**. Then, if D_T is the diagonalizable part of T, and $g(x)$ is any complex polynomial, prove that the diagonalizable part of $g(T)$ is $g(D_T)$.

6.6 Nilpotent Operators and T-cyclic Subspaces

We have seen that any linear operator T on a complex vector space V of finite dimension can be expressed in the form $T = D_T + N_T$, where D_T and N_T are operators that are diagonalizable and nilpotent, respectively. For the matrix representations of T, this means that T can be represented by a matrix A, with $A = B + C$, where B and C are matrices that are respectively diagonal and nilpotent. The diagonal entries of B are the eigenvalues of T, each repeated according to its multiplicity in the characteristic polynomial, and so B is completely determined by this polynomial except for the order of appearance of the diagonal entries. On the other hand, the nilpotent matrix C is not so uniquely determined: Its entries are dependent on the choice of bases for the null spaces of $(T - c_i 1)^{r_i}$, where $m(x) = (x - c_1)^{r_1}(x - c_2)^{r_2} \cdots (x - c_k)^{r_k}$ is the prime power factorization of the minimum polynomial of T. It is the purpose of this section to develop some preliminary ideas, which will be used in Sec. 6.7 to show that a basis for V exists, with respect to which the nilpotent component C—and so also A—is in a very simple form.

For any vector $\alpha \in V$, there clearly exists a smallest T-invariant subspace that contains α; in fact, it can be defined as the intersection of *all* subspaces of V that are invariant under T and contain α. However, it is more useful to observe (Prob. 10) that any T-invariant subspace that contains α must also contain $T\alpha, T^2\alpha, T^3\alpha, \ldots$ as well as $[g(T)]\alpha$, for any polynomial $g(x)$ over the field of scalars (**R** or **C**). Since the set of these "polynomial" vectors constitutes a T-invariant subspace (Prob. 11), it must be the smallest T-invariant subspace that contains α. This type of subspace is the key idea in this section, and so we formalize it with a definition.

Definition *If* T *is a linear operator on a vector space* V, *and* α $(\neq 0)$ *is a vector of* V, *the set of all vectors* $[g(T)]\alpha$, *for arbitrary polynomials* $g(x)$ *over the field of scalars, is called the* T-*cyclic subspace generated by* α. *A subspace of* V *is said to be* T-*cyclic if it has a ("*T-*cyclic") generator of this kind.*

For any linear operator T on V the subspace $\mathcal{O}$ can be regarded as a T-cyclic subspace with 0 as its T-cyclic generator. If α is an eigenvector of T, the

T-cyclic subspace generated by α is clearly one-dimensional; if T = I, all T-cyclic subspaces must have dimension 1. Except for these special cases, however, the dimension of a T-cyclic subspace is at least 2.

EXAMPLE 1

Let us examine the T-cyclic subspace of $V_3(\mathbf{R})$, generated by $X = (1,1,1)$, where $T(x,y,z) = (x,-y,z)$ for any $(x,y,z) \in V_3(\mathbf{R})$. We see that $TX = (1,-1,1)$ and $T^2 X = (1,1,1) = X$, so that the subspace has dimension 2. A basis for the subspace is $\{(1,1,1), (1,-1,1)\}$.

EXAMPLE 2

Let the matrix of a linear operator T on $V_3(\mathbf{R})$, relative to the $\{E_i\}$ basis, be

$$\begin{bmatrix} 1 & 0 & 1 \\ -1 & 1 & 0 \\ 0 & 1 & 0 \end{bmatrix}$$

If we examine the T-cyclic subspace generated by $(1,1,1)$, we discover that $TX = (2,0,1)$ and $T^2 X = (3,-2,0)$. In this case, the three vectors X, TX, and $T^2 X$ are linearly independent and so make up a basis for the whole space $V_3(\mathbf{R})$. This space is then T-cyclic, for this particular operator T, and is generated by $(1,1,1)$.

Lemma 1 Let W be a T-cyclic subspace of a finite-dimensional vector space V. Then there exists a vector $\alpha \in W$ and a positive integer t such that $\{\alpha, T\alpha, T^2\alpha, \ldots, T^{t-1}\alpha\}$ is a basis for W, and the associated matrix of the operator induced on W by T (that is, T restricted to W) has the form

$$\begin{bmatrix} 0 & 0 & 0 & \cdots & 0 & -a_0 \\ 1 & 0 & 0 & \cdots & 0 & -a_1 \\ 0 & 1 & 0 & \cdots & 0 & -a_2 \\ \cdots & \cdots & \cdots & \cdots & \cdots & \cdots \\ 0 & 0 & 0 & \cdots & 0 & -a_{t-2} \\ 0 & 0 & 0 & \cdots & 1 & -a_{t-1} \end{bmatrix}$$

Proof
Since W is a T-cyclic subspace, there exists $\alpha \in W$ such that W is generated by the set of vectors $\{\alpha, T\alpha, T^2\alpha, \ldots\}$. However, because V has finite

dimension, there must exist an integer $t > 0$ such that $\{\alpha, T\alpha, \ldots, T^{t-1}\alpha\}$ is linearly independent, whereas $T^t\alpha \in \mathcal{L}\{\alpha, T\alpha, \ldots, T^{t-1}\alpha\}$. Since $T^m\alpha \in \mathcal{L}\{\alpha, T\alpha, \ldots, T^{t-1}\alpha\}$, for any integer $m \geq t$, it is clear that the vectors $\alpha, T\alpha, \ldots, T^{t-1}\alpha$ make up a basis for W. The linear dependence of $\{\alpha, T\alpha, \ldots, T^{t-1}\alpha, T^t\alpha\}$ requires that $a_0\alpha + a_1(T\alpha) + a_2(T^2\alpha) + \cdots + a_{t-2}(T^{t-2}\alpha) + a_{t-1}(T^{t-1}\alpha) + a_t(T^t\alpha) = 0$, and there is no loss in generality if we assume that $a_t = 1$. But then $T^t\alpha = -a_0\alpha - a_1(T\alpha) - a_2(T^2\alpha) \cdots - a_{t-2}(T^{t-2}\alpha) - a_{t-1}(T^{t-1}\alpha)$, and it is easy to see that the matrix of T, relative to the basis $\{\alpha, T\alpha, \ldots, T^{t-1}\alpha\}$, is the one shown. ∎

The matrix displayed in Lemma 1 is called the *companion matrix* of the polynomial $a_0 + a_1 x + \cdots + a_{t-1}x^{t-1} + x^t$. It is not difficult to see (Prob. 12) that this polynomial is both the characteristic and minimum polynomial of its companion matrix and so, also, of any linear operator that this matrix represents.

Lemma 1 is valid for any linear operator T on a finite-dimensional vector space. However, our present interest is in nilpotent operators and, in this case, the matrix representation of T is simpler. If T is a nilpotent operator, such that $T^p = 0$ but $T^{p-1} \neq 0$, we may speak of p as the *index of nilpotency* of T or say that T is *nilpotent of index p*. We now rephrase Lemma 1 for a direct application to nilpotent operators.

Lemma 2   *Let* T *be a nilpotent linear operator on a finite-dimensional vector space* V. *Then, if* W *is a* T*-cyclic subspace of* V, *there exists a vector* $\alpha \in W$, *and a positive integer* t, *such that* $\{\alpha, T\alpha, \ldots, T^{t-1}\alpha\}$ *is a basis of* W, *and the* $t \times t$ *matrix of the operator induced by* T *on* W, *relative to this basis, is*

$$\begin{bmatrix} 0 & 0 & 0 & \cdots & 0 & 0 \\ 1 & 0 & 0 & \cdots & 0 & 0 \\ 0 & 1 & 0 & \cdots & 0 & 0 \\ \cdots & \cdots & \cdots & \cdots & \cdots & \cdots \\ 0 & 0 & 0 & \cdots & 0 & 0 \\ 0 & 0 & 0 & \cdots & 1 & 0 \end{bmatrix}$$

Proof

The existence of the basis $\{\alpha, T\alpha, \ldots, T^{t-1}\alpha\}$, for some positive integer t, is established as in Lemma 1. If we use T_W to denote the operator induced by T on W, the nilpotency of T implies that T_W is also nilpotent. Moreover, the minimum polynomial of the induced operator T_W is x^t, where $t = \dim W$. Hence $T_W^t = 0$, and this implies, in particular, that $(T_W)\alpha = T^t\alpha = 0$. The special form of the matrix of T_W, relative to the basis generated by α, is now an immediate consequence of the latter result. It may be of interest to observe that, if the subspace W happens to coincide with the whole space V, the displayed matrix is the matrix of T on V.

If a linear operator T on a vector space V is nilpotent of index p, it is clear (Prob. 13) that there exists a T-cyclic subspace of dimension p. The proof that this subspace is a direct summand of V and that its complementary subspace is also T-invariant is somewhat difficult, however, and we shall postpone it until the next section. Once this result—and its inductive consequence—has been established, it will be an easy matter to obtain the desired Jordan canonical matrix for any linear operator on a finite-dimensional vector space.

EXAMPLE 3

If we know that $T^3\alpha = T^2\alpha - 2(T\alpha) + \alpha$, for a linear operator T on a vector space V, for α ($\neq 0$) $\in V$, and no similar equality exists with lower powers of T, it is easy to write a matrix to represent the operator induced by T on the T-cyclic subspace generated by α. A basis of this subspace is $\{\alpha, T\alpha, T^2\alpha\}$ and so, relative to this basis, the representing matrix is the companion matrix of the polynomial $x^3 - x^2 + 2x - 1$. This matrix is

$$\begin{bmatrix} 0 & 0 & 1 \\ 1 & 0 & -2 \\ 0 & 1 & 1 \end{bmatrix}$$

EXAMPLE 4

Find a matrix to represent an operator T, which is nilpotent of index 4, on a vector space V of dimension 4.

Solution
Since the index of nilpotency of T is 4, there must exist a vector $\alpha \in V$ such that $T^3\alpha \neq 0$ but $T^4\alpha = 0$. Since the dimension of the space is 4, it follows that $\{\alpha, T\alpha, T^2\alpha, T^3\alpha\}$ is a basis, and, relative to this basis, the matrix of T is

$$\begin{bmatrix} 0 & 0 & 0 & 0 \\ 1 & 0 & 0 & 0 \\ 0 & 1 & 0 & 0 \\ 0 & 0 & 1 & 0 \end{bmatrix}$$

Problems 6.6

1. Express each of the following matrices as the sum of a diagonal matrix and a matrix with all diagonal entries 0:

    (a) $\begin{bmatrix} 2 & 1 & 0 \\ -2 & 1 & 1 \\ 0 & 2 & 1 \end{bmatrix}$
    (b) $\begin{bmatrix} 1 & 1 & 0 & 1 \\ 0 & 1 & -1 & 1 \\ 2 & -1 & 1 & 0 \\ 3 & 1 & 0 & 1 \end{bmatrix}$

    (c) diag $\{A_1, A_2\}$

    where $A_1 = \begin{bmatrix} 2 & 1 \\ 1 & 3 \end{bmatrix}$ and $A_2 = \begin{bmatrix} 1 & 0 & 1 \\ 2 & -1 & 1 \\ 3 & 1 & 2 \end{bmatrix}$.

2. Check whether the nondiagonal matrices in the decompositions in Prob. 1 are nilpotent.

3. If $T(x_1, x_2, x_3, x_4) = (x_1 - x_2, x_2, x_3, x_4 + x_2)$ defines the operator T on $V_4(\mathbf{R})$, determine the T-cyclic subspace that is generated by $(1, 0, 0, 1)$.

4. Discover some linear operator T and a T-cyclic generator for the space $V_3(\mathbf{C})$.

5. Write the companion matrix for the polynomial $x^5 - 2x^4 + 5x^3 - x^2 - 3x + 1$.

6. If T is a nilpotent operator of index 3 on a vector space of dimension 5, give a matrix that will represent T relative to some basis. If the space has dimension 3, describe a basis for the space and display the associated matrix of T.

7. If

$$A = \begin{bmatrix} 2 & 0 & -1 \\ 1 & 0 & 1 \\ 0 & 1 & 0 \end{bmatrix}$$

is the matrix of a linear operator T on V_3, find the T-cyclic subspace generated by the vector $(1, -1, 1)$.

8. Let D be the differentiation operator on the space P_4 of complex polynomials of degree less than 4 in a variable x. Find a D-cyclic generator α for the space and give the matrix representation of D, relative to the basis $\{\alpha, D\alpha, D^2\alpha, D^3\alpha\}$.

9. Let R be the operator that rotates each vector of $V_2(\mathbf{R})$ about the origin through an angle of $\pi/2$ rad. If $X = (1,1)$, check that $\{X, TX\}$ is a basis of $V_2(\mathbf{R})$, and find the matrix that represents R with respect to this basis.

10. Let T be a linear operator on a vector space V. Then explain why the intersection of all T-invariant subspaces of V, that contain a vector $\alpha \in V$, is the same as the subspace generated by $\{\alpha, T\alpha, T^2\alpha, \ldots\}$.
11. Explain why any T-cyclic subspace must be T-invariant.
12. Prove that the polynomial $a_0 + a_1 x + \cdots + a_{t-1} x^{t-1} + x^t$ is both the minimum and characteristic polynomial of its companion matrix.
13. If T is any nilpotent linear operator of index p on a vector space V, explain why there exists a T-cyclic subspace of V of dimension p.

14. Prove that any matrix that is similar to a nilpotent matrix of index t is also nilpotent of index t.
15. If U_1 and U_2 are linear operators on a finite-dimensional vector space V, and we let $T = U_1 U_2 - U_2 U_1$, prove that $I - T$ is not nilpotent on V.
16. If V is a T^2-cyclic vector space, for a linear operator T on V, prove that V is also T-cyclic.
17. Let
$$A = \begin{bmatrix} 3 & -1 & -1 & -2 \\ 1 & 1 & -1 & -1 \\ 1 & 0 & 0 & -1 \\ 0 & -1 & 1 & 1 \end{bmatrix}$$
be the matrix representation of a linear operator T on a vector space V. Then, if $(1, -1, -1, 0)$ is the coordinate vector of $\alpha \in V$, verify that α is the generator of a T-cyclic subspace of V of dimension 3.
18. If
$$A = \begin{bmatrix} 2 & 0 & 0 \\ 0 & 2 & 0 \\ 0 & 0 & -1 \end{bmatrix}$$
is the matrix representation of a linear operator T on a vector space V, prove that V is not T-cyclic. Find the T-cyclic subspace generated by $\alpha \in V$, where the coordinate vector of α, with respect to the basis understood for V, is $(1, 1, -1)$.
19. If
$$A = \begin{bmatrix} 1 & i & 1 \\ 0 & -1 & i \\ 1 & 0 & 1 \end{bmatrix}$$
represents a linear operator T on a complex vector space V, decide whether V is T-cyclic.

20. The matrix of a linear operator T on a vector space V, relative to some basis $\{\alpha_1, \alpha_2, \alpha_3, \alpha_4\}$, is

$$\begin{bmatrix} -2 & 4 & 1 & 1 \\ 0 & 1 & 0 & 0 \\ 0 & 0 & 1 & 0 \\ -1 & 0 & -1 & 0 \end{bmatrix}$$

By trial and error, discover a decomposition of V into the direct sum of two T-cyclic subspaces.

21. If
$$A = \begin{bmatrix} 0 & 1 & 2 \\ 0 & 0 & 1 \\ 0 & 0 & 0 \end{bmatrix}$$

is the matrix of a linear operator T on a vector space V, relative to a basis $\{\alpha_1, \alpha_2, \alpha_3\}$, find the rank, null space, and index of nilpotency of T.

22. If T is a diagonalizable linear operator on an n-dimensional vector space V, prove that (a) T has n distinct eigenvalues if V is T-cyclic; (b) V is T-cyclic if T has n distinct eigenvalues. [*Hint:* If $\{\alpha_1, \alpha_2, \ldots, \alpha_n\}$ is a basis of eigenvectors for V, show that $\alpha = \alpha_1 + \alpha_2 + \cdots + \alpha_n$ is a T-cyclic generator of V.]

6.7 The Jordan Canonical Form

Now that the necessary preliminary lemmas have been derived in Sec. 6.6, we may proceed to the results which are the ultimate objective of this chapter. The proof of Theorem 7.1 is somewhat difficult, and the one we give capitalizes on some of the notions introduced in Chap. 2 but not used very extensively in subsequent discussions. We refer, in particular, to the following notions, which the reader is urged to review before going further: the space V^* that is *dual* to a given vector space V; the operator T' that is the *transpose* of a linear operator T; and the subspace W^0 of V^* that is the *annihilator* of a subset W of V.

Theorem 7.1   Let T be a *nilpotent operator of index p on an n-dimensional (real or complex) vector space V. Then $V = W_1 \oplus W_2 \oplus \cdots \oplus W_s$, where W_i is a T-cyclic subspace of dimension p_i, $i = 1, 2, \ldots, s$, such that $p = p_1 \geq p_2 \geq \cdots \geq p_s$ and $\sum_{i=1}^{s} p_i = n$.*

Proof
Since T is nilpotent of index p on V, there exists a T-cyclic subspace W of V, with basis $\{\alpha, T\alpha, \ldots, T^{p-1}\alpha\}$, for some $\alpha \in W$. If $W = V$, the proof is complete; otherwise, $W \neq V$ and the proof continues. We shall prove that W is a direct summand of V, with a nonzero T-invariant complementary subspace, by first showing that the dual space V^* can be decomposed into the direct sum of two nontrivial subspaces, one of which is invariant under the transpose T' of T.

Any basis $\{\alpha_1, \alpha_2, \ldots, \alpha_p\}$ of W can be extended to a basis $\{\alpha_1, \alpha_2, \ldots, \alpha_n\}$ of V, and we let $\{f_1, f_2, \ldots, f_n\}$ be the associated dual basis of V^*. We leave it for the reader to verify (Prob. 14) that $\{f_{p+1}, f_{p+2}, \ldots, f_n\}$ is a basis for the annihilator W^0 of W, so that dim $W^0 = n - p$. The definition of the transpose operator T' implies (Prob. 15) that $(T')^m = (T^m)'$, for any integer $m \geq 1$; and, since $T^p = 0$, we conclude that $(T')^p = 0$. However, there must exist a function $f \in V^*$ such that $(T')^{p-1}f \notin W^0$. For, otherwise, $0 = [(T')^{p-1}f]\alpha = f(T^{p-1}\alpha)$, for all $f \in V^*$, so that $T^{p-1}\alpha = 0$, contrary to our definition of α. We now use U_1^0 to designate the p-dimensional (Prob. 12) subspace of V^*, which is generated by $f, T'f, \ldots, (T')^{p-1}f$, our notation implying that we are to be interested later in the subspace U_1 of *all* vectors in V that are annihilated by U_1^0. In order to see that $U_1^0 \cap W^0 = \emptyset$, we suppose that $a_0 f + a_1(T'f) + \cdots + a_{p-1}(T')^{p-1}f \in W^0$, where $a_0, a_1, \ldots, a_{p-1}$ are scalars with a_k the nonzero scalar of lowest index in the set. This means that $a_k(T')^k f + \cdots + a_{p-1}(T')^{p-1}f \in W^0$; and, on applying the operator $(T')^{p-1-k}$ to this element [recalling that $(T')^p = 0$] we find that $a_k(T')^{p-1}f \in W^0$. Since $a_k \neq 0$, this is contrary to our choice of f, and we must conclude that $U_1^0 \cap W^0 = \emptyset$. Hence $V^* = W^0 \oplus U_1^0$, and we note [since $(T')^p = 0$] that U_1^0 is a subspace of V^* that is invariant under T'. This is the decomposition of V^* to which we referred above.

Our proof will be completed by showing that $V = W \oplus U_1$, and our first observation is that U_1 is a T-invariant subspace of V, because (Prob. 13) U_1^0 is a T'-invariant subspace of V^*. If $\beta \in W \cap U_1$, it is a consequence of the decomposition of V^* obtained above that $h(\beta) = 0$, for all $h \in V^*$. Thus $\beta = 0$, so that $W \cap U_1 = \emptyset$. We already know that dim $U_1^0 = p = p_1$, and so it follows from Theorem 5.1 (Chap. 2) that dim $U_1 = n - p_1$. But now, dim W + dim $U_1 = p_1 + (n - p_1) = n$, whence $V = W \oplus U_1$, where $W = W_1$ is T-cyclic and U_1 is T-invariant. We now observe that T is nilpotent of index $p_2 \leq p_1$ on U_1, and dim U_1 < dim V. A similar argument to the above can now be applied to T as a nilpotent operator on U_1, with the result that $U_1 = W_2 \oplus U_2$ where W_2 is T-cyclic of dimension p_2 and U_2 is T-invariant. If $U_2 \neq \emptyset$, the argument can be repeated and, in any case, the fact that V is of finite dimension leads ultimately to the conclusion of the theorem. ∎

It is sometimes useful to restate Theorem 7.1 in the following form, which we list as a corollary.

Corollary   Let T *be a nilpotent operator of index p on an n-dimensional complex vector space. Then there exists a positive integer* s, *a set of s distinct vectors* $\{\alpha_1, \alpha_2, \ldots, \alpha_s\}$, *and s integers* $p = p_1 \geq p_2 \geq \cdots \geq p_s$, *such that the following is true:*

(1) $V = W_1 \oplus W_2 \oplus \cdots \oplus W_s$, *where* W_i *is the T-cyclic subspace of V generated by* α_i ($i = 1, 2, \ldots, s$).

(2) *The set* $\{\alpha_1, T\alpha_1, \ldots, T^{p_1-1}\alpha_1; \alpha_2, T\alpha_2, \ldots, T^{p_2-1}\alpha_2;$
$\cdots ; \alpha_s, T\alpha_s, \ldots, T^{p_s-1}\alpha_s\}$ *is an ordered basis of V, so that* $\sum_{i=1}^{s} p_i = n$.

(3) *The matrix of* T, *relative to the ordered basis in* (2) *is* $A = \mathrm{diag}\{A_1, A_2, \ldots, A_s\}$, *where the order of* A_i *is* p_i ($i = 1, 2, \ldots, s$), *and each of these s matrices is of the form shown in Lemma 2 of Sec. 6.6.*

With this important result established, we are now able to combine it with Theorem 5.1 and its corollary to obtain our major result.

Theorem 7.2   Let T *be a linear operator on an n-dimensional vector space over* **C**, *with* $f(x) = (-1)^n(x - c_1)^{t_1}(x - c_2)^{t_2} \cdots (x - c_k)^{t_k}$ *the characteristic polynomial of* T. *Then there exists a basis of V such that the related matrix of* T *is* $J = \mathrm{diag}\{J_1, J_2, \ldots, J_k\}$, *where* J_i *is a square matrix of order* t_i ($i = 1, 2, \ldots, k$) *of the form* $\mathrm{diag}\{J_{i1}, J_{i2}, \ldots, J_{is(i)}\}$; *and* J_{ij} *is the square matrix of order* p_{ij}, *with* $p_{i1} \geq p_{i2} \geq \cdots \geq p_{is(i)}$ *and* $\sum_{j=1}^{s(i)} p_{ij} = t_i$, *defined by*

$$J_{ij} = \begin{bmatrix} c_i & 0 & 0 & \cdots & 0 & 0 \\ 1 & c_i & 0 & \cdots & 0 & 0 \\ 0 & 1 & c_i & \cdots & 0 & 0 \\ \cdots & \cdots & \cdots & \cdots & \cdots & \cdots \\ 0 & 0 & 0 & \cdots & c_i & 0 \\ 0 & 0 & 0 & \cdots & 1 & c_i \end{bmatrix}$$

Proof
As usual, we let $m(x) = (x - c_1)^{r_1}(x - c_2)^{r_2} \cdots (x - c_k)^{r_k}$ be the minimum polynomial of T on V, with $r_1 \leq t_1, r_2 \leq t_2, \ldots, r_k \leq t_k$. It now follows from Theorem 5.1 and its corollary that $V = W_1 \oplus W_2 \oplus \cdots \oplus W_k$, where W_i is the null space of $(T - c_i 1)^{r_i}$, and $\dim W_i = t_i, i = 1, 2, \ldots, k$.

Since we may regard $T - c_i\mathbf{I}$ as a nilpotent operator on W_i, the corollary to Theorem 7.1 implies the existence of a basis for W_i, with respect to which the matrix B_i of this operator has the form described for A in the corollary. If J_i is the matrix (relative to this same basis) of the operator induced on W_i by T, it follows that $B_i = J_i - c_i\mathbf{I}$ and $J_i = c_i\mathbf{I} + B_i$. On combining the bases for W_i, $i = 1, 2, \ldots, k$, we obtain a basis for V, and the associated matrix of T has the form detailed in the statement of the theorem. ∎

The matrix J of T, as described in Theorem 7.2, is known as the *Jordan canonical* matrix of T, and the submatrices J_{ij} are called its component *Jordan blocks*. The matrix J is said to be in Jordan canonical *form*. It may be observed that there are $s(i)$ Jordan blocks, that are associated with the eigenvalue c_i, and that their orders are $p_{i1}, p_{i2}, \ldots, p_{is(i)}$, where $p_{i1} \geq p_{i2} \geq \cdots \geq p_{is(i)}$, $i = 1, 2, \ldots, k$. In the notation of the theorem, it is clear (Prob. 16) that $(x - c_i)^{p_{i1}}$ is the minimum polynomial of J_i, and the minimum polynomials of all Jordan blocks associated with c_i divide this. Hence the minimum polynomial of T is $(x - c_1)^{p_{11}}(x - c_2)^{p_{21}} \cdots (x - c_k)^{p_{k1}}$, so that $p_{i1} = r_i$, $i = 1, 2, \ldots, k$. In other words, the minimum polynomial of T is the product of the minimum polynomials of the Jordan blocks of *largest* orders associated with each of the k distinct eigenvalues. The minimum polynomials of the various Jordan blocks in the Jordan matrix of T are also called the *elementary divisors* of T, and so the minimum polynomial of T can then be described as the product of the elementary divisors of highest degrees, one associated with each of the distinct eigenvalues. While a matrix in Jordan canonical form is more complicated than a diagonal matrix, it is still relatively easy to perform algebraic operations on it. For a discussion of the computation of a polynomial in a Jordan matrix, the reader is referred to the book by Gel'fand (pp. 135–137) listed at the end of the chapter.

Inasmuch as a Jordan matrix is completely determined by its eigenvalues and the orders of the Jordan blocks that are associated with them, these are sometimes given to describe a Jordan matrix. In the notation that we are using, such a matrix is completely described by the spectrum $\{c_1, c_2, \ldots, c_k\}$ and an ordered listing of the orders of the Jordan blocks, which may be abbreviated by $[(p_{11}, p_{12}, \ldots, p_{1s(1)})(p_{21}, p_{22}, \ldots, p_{2s(2)}) \cdots (p_{k1}, p_{k2}, \ldots, p_{ks(k)})]$. This numerical abbreviation is called the *Segre characteristic* of the operator represented by the Jordan matrix. For example, let us suppose that the spectrum of a linear operator T is $\{a, b, c\}$, and its Segre characteristic is $[(311)(22)(1)]$. It is now apparent that the minimum polynomials of the ordered blocks in the associated Jordan matrix are $(x - a)^3$, $(x - a)$, $(x - a)$, $(x - b)^2$, $(x - b)^2$, $(x - c)$. Moreover, the minimum polynomial $m(x) = (x - a)^3(x - b)^2(x - c)$ and the characteristic polynomial $f(x) = (x - a)^5(x - b)^4(x - c)$.

Finally, a brief comment is in order on the uniqueness of the Jordan matrix

SEC. 6.7 THE JORDAN CANONICAL FORM

and other possible invariants of the similarity equivalence class of a given square matrix. What we have shown is that any linear operator T can be represented by a Jordan matrix, or, equivalently, that any square matrix A is equivalent to a matrix in Jordan form. Although we have not given a proof that there can be only one Jordan matrix (for a given arrangement of eigenvalues on its diagonal) in any given equivalence class under similarity, this is not difficult to do (Prob. 17), and so we may refer to such a matrix as *canonical*. We have already noted that a Jordan matrix is determined by its *spectrum* and *Segre characteristic* or, equivalently, by its *elementary divisors;* and so these, along with the *order* of the matrices, may be considered to constitute the invariants of a similarity equivalence class of matrices. The same invariants may be associated with a linear operator on a vector space, if we replace the *order* of the matrix by the *dimension* of the vector space. We leave unanswered the question of how to find an algorithm for putting a given square matrix in Jordan form: that is, how to find a nonsingular matrix P such that $P^{-1}AP$ is the Jordan canonical matrix that is similar to the matrix A. This is a matter of discovering a basis of the correct kind for the vector space, but, although algorithms exist for this purpose, a knowledge of the *existence* of such a basis is quite often sufficient for most purposes. Accordingly, we leave the development of such an algorithm to other more detailed treatises on linear algebra.

EXAMPLE 1

The spectrum of a linear operator T is $\{-1, 1, 2\}$, and its related Segre characteristic is $[(31)(211)(2)]$. Write the Jordan matrix that represents T, relative to some basis.

Solution

The dimension of the vector space on which T operates must be 10, and the Jordan matrix is the following:

$$\begin{bmatrix} -1 & 0 & 0 & 0 & 0 & 0 & 0 & 0 & 0 & 0 \\ 1 & -1 & 0 & 0 & 0 & 0 & 0 & 0 & 0 & 0 \\ 0 & 1 & -1 & 0 & 0 & 0 & 0 & 0 & 0 & 0 \\ 0 & 0 & 0 & -1 & 0 & 0 & 0 & 0 & 0 & 0 \\ 0 & 0 & 0 & 0 & 1 & 0 & 0 & 0 & 0 & 0 \\ 0 & 0 & 0 & 0 & 1 & 1 & 0 & 0 & 0 & 0 \\ 0 & 0 & 0 & 0 & 0 & 0 & 1 & 0 & 0 & 0 \\ 0 & 0 & 0 & 0 & 0 & 0 & 0 & 1 & 0 & 0 \\ 0 & 0 & 0 & 0 & 0 & 0 & 0 & 0 & 2 & 0 \\ 0 & 0 & 0 & 0 & 0 & 0 & 0 & 0 & 1 & 2 \end{bmatrix}$$

In our abbreviated partitioned "diagonal" form, this could also be given as diag $\{J_{11}, J_{12}, J_{21}, J_{22}, J_{23}, J_{31}\}$, where

$$J_{11} = \begin{bmatrix} -1 & 0 & 0 \\ 1 & -1 & 0 \\ 0 & 1 & -1 \end{bmatrix} \quad J_{12} = [-1]$$

$$J_{21} = \begin{bmatrix} 1 & 0 \\ 1 & 1 \end{bmatrix} \quad J_{22} = [1] = J_{23} \quad J_{31} = \begin{bmatrix} 2 & 0 \\ 1 & 2 \end{bmatrix}$$

EXAMPLE 2

Write the Jordan matrix that represents a linear operator T, given that the elementary divisors of T are $(x+i)^2$, $x+i$, $(x-1)^2$, $(x-i)^2$.

Solution

The desired Jordan matrix is the seven-dimensional matrix J, where

$$J = \begin{bmatrix} -i & 0 & 0 & 0 & 0 & 0 & 0 \\ 1 & -i & 0 & 0 & 0 & 0 & 0 \\ 0 & 0 & -i & 0 & 0 & 0 & 0 \\ 0 & 0 & 0 & 1 & 0 & 0 & 0 \\ 0 & 0 & 0 & 1 & 1 & 0 & 0 \\ 0 & 0 & 0 & 0 & 0 & i & 0 \\ 0 & 0 & 0 & 0 & 0 & 1 & i \end{bmatrix}$$

Problems 6.7

1. Write, from inspection, the characteristic and minimum polynomials of each of the following Jordan matrices:

   (a) diag $\{J_{11}, J_{12}, J_{21}, J_{22}\}$, where $J_{11} = J_{12} = \begin{bmatrix} 2 & 0 \\ 1 & 2 \end{bmatrix}$ and $J_{21} = J_{22} = [3]$.

   (b) diag $\{J_{11}, J_{12}, J_{13}, J_{21}\}$, where

   $$J_{11} = \begin{bmatrix} 1 & 0 & 0 \\ 1 & 1 & 0 \\ 0 & 1 & 1 \end{bmatrix} \quad J_{12} = \begin{bmatrix} 1 & 0 \\ 1 & 1 \end{bmatrix}$$

   $$J_{13} = [1] \quad \text{and} \quad J_{21} = \begin{bmatrix} 2 & 0 \\ 1 & 2 \end{bmatrix}$$

2. Give the form of a Jordan matrix whose Segre characteristic is (a) [(32)(221)]; (b) [(211)(11)(3)]. Express the characteristic and minimum polynomial of each matrix.

3. List the elementary divisors of both matrices in Prob. 1.
4. Write the Jordan matrix whose elementary divisors are (a) $(x + 2)^2$, $(x - 1)^2$, $(x - 2)$; (b) $(x + i)^2$, $(x - i)^2$, x; (c) $(x + 1 + i)^2$, $(x + 1 - i)^2$, x.
5. What are the possible Jordan matrices of order 5 whose minimum polynomials are (a) $(x + 2)^3$; (b) $(x - 2)^4$; (c) $(x - 1)^5$?
6. Find the Jordan matrix that is similar to (a) $\begin{bmatrix} 1 & 2 \\ 1 & -1 \end{bmatrix}$; (b) $\begin{bmatrix} 2 & 1 \\ -1 & 1 \end{bmatrix}$; (c) $\begin{bmatrix} 1 & -1 \\ 2 & -1 \end{bmatrix}$.
7. Find the Jordan matrix that is similar to

(a) $\begin{bmatrix} 1 & 0 & 0 \\ 2 & 2 & 0 \\ -1 & 0 & -3 \end{bmatrix}$ (b) $\begin{bmatrix} -4 & 3 & 1 \\ -2 & 3 & 0 \\ 0 & 0 & 1 \end{bmatrix}$

8. Describe the Segre characteristic of any diagonalizable operator.
9. What is the maximum number of nonsimilar 3×3 matrices that have $(x - 1)^3$ as their characteristic polynomial?
10. Let $f(x)$ and $m(x)$ designate the characteristic and minimum polynomials, respectively, of

$$A = \begin{bmatrix} 2 & 0 & 0 \\ a & 2 & 0 \\ b & c & 1 \end{bmatrix}$$

for complex numbers a, b, c.
(a) Find a necessary and sufficient condition that $m(x) = (x - 2)(x - 1)$, and give the Jordan matrix that is then similar to A.
(b) Find a necessary and sufficient condition for $f(x)$ to equal $m(x)$, and give the Jordan matrix that is then similar to A.
11. Give a complete classification of all 2×2 matrices under similarity, by listing all possible Jordan matrices of order 2.

Note Problems 12 through 15 refer to the proof of Theorem 7.1.
12. Explain why dim $U^0 = p$.
13. Why does the fact that U^0 is T'-invariant imply that U is T-invariant?
14. Explain why $\{f_{p+1}, f_{p+2}, \ldots, f_n\}$ is a basis for the annihilator W^0 of W.

15. Prove that $(T')^m = (T^m)'$ for any linear operator on a finite-dimensional vector space.
16. In the notation of Theorem 7.2, explain why $(x - c_i)^{p_{i1}}$ is the minimum polynomial of J_i.
17. Prove that the Jordan matrix that is similar to a given square matrix A is unique for any given arrangement of its eigenvalues along the diagonal.
18. If $c_1, c_2, \ldots, c_n$ are the eigenvalues of an $n \times n$ matrix A, use the Jordan matrix similar to A to prove that $\det A = c_1 c_2 \cdots c_n$.
19. Use the Jordan canonical form to prove that any square matrix is similar to its transpose.
20. Use the Jordan canonical form to (a) solve Prob. 23 of Sec. 6.5; (b) show that similar matrices have equal traces (see Prob. 20 of Sec. 2.10).
21. The characteristic and minimum polynomials of a matrix A are, respectively, $(x - 1)^3(x + 2)^2$ and $(x - 1)^2(x + 2)^2$. Find the Jordan matrix that is similar to A.
22. Classify all complex matrices of order 3 by listing all possible Jordan matrices to which they may be similar.
23. Use the Jordan canonical form to prove the Cayley-Hamilton theorem.
24. The Segre characteristic of a linear operator T on a vector space V is $[(22)(321)(2)(2)]$, and its associated spectrum is $\{-1,1,2,3\}$. Find (a) the dimension of V; (b) the rank, elementary divisors, and minimum polynomial of T; and (c) the Jordan matrix that may represent T, assuming the order of listing of its eigenvalues.
25. Let

$$A = \begin{bmatrix} 3 & 0 & 0 & 0 \\ 1 & 3 & 0 & 0 \\ 0 & 0 & 3 & 0 \\ 0 & 0 & a & 3 \end{bmatrix}$$

Note that A is a Jordan matrix, if either $a = 0$ or $a = 1$, and that both matrices have the same characteristic and the same minimum polynomials. Explain why these two matrices are not similar.
26. If the characteristic polynomial of a linear operator T is $x^4 - 9$, explain why this is also the minimum polynomial of T.
27. Prove that any two nilpotent matrices of order 3 are similar if and only if they have the same minimum polynomial.

28. Find the Jordan matrix that represents the differentiation operator D on the space P_4 of all polynomials of degree 3 or less in one variable.
29. If N is a nilpotent matrix of order n, use the binomial expansion of $(1+x)^{\frac{1}{2}}$ to obtain a matrix A such that $A^2 = I + N$.
30. Use Prob. 29 and the Jordan canonical form to obtain, for any given nonsingular complex matrix B, a matrix A such that $A^2 = B$.

APPENDIX

The Real Case: A Variant of the Jordan Canonical Form

There does exist a *real* Jordan canonical matrix to represent a linear operator on a finite-dimensional *real* vector space, but even though discussions of this type of matrix are given in many texts on linear algebra, we have decided not to do so here. Our decision was based on two considerations: the theory for the real case is relatively complicated, as compared with the complex case, even though the same basic ideas are involved, and the real Jordan matrix does not seem to be used very much in the applications of linear algebra to other subject areas. The complexity of the development leading to the real Jordan matrix can be attributed to the fact that an irreducible polynomial with coefficients in **R** may be of the second degree, whereas irreducible complex polynomials are always linear. There is, however, a variant of the complex Jordan canonical form of a matrix which happens to be very useful for real vector spaces—at least in a study of certain differential equations. This result is due to **S**. Lefschetz, and a discussion of it can be found in his book "Differential Equations: Geometric Theory," 2d ed., pp. 352–354, Interscience Publishers, New York, 1963. We shall include a brief survey of the Lefschetz theory here, because it does not seem to be found elsewhere in the literature—except in his book. But first, let us illustrate one of the important aspects of the result with a simple example from the theory of differential equations.

The differential equation $d^3y/dx^3 - d^2y/dx^2 + dy/dx - y = 0$ is a very elementary homogeneous one which, in our familiar symbolism with D denoting the differential operator, assumes the form $(D^3 - D^2 + D - 1)y = (D^2 + 1)(D - 1)y = 0$. It is well known from the theory associated with such equations that the general solution of this one is

$$y = a_1 \cos x + a_2 \sin x + a_3 e^x$$

where a_1, a_2, a_3 are arbitrary real or complex numbers, depending upon whether the real or complex solutions of the equation are desired. Even if our interest lies only in the real solutions, it is a simple matter to imbed these in the complex solution space—and this can be done either by using the same (real) basis $\{\cos x, \sin x, e^x\}$ or with the use of some nonreal basis. In the past, we have always referred to a vector as "real" or "complex" according as the scalars of the space to which it belongs are real or complex. In the present context of functions, however, it is convenient to use the familiar terminology of analysis and to speak of the elements of a vector space of functions as *real* (or *complex*) *vectors* if they are *real* (or *complex*) as *func-*

tions. The set $\{\cos x, \sin x, e^x\}$ for the complex solution space just discussed (and with x understood to be a real variable) is then a *real* basis, and another basis for the same space is known to be $\{e^{ix}, e^{-ix}, e^x\}$ in which it should be noted that two members are not real. Since $\cos x = (e^{ix} + e^{-ix})/2$ and $\sin x = (e^{ix} - e^{-ix})/2i$, it follows that the general *real* solution $y = a_1 \cos x + a_2 \sin x + a_3 e^x$, with $a_1, a_2, a_3 \in \mathbf{R}$, can also be expressed in the form

$$y = \frac{a_1 - a_2 i}{2} e^{ix} + \frac{a_1 + a_2 i}{2} e^{-ix} + a_3 e^x$$

Two observations are now in order in connection with this expression of the real vector y in terms of the basis $\{e^{ix}, e^{-ix}, e^x\}$:

1. The two nonreal basis vectors are complex conjugates of each other, assuming our usual understanding of what this means in analysis.

2. The coefficients of the nonreal basis vectors are conjugate complex numbers, and the coefficient of the real basis vector is a real number.

This example may be considered related to the Jordan decomposition, relative to the operator D with eigenvalues $i, -i, 1$, of the complex solution space of the differential equation. Although this example is almost trivial, it is part of our present objective to demonstrate that the result in the example—in particular the two points itemized above—are not at all unusual, but that a like situation exists for any *real function space* of finite dimension on which a *real* operator has been defined. We now proceed to a brief discussion of this more general situation.

The setting for our discussion is then some n-dimensional space $V = V(\mathbf{R})$ of real functions, with T a real linear operator on V. In other words, *we are considering a real vector space, whose basis elements are real functions and any matrix of the operator* T *is an* $n \times n$ *real matrix*. In our earlier example, we regarded the functions e^{ix} and e^{-ix} as "conjugate" in a way that was assumed to be familiar, but let us make the notion precise in our present, more general setting. The real space $V = V(\mathbf{R})$ is imbedded as a *subset* in the complex space having the same *real* basis that has been associated with the real space. Thus, if $\{\beta_1, \beta_2, \ldots, \beta_n\}$ is a (real) basis of $V(\mathbf{R})$, these same vectors will generate over $\mathbf{C}$ the complex space $V(\mathbf{C})$ and $V(\mathbf{R}) \subset V(\mathbf{C})$. With any vector $\beta = c_1\beta_1 + c_2\beta_2 + \cdots + c_n\beta_n$ in $V(\mathbf{C})$, we may then associate the vector $\bar{\beta} = \bar{c}_1\beta_1 + \bar{c}_2\beta_2 + \cdots + \bar{c}_n\beta_n$, where $\bar{c}_i$ is the complex conjugate of c_i ($i = 1, 2, \ldots, n$), and we refer to $\bar{\beta}$ as the *conjugate* of β. The operator T can, of course, be considered a linear operator on the expanded space $V(\mathbf{C})$ in the natural way, and if A is the *real* matrix of T on $V(\mathbf{R})$, relative to some original real basis, we first wish to compare the matrices that represent T on $V(\mathbf{C})$, relative to bases with the property that the members of one are the conjugates of those in the other. If $P = [p_{ij}]$ is the matrix of transition from the real basis to one of these nonreal bases, the associated matrix of T is $P^{-1}AP$. However, if we recall that the columns of a transition matrix P are the coefficients of the new basis vectors in terms of the old, it is clear that the corresponding matrix of transition to the other nonreal basis of conjugate vectors is $\bar{P}$, where $\bar{P} = [\bar{p}_{ij}]$ is the matrix *conjugate* to P. It follows that the matrix of T, relative to the second basis, is $(\bar{P})^{-1}A\bar{P} = \overline{P^{-1}AP}$, inasmuch as $A = \bar{A}$ is real. Thus *the matrices of* T, *relative to conjugate bases, are conjugate to each other*.

Because the real matrix A represents T on $V(\mathbf{C})$, relative to the original real basis, the characteristic polynomial of T is real, and so its eigenvalues are either real or pairs of conjugate complex numbers. Let us suppose that, of the k distinct eigenvalues, $2p$ of them are pairs of complex conjugates while q of them are real numbers. We now label them so that $\{c_1, c_3, \ldots, c_{2p-1}\}$ is one complex set, $\{c_2, c_4, \ldots, c_{2p}\}$ is the ordered set of associated conjugates, and $\{c_{2p+1}, \ldots, c_{2p+q}\}$ is the set of real eigenvalues. The space $V(\mathbf{C})$ has, of course, the primary decomposition

$$V(\mathbf{C}) = W_1 \oplus W_2 \oplus \cdots \oplus W_{2p+q}$$

and our theory related to this type of decomposition yields the following results:

$$(\mathsf{T} - c_i \mathsf{l})^r W_i = 0 \quad \text{for } some\ r \in \mathbf{Z}$$
$$(\mathsf{T} - c_i \mathsf{l})^r W_j = W_j \quad i \neq j \quad \text{for } every\ r \in \mathbf{Z}$$

If we now let $W_1 \oplus W_3 \oplus \cdots \oplus W_{2p-1} = W'$, $W_2 \oplus W_4 \oplus \cdots \oplus W_{2p} = W''$, and $W_{2p+1} \oplus \cdots \oplus W_{2p+q} = W'''$, we see that $V(\mathbf{C})$ can be expressed as

$$V(\mathbf{C}) = W' \oplus W'' \oplus W'''$$

A basis of W' in this decomposition consists, in general, of nonreal vectors, and it can be seen from the *real* nature of T that a basis of W''' can be chosen to be real. An elementary argument involving the linear independence of vectors can now be used (see book by Lefschetz) to show that the sets of real and complex vectors just mentioned, when augmented by the conjugates of those in the basis selected for W', constitute a basis for $V(\mathbf{C})$. If we now express W' in its Jordan decomposition as a direct sum of irreducible subspaces and take the union of the bases of these component subspaces as the basis of W', it can be shown that W'' has a similar decomposition. Since the bases of the irreducible subspaces of W' and W'' are conjugates of each other, it follows from the argument leading to the result obtained in the first part of this general discussion that the associated blocks in the related matrix of T are conjugate submatrices. The block submatrices that are associated with the irreducible subspaces of W''' are, of course, real, and so we have obtained the following result:

> Let T be a linear operator on a real finite-dimensional vector space $V = V(\mathbf{R})$ of real functions, imbedded as a subset in the complex space $V(\mathbf{C})$ with the same (real) basis as the one selected for $V(\mathbf{R})$. It is then possible to decompose $V(\mathbf{C})$ into a direct sum of indecomposable subspaces W_i such that the diagonal block submatrices of the corresponding matrix of T are related to W_i as follows: if the eigenvalue c_i of T is real, the associated subspace W_i and the related block submatrix B_i are both real; if c_j and c_k are complex conjugate eigenvalues, the associated subspaces W_j, W_k and their related block submatrices B_j, B_k are such that the bases of W_j and W_k are conjugate vectors while B_j and B_k are conjugate submatrices.

The interesting application of this result—and the one illustrated in the introductory example—may now be obtained in general. If the matrix of the operator T on $V(\mathbf{C})$ is in the normal form just described, the associated basis of the complex space is $\{\alpha_1, \bar{\alpha}_1, \ldots, \alpha_p, \bar{\alpha}_p, \alpha_{2p+1}, \ldots, \alpha_{2p+q}\}$, in which the first $2p$ vectors are complex conjugate pairs, and the final q vectors are real. An arbitrary vector $\alpha \in V(\mathbf{C})$ can be expressed in the form $\alpha = x_1\alpha_1 + y_1\bar{\alpha}_1 + \cdots + x_p\alpha_p + y_p\bar{\alpha}_p + x_{2p+1}\alpha_{2p+1} + \cdots + x_{2p+q}\alpha_{2p+q}$, for complex numbers $x_1, y_1, \ldots, x_p, y_p, x_{2p+1}, \ldots, x_{2p+q}$. In particular, if $\alpha = \bar{\alpha}$ is a real vector—and so in $V(\mathbf{R})$—we have

$$\alpha = \sum_{1 \leq i \leq p} (x_i\alpha_i + y_i\bar{\alpha}_i) + \sum_{j=1}^{q} x_{2p+j}\alpha_{2p+j} = \sum_{1 \leq i \leq p} (\bar{x}_i\bar{\alpha}_i + \bar{y}_i\alpha_i) + \sum_{j=1}^{q} \bar{x}_{2p+j}\alpha_{2p+j}$$

But now the linear independence of basis vectors implies that $y_i = \bar{x}_i$ and $\bar{x}_{2p+j} = x_{2p+j}$ for all i and j occurring in the expressions. This means that, if a basis of the kind just described is assumed for $V(\mathbf{C})$, any *real* vector of this space has real coefficients associated with real basis vectors and any coefficient associated with a complex basis vector is the complex conjugate of the coefficient of the conjugate vector of the basis. It perhaps should be pointed out that the decomposition described in this appendix is a decomposition of the complex space $V(\mathbf{C})$ in which $V(\mathbf{R})$ has been imbedded, and is not to be construed as a decomposition of the original real space $V = V(\mathbf{R})$ into a direct sum of subspaces. Our major interest, however, is with *real* vectors (of V) that are imbedded in a natural way in the complex space $V(\mathbf{C})$, and the components of the real vectors comprise *subsets*—but in general not subspaces—of either $V(\mathbf{C})$ or $V(\mathbf{R})$.

Selected Readings

Blumerson, L.: Diagonalization of Quadratic Forms with Matrix Coefficients, *Amer. Math. Monthly*, **70**(1): 53–55 (January, 1963).

Brand, L.: Applications of the Companion Matrix, *Amer. Math. Monthly*, **75**(2): 146–152 (February, 1968). [See also **References** in Article.]

Cater, S.: An Elementary Development of the Jordan Canonical Form, *Amer. Math. Monthly*, **69**(5): 391–393 (May, 1962).

Cullen, C. G.: A Note on Normal Matrices, *Amer. Math. Monthly*, **72**(6): 643–644 (June–July, 1965).

Gel'fand, I. M.: "Lectures on Linear Algebra," Interscience Publishers, Inc., New York, 1961.

Hausner, Alvin: Uniqueness of Polar Decomposition, *Amer. Math. Monthly*, **74**(3): 303–304 (March, 1967).

Hochstadt, H.: Laplace Transforms and Canonical Matrices, *Amer. Math. Monthly*, **71**(7): 728–736 (August–September, 1964).

Lotkin, Mark: The Diagonalization of Certain Normal Matrices, *Amer. Math. Monthly*, **67**(9): 861–865 (November, 1960).

Moppert, C. F.: Construction of a Characteristic Basis for a Matrix, *Amer. Math. Monthly*, **73**(10): 1062–1069 (December, 1966).

Putzer, E. J.: Avoiding the Jordan Canonical Form in the Discussion of Linear Systems with Constant Coefficients, *Amer. Math. Monthly*, **73**(1): 2–7 (January, 1966).

Wallach, Sylvan: A Simple Derivation of Jordan Canonical Matrices over Arbitrary Fields, *Amer. Math. Monthly*, **72**(6): 614–618 (June–July, 1965).

Zadeh, L. A., and C. A. Desoer: "Linear System Theory," McGraw-Hill Book Company, New York, 1963.

GLOSSARY OF SYMBOLS

Although there may be some exceptions, the following symbols have been used with the indicated meanings throughout the text.

R   real numbers
C   complex numbers
V vector space
V_n or $V_n(\mathbf{R})$ real coordinate space of n-tuples
$V_n(\mathbf{C})$ complex coordinate space of n-tuples
E_i n-tuple of 0s, except for 1 in the ith position
$X, Y, Z, \ldots$ coordinate vectors
$\alpha, \beta, \gamma, \ldots$ vectors of an abstract vector space
$a, b, c, \ldots$ scalars
$A, B, C, \ldots$ matrices
I identity matrix
O zero matrix
A' the transpose (matrix) of A
A^* the matrix adjoint to A
adj A the (classical) adjoint of A
det A or $|A|$ determinant of A
diag $\{a_1, a_2, \ldots, a_n\}$ diagonal matrix with $a_1, a_2, \ldots, a_n$ on diagonal
$\mathcal{L}\{\alpha, \beta, \gamma, \ldots\}$ space of all linear combinations of $\alpha, \beta, \gamma, \ldots$
T or $\mathbf{T}$ linear transformation or operator
I   identity operator
0   zero operator, number zero, or *abstract* zero vector
T'   the transpose (operator) of T
T*   the operator adjoint to T
(,)   inner product
N_T null space or kernel of T
R_T range (space) of T
W^0 annihilator of W
$W^\perp$ orthogonal complement of the subspace W
$\mathcal{O}$ zero space or subspace
$\mathit{0}$ zero coordinate vector
$B(X, Y)$ bilinear form
$Q(X)$ quadratic form
S_n symmetric group (of permutations) on n symbols
P_n space of polynomials of degree less than n in an indeterminate or real variable

ANSWERS

Chapter 1

Sec. 1.1
1. For **7**: The sum of two continuous functions is a continuous function; any scalar multiple of a continuous function is a continuous function. Similarly, for **8**.
2. (a) $(3,2,-3,2)$; (b) $(-1,0,-1,2)$; (c) $(2,2,-4,4)$; (d) $(-5,-1,-2,6)$.
3. (a) $2x^2 - 2x - 3$; (b) $x^2 + 4x + 1$; (c) $3x^2 - 6x - 6$.
4. (a) $(4,1,0)$; (b) $(4,1,1)$; (c) $(1,2,-3)$.
6. **R** is a real vector space over **R**; **R** is also a vector space over **Q**, but **Q** is not a vector space over **R**.
7. $\gamma - \alpha = (\alpha + \beta) - \alpha = [\alpha + (-\alpha)] + \beta = (\alpha - \alpha) + \beta = \beta$.

Sec. 1.2
1. No. To each geometric vector, there corresponds infinitely many distinct directed line segments.
2. The equality or inequality of two geometric vectors is independent of the initial points of any representing line segments.
6. No. It is a consequence of our definitions of $-\alpha$ and multiplication by scalars, for any geometric vector α.

Sec. 1.3
1. $(x_1,x_2) + (y_1,y_2) = (x_1 + y_1, x_2 + y_2)$; $r(x_1,x_2) = (rx_1,rx_2)$. With each pair (x_1,x_2), there is associated one geometric vector in the plane but infinitely many directed line segments of equal length and in the same direction.
2. The elements associated with AB, BC, and AC are $(-3,4,3)$, $(5,1,-11)$, and $(2,5,-8)$, respectively; $(-3,4,3) + (5,1,-11) = (2,5,-8)$.
4. (a) If A, B, C are the points $(0,0,1)$, $(1,-2,4)$, $(1,1,2)$, respectively, then $AB + BC = AC$.
   (b) If O, A, B are $(0,0,0)$, $(1,3,-4)$, $(2,6,-8)$, respectively, then $2(OA) = OB$.
   (c) If O, A, B, C, P are $(0,0,0)$, $(1,0,0)$, $(0,1,0)$, $(0,0,1)$, $(2,3,-2)$ respectively, then $OP = 2(OA) + 3(OB) - 2(OC)$.
6. $(x_1,x_2,0) + (y_1,y_2,0) = (x_1 + y_1, x_2 + y_2, 0)$; $r(x_1,x_2,0) = (rx_1,rx_2,0)$.

Hence the subset is a vector space. The isomorphism is $(x_1,x_2,0) \to (x_1,x_2)$.

8. (a) All real polynomials in the indeterminate or real variable t; (b) all complex numbers; (c) all real-valued functions on **R**; (d) all solutions of $y'' + y' + y = 0$; (e) all real polynomials of degree less than n in the indeterminate or real variable t.

Sec. 1.4
1. Yes, both are subspaces.
2. If S and T are subspaces of V, then $0 \in S \cap T$. Also, if $\alpha, \beta \in S \cap T$, then $\alpha + \beta \in S$ and $\alpha + \beta \in T$, so that $\alpha + \beta \in S \cap T$; $r\alpha \in S$, $r\alpha \in T$, so that $r\alpha \in S \cap T$ for $r \in$ **R**.
3. P_3.
4. (a) Yes; (b) Yes; (c) Yes; (d) No; (e) Yes; (f) Yes; (g) No; (h) No.
5. $X_3 = X_1 + 2X_2$, $X_4 = -X_1 + X_2$, so that $\mathcal{L}\{X_3,X_4\} \subset \mathcal{L}\{X_1,X_2\}$; $X_1 = \frac{1}{3}X_3 - \frac{2}{3}X_4$, $X_2 = \frac{1}{3}X_3 + \frac{1}{3}X_4$, so that $\mathcal{L}\{X_1,X_2\} \subset \mathcal{L}\{X_3,X_4\}$.
6. As in Prob. 5, noting that $1 = [(1-\sqrt{3}) + (1+\sqrt{3})]/2$ and $\sqrt{3} = [(1+\sqrt{3}) - (1-\sqrt{3})]/2$.
7. (a) R, $\mathcal{O}$; (b) V_2, $\mathcal{L}\{X\}$, $\mathcal{O}$, where $X (\neq 0) \in V_2$.
8. The whole 3-space; any plane through the origin; any line through the origin; the origin.

Sec. 1.5
1. $(2,3,-1,0)$, $(1,-1,2,0)$, $(1,2,-1,0)$; $(2,1,1)$, $(3,-1,2)$, $(-1,2,-1)$, $(0,0,0)$.
2. $a_{i1}x_1 + a_{i2}x_2 + \cdots + a_{in}x_n = 0$, $i = 1, 2, \ldots$, in if $x_1 = x_2 = \cdots = x_n = 0$.
3. $x + y = 1$, $x + y = 2$.
7. $[2,-3,1,4]$; $x = 2 + \frac{3}{2}c_1 - \frac{1}{2}c_2$, $y = c_1$, $z = c_2$.

Sec. 1.6
1. $\pi(i)$ is always written directly below i, regardless of where i is.
2. (a) $a_{11}x_1 + a_{21}x_1 + a_{31}x_1 + a_{12}x_2 + a_{22}x_2 + a_{32}x_2 + a_{13}x_3 + a_{23}x_3 + a_{33}x_3$.
   (b) $a_{11}x_1^2 + a_{21}x_1x_2 + a_{31}x_1x_3 + a_{12}x_2x_1 + a_{22}x_2^2 + a_{32}x_2x_3$.
3. (a) $\begin{pmatrix} 1 & 2 & 3 & 4 \\ 4 & 3 & 1 & 2 \end{pmatrix}$; (b) $\begin{pmatrix} 1 & 2 & 3 & 4 & 5 \\ 4 & 3 & 2 & 5 & 1 \end{pmatrix}$
4. $\pi(1) = 4$, $\pi(3) = 1$, $\pi(5) = 6$, $\pi(6) = 5$.
5. $\pi_1\pi_2 = \begin{pmatrix} 1 & 2 & 3 & 4 & 5 & 6 \\ 2 & 4 & 3 & 5 & 6 & 1 \end{pmatrix}$; $\pi_2\pi_1 = \begin{pmatrix} 1 & 2 & 3 & 4 & 5 & 6 \\ 5 & 1 & 6 & 4 & 3 & 2 \end{pmatrix}$
6. $\pi_1^{-1} = \begin{pmatrix} 1 & 2 & 3 & 4 & 5 & 6 \\ 2 & 1 & 4 & 5 & 3 & 6 \end{pmatrix}$; $\pi_2^{-1} = \begin{pmatrix} 1 & 2 & 3 & 4 & 5 & 6 \\ 1 & 6 & 4 & 3 & 2 & 5 \end{pmatrix}$
7. (a) 6; (b) 7; (c) 5.
8. 3.
9. (a) -3; (b) 6; (c) 13; (d) 3.

344 ANSWERS

10. $|A| = 1(-3) + 2(-17) + 1(13) = -24$.
11. $|A| = 2(-17) + 1(6) + 4(1) = -24$.
12. $|A| = 2(2) + 3(1) - 5(-6) = 37$.

Sec. 1.7
1. If $a_1\alpha_1 + a_2\alpha_2 + \cdots + a_r\alpha_r = 0$ for $r \leq n$, then also $a_1\alpha_1 + \cdots + a_r\alpha_r + 0\alpha_{r+1} + \cdots + 0\alpha_n = 0$.
2. $2x - 2y - z = 0$.
   $x + 3y + z = 0$.
   $x + 2z = 0$.
3. Any vector in $\mathcal{L}\{(1,0,0),(0,1,1)\}$ is of the form (a,b,b).
4. (a) Linearly independent; (b) $(2,2,2,6) = 2(1,1,1,3)$; (c) $(0,0,0) = 0(2,1,0) + 0(1,1,-2)$; (d) $(0,2,0,0) = 2(0,1,0,0) + 0(1,0,1,0)$.
5. (a) $(-4,5,-4,-1) = 2(1,1,-2,1) - 3(2,-1,0,1)$; (b) $(1,-1,-2,-1) = (2,0,-1,0) - (1,1,1,1)$; (c) linearly independent; (d) $(-1,1) = -\frac{4}{5}(2,-1) + \frac{1}{5}(3,1)$.

Sec. 1.8
1. (a) $\{(1,1),(0,1)\}$; (b) $\{(-1,0),(0,1)\}$; (c) $\{(-2,-3),(0,1)\}$.
2. $(1,1) = (0,1) - (-1,0)$, $(0,1) = (0,1) + 0(-1,0)$; $(-1,0) = (0,1) - (1,1)$, $(0,1) = (0,1) + 0(1,1)$.
3. (a) $\{(1,1,0),(0,1,0),(0,0,1)\}$; (b) $\{(1,1,1),(0,-1,1),(0,0,1)\}$.
4. (a) $(1,1,0) = (1,0,0) + (0,1,0) + 0(0,0,1)$, $(0,1,0) = 0(1,0,0) + (0,1,0) + 0(0,0,1)$, $(0,0,1) = 0(1,0,0) + 0(0,1,0) + (0,0,1)$; (b) $(1,1,1) = (1,0,0) + (0,1,0) + (0,0,1)$, $(0,-1,1) = 0(1,0,0) - (0,1,0) + (0,0,1)$.
5. $\{(1,1,0),(-1,2,1),(1,0,0)\}$.

Sec. 1.9
1. 3, 2, or 1.
2. $(2,1,0)$ and $(-1,1,-1)$, respectively.
3. V is a subspace of V_3 of dimension 3.
4. If $f: V_1 \to V$ and $g: V \to V_2$ are isomorphic mappings (see Sec. 1.3), then $gf: V_1 \to V_2$ is also an isomorphism.
5. Map $ax^2 + bx + c$ onto $(a,b,c) \in V_3$.
6. (a) $(2,2,0)$ and $(0,-1,-1)$; (b) $(-2,-2,0)$ and $(0,1,1)$; (c) $(3,3,0)$ and $(0,-2,-2)$.
7. (a) $\alpha \in T$; (b) $\alpha \in S$.

Chapter 2

Sec. 2.1
1. Any property of the space that is left unchanged by the operator.
2. With each $x \in \mathbf{R}$, there is associated a unique $y \in \mathbf{R}$, and so $x \to y$ is a mapping; not an "onto" mapping.
5. (a) No; (b) Yes; (c) No.
6. (a) Yes; (b) No; (c) No.

Sec. 2.2
1. Use induction on n.
2. $N_T = V$, $R_T = \emptyset$.

ANSWERS   345

3. $T\alpha = (x_1 + x_2 + x_3, 2x_1 - x_2 + x_3)$.
4. $N_T = \mathcal{L}\{(2,1,-3)\}$, $R_T = V_2$.
5. (a) 6; (b) 5; (c) 4; (d) 0.

Sec. 2.3
1. (a) $(3x_1 - x_2 - x_3, x_1 + x_2 + x_3, x_2 - x_3)$; (b) $(2x_1 - x_2 - 3x_3, x_1 - x_3, x_2 + x_3)$; (c) $(2x_1 - 2x_2 + x_3, x_1 + x_2 - x_3, 0)$; (d) $(4x_1 - 2x_2, 2x_1, 2x_2 - 2x_3)$; (e) $(7x_1 - 2x_2 - 3x_3, 2x_1 + 3x_2 + 3x_3, 2x_2 - 2x_3)$.
2. (a) $(3x_1 + 3x_2 + x_3, x_1 + x_2 - x_3, -x_1 - x_2 + x_3)$; (b) $(x_1 + 3x_2 + 2x_3, 2x_1 + 2x_2, -2x_1 - 6x_2 - 4x_3)$; (c) $(4x_1 - 4x_3, -x_1 - 3x_2 - 2x_3, x_1 - x_2 - 2x_3)$.
3. (a) $(f + g)(x_1,x_2) = 5x_1 - x_2$; (b) $(f - g)(x_1,x_2) = -x_1 + 3x_2$; (c) $(2f + g)(x_1,x_2) = 7x_1$; (d) $(g - 3f)(x_1,x_2) = -3x_1 - 5x_2$.
4. The range of T_1 is contained in the domain of T_2, but the range of T_2 is not contained in the domain of T_1.
5. $(T_1 + T_2)(x_1,x_2) = (x_1 + r_1x_1 + r_2x_2, x_2 + r_1x_2) = (T_2 + T_1)(x_1,x_2)$, for $r_1 > 1$ and $r_2 \neq 0$.
6. $T_1T_2(x_1,x_2) = (r_1x_1 + r_1r_2x_2, r_1x_2) = T_2T_1(x_1,x_2)$. A dilation followed by a shear has the same effect on a geometric vector as the shear followed by the dilation.

Sec. 2.4
1. $f(a + b) = a + b + 1$, $f(a) = a + 1$, $f(b) = b + 1$, so that $f(a + b) \neq f(a) + f(b)$.
2. No.
3. (a) 1; (b) 3; (c) 3.
4. (a) Yes; (b) No; (c) No; (d) Yes.
5. Yes. If $f(\alpha) = k \neq 0$, then $f(a\alpha/k) = a$.
6. $f(x_1,x_2,x_3) = rx_3$, $r \neq 0$.
7. There exists a mapping $f: V_1 \times V_2 \to \mathbf{R}$, such that
    (i) $f(x_1,x_2) = 0$, for all $x_1 \in V_1$ implies that $x_2 = 0$; $f(x_1,x_2) = 0$, for all $x_2 \in V_2$ implies that $x_1 = 0$.
    (ii) For $x_1,y_1 \in V_1, x_2,y_2 \in V_2$, and $a,b \in \mathbf{R}$: $f(ax_1 + by_1, x_2) = af(x_1,x_2) + bf(y_1,x_2)$; $f(x_1, ax_2 + by_2) = af(x_1,x_2) + bf(x_1,y_2)$.
8. $\{f_1,f_2\}$, where $f_1(x_1,x_2) = \frac{2}{3}x_1 + \frac{1}{3}x_2$, $f_2(x_1,x_2) = \frac{1}{3}x_1 - \frac{1}{3}x_2$.

Sec. 2.5
2. $f(0) = 0$, for all $f \in V^*$.
3. $f(x) = 0$, for all $x \in V$, implies that $f = 0$.
4. If $\alpha_1 (\neq 0) \in S$, extend $\{\alpha_1\}$ to a basis for V, and let $\{f_i\}$ be the corresponding dual basis of V^*. Then $f_1 \notin S^0$.
5. 3.
6. $f \in S_2^0$ implies that $f(\alpha) = 0$, for all $\alpha \in S_2$. Since $S_1 \subset S_2$, $f(\alpha) = 0$ for all $\alpha \in S$, and so $f \in S_1^0$.

Sec. 2.6
1. $TT^{-1} = I_W$, $T^{-1}T = I_V$, where $T: V \to W$, but $I_W \neq I_V$ unless $V = W$.
2. A one-to-one mapping need not be "onto," whereas a one-to-one

correspondence is always a one-to-one mapping that is "onto" (either way). Example: $f(x_1,x_2) = (x_1,x_2,0)$ defines f as a one-to-one mapping of V_2 into V_3.

3. T is one-to-one if and only if $N_T = \mathcal{O}$ if and only if dim R_T = dim V.
4. See example in Prob. 2.
5. $T^{-1}(x_1,x_2) = \left(\dfrac{x_1 - x_2}{2}, \dfrac{x_1 + x_2}{2}\right)$.
6. The null space of $T_1 T_2 \cdots T_r$ must be $\mathcal{O}$; otherwise there would be a nonzero null space for at least one of $T_1, T_2, \ldots, T_n$, a contradiction. Verification of the inverse, as in Theorem 6.4 (Chap. 2).
7. If T is invertible, $N_T = \mathcal{O}$, and so T is nonsingular. If T is nonsingular, it follows from $N_T = \mathcal{O}$ that dim R_T = dim V = dim W, so that $R_T = W$. Hence T is one-to-one and onto, and so T is invertible.

Sec. 2.7

1. (a) $\begin{bmatrix} 0 & 7 & 4 \\ 4 & 5 & 0 \\ 0 & 7 & 10 \end{bmatrix}$ (b) $\begin{bmatrix} 2 & -1 & -8 \\ 0 & -5 & 4 \\ -2 & 1 & -8 \end{bmatrix}$

   (c) $\begin{bmatrix} -1 & 18 & 14 \\ 10 & 15 & -2 \\ 1 & 17 & 29 \end{bmatrix}$ (d) $\begin{bmatrix} 5 & -6 & -22 \\ -2 & -15 & 10 \\ -5 & -1 & -25 \end{bmatrix}$

2. (a) $\begin{bmatrix} 4 & 2 & -2 \\ 2 & 4 & 2 \\ 2 & -4 & 4 \end{bmatrix}$ (b) $\begin{bmatrix} -2 & -1 & 1 \\ -1 & -2 & -1 \\ -1 & 2 & -2 \end{bmatrix}$

3. $\begin{bmatrix} 5 & 6 & -3 \\ 3 & 11 & 6 \\ -6 & 0 & 5 \end{bmatrix}$

4. Consider *definition* of the matrix of a linear transformation.
5. Use argument similar to that in Prob. 4.
6. (a) $\begin{bmatrix} 2 & -1 & 0 \\ 0 & 1 & 1 \end{bmatrix}$ (b) $\begin{bmatrix} 1 & 0 & 0 \\ 0 & 1 & -1 \\ 0 & 1 & 2 \end{bmatrix}$ (c) $\begin{bmatrix} 1 & 1 & 0 \\ 0 & 1 & 2 \\ 1 & 0 & 3 \end{bmatrix}$

7. (a) $\begin{bmatrix} 1 & 0 \\ 0 & 1 \end{bmatrix}$ (b) $\begin{bmatrix} 0 & 0 & 0 \\ 0 & 0 & 0 \\ 0 & 0 & 0 \end{bmatrix}$ (c) $\begin{bmatrix} 1 & 1 & 0 & 0 \\ 0 & 0 & 1 & 0 \\ 0 & 1 & 0 & 1 \end{bmatrix}$

8. $\begin{bmatrix} 1 & -2 & 0 \\ 1 & 0 & 3 \\ 0 & 1 & -1 \end{bmatrix}$

9. $\begin{bmatrix} -\frac{1}{2} & 3 & \frac{1}{2} \\ -\frac{1}{2} & -2 & -\frac{5}{2} \\ \frac{3}{2} & 1 & \frac{5}{2} \end{bmatrix}$

10. (a) $\begin{bmatrix} \frac{1}{2} & \frac{7}{2} \\ \frac{1}{2} & \frac{1}{2} \end{bmatrix}$, relative to $\{(1,1),(-1,1)\}$; $\begin{bmatrix} -\frac{3}{2} & \frac{3}{2} \\ -\frac{3}{2} & \frac{5}{2} \end{bmatrix}$, relative to the $\{E_i\}$ basis.

(b) $\begin{bmatrix} \frac{1}{2} & \frac{4}{3} \\ -\frac{1}{2} & -\frac{5}{3} \end{bmatrix}$, relative to $\{(4,1),(2,-1)\}$; $\begin{bmatrix} \frac{1}{2} & -1 \\ \frac{2}{3} & -\frac{5}{3} \end{bmatrix}$, relative to the $\{E_i\}$ basis.

(c) $\begin{bmatrix} -1 & -2 \\ 4 & 4 \end{bmatrix}$, relative to $\{(1,2),(1,0)\}$; $\begin{bmatrix} 2 & \frac{1}{2} \\ -4 & 1 \end{bmatrix}$, relative to the $\{E_i\}$ basis.

Sec. 2.8 2. (a) $\begin{bmatrix} 7 & 10 & -14 \\ 0 & 6 & -11 \end{bmatrix}$ (b) $\begin{bmatrix} -9 & -8 & 16 \\ 5 & 10 & -15 \\ -2 & 6 & -7 \end{bmatrix}$

3. $AB = \begin{bmatrix} 31 & 27 \\ 26 & 42 \end{bmatrix}$, $BA = \begin{bmatrix} 16 & 28 & 25 & 32 \\ 7 & 12 & 11 & 13 \\ 12 & 16 & 20 & 4 \\ 11 & 20 & 17 & 25 \end{bmatrix}$

4. (a) (7,8); (b) (2,−13); (c) (7,2).
5. (a) (−11,−4,8,−10); (b) (15,7,20,9); (c) (4,1,−12,5).
6. (a) (156,66,72,114); (b) (−152,−66,−104,−106); (c) (−70,−29,−20,−53).
7. $A^3 + ABA + BA^2 + B^2A + A^2B + AB^2 + BAB + B^3$.
8. (a) $\begin{bmatrix} -\frac{1}{11} & \frac{3}{11} \\ \frac{4}{11} & -\frac{1}{11} \end{bmatrix}$ (b) $\begin{bmatrix} -3 & 2 \\ 8 & -5 \end{bmatrix}$ (c) $\begin{bmatrix} -9 & 2 \\ 5 & -1 \end{bmatrix}$

(d) $\begin{bmatrix} -\frac{1}{10} & \frac{1}{5} \\ -\frac{3}{10} & \frac{1}{10} \end{bmatrix}$

10. $AB = I$, but

$$BA = \begin{bmatrix} 0 & -1 & 1 \\ 1 & 2 & -1 \\ 1 & 1 & 0 \end{bmatrix}$$

A transformation represented by A would not be one-to-one, and a transformation represented by B would not be onto.

Sec. 2.9 1. $(AB)' = B'A' = \begin{bmatrix} 1 & -1 & -1 \\ 3 & 0 & 1 \\ 4 & -3 & 0 \end{bmatrix}$

2. The n-tuple *matrix* can be multiplied on the right by an $n \times n$ matrix, but this is not possible for the n-tuble *vector*.

3. $\begin{bmatrix} 1 & 0 & 2 \\ 1 & 1 & 1 \\ 0 & 1 & -2 \end{bmatrix}$

4. $\begin{bmatrix} 1 & 2 & 0 \\ 2 & 0 & -1 \\ -1 & 5 & 2 \end{bmatrix}$

5. $\beta_1 = 2\alpha_1 + 3\alpha_2 + \alpha_3, \beta_2 = \alpha_1 + \alpha_2 + 2\alpha_3, \beta_3 = -2\alpha_1 + 4\alpha_2 + \alpha_3$.

6. (a) (3,1,3); (b) (2,7,1); (c) (0,1,5).

7. $(A')^{-1} = (A^{-1})' = \begin{bmatrix} \frac{3}{5} & \frac{1}{5} \\ -\frac{2}{5} & \frac{1}{5} \end{bmatrix}$

Sec. 2.10 1. (a) $\begin{bmatrix} 2 & 4 \\ 1 & 1 \end{bmatrix}$ (b) $\begin{bmatrix} 1 & 1 & 3 \\ 2 & 2 & 1 \\ -1 & 3 & 5 \end{bmatrix}$ (c) $\begin{bmatrix} 2 & -1 & 3 \\ -1 & 4 & 1 \\ 3 & 1 & 5 \end{bmatrix}$

2. Since (1,7) and (3,−2) are linearly independent, the column rank is 2; similarly for the row rank.

3. Row rank is 2 and, since (2,1,0) and (1,2,0) are linearly independent while (2,1,0) − (1,2,0) = (1,−1,0), the column rank is also 2.

4. $T'g(x,y) = -3y$.

5. $T'g(x,y) = 5x + y$.

6. $T'g(x,y) = 3x - 3y$.

7. Row space is $\mathcal{L}\{(1,-1,2),(2,0,1)\}$; column space is $\mathcal{L}\{(-1,0,-1),(2,1,2)\}$.

8. $A' = \begin{bmatrix} 1 & 1 & 2 \\ 4 & 2 & -1 \\ -1 & 1 & 3 \\ 1 & 0 & 1 \end{bmatrix}$

is the matrix of T', relative to the respective dual bases $\{f_i\}$ and $\{g_i\}$ of V^* and W^*. The operator T' is nonsingular, but not an "onto" mapping.

Chapter 3

Sec. 3.1 1. $1(1) + 2(0) - 1(-1) = 2, 3(1) + 1(0) + 2(-1) = 1, 2(1) - 1(0) + 1(-1) = 1$.

2. $2(2) + 2(1) - 3(2) = 0, -1(2) + 4(1) - 1(2) = 0, 3(2) + 2(1) - 4(2) = 0$.

3. $c[2 \quad 1 \quad 2]' + [1 \quad 1 \quad -1]' = [2c+1 \quad c+1 \quad 2c-1]'$; $2(2c+1) + 2(c+1) - 3(2c-1) = 7, -1(2c+1) + 4(c+1) - 1(2c-1) = 4, 3(2c+1) + 2(c+1) - 4(2c-1) = 9$.

4. (a) $\begin{bmatrix} 2 & -3 & 1 \\ 1 & -2 & 4 \\ 3 & 1 & 1 \end{bmatrix} \begin{bmatrix} x \\ y \\ z \end{bmatrix} = \begin{bmatrix} -1 \\ 2 \\ 9 \end{bmatrix}$

(b) Rank of coefficient matrix is 3, and so the coefficient and augmented matrices have the same rank; hence the system is solvable.

5. $[2 \quad -3 \quad 1] \begin{bmatrix} x \\ y \\ z \end{bmatrix} = [4]$

Three solutions are $(2,0,0)$, $(0,-\frac{4}{3},0)$, $(0,0,4)$, but there are infinitely many solutions of the form $[(4 + 3c_1 - c_2)/2, \; c_1, c_2]$.

Sec. 3.2

1. $\begin{bmatrix} -7 & 2 & 3 \\ 3 & -1 & -1 \\ 5 & -1 & -2 \end{bmatrix}$

2. $\begin{bmatrix} \frac{1}{2} & \frac{1}{2} & -2 \\ -1 & 0 & 2 \\ \frac{1}{2} & -\frac{1}{2} & 1 \end{bmatrix}$

3. $\begin{bmatrix} 1 & 0 & 0 & 0 \\ -3 & 3 & -2 & -1 \\ 1 & -1 & 1 & 0 \\ 0 & -1 & 1 & 1 \end{bmatrix}$

4. $\begin{bmatrix} 1 & 0 & 0 \\ 1 & 1 & 0 \\ 0 & 0 & 1 \end{bmatrix} \begin{bmatrix} 1 & 0 & 0 \\ 0 & 1 & 0 \\ 2 & 0 & 1 \end{bmatrix} \begin{bmatrix} 1 & 0 & 0 \\ 0 & 0 & 1 \\ 0 & 1 & 0 \end{bmatrix} \begin{bmatrix} 1 & 1 & 0 \\ 0 & 1 & 0 \\ 0 & 0 & 1 \end{bmatrix} \begin{bmatrix} 1 & 0 & 0 \\ 0 & 1 & 0 \\ 0 & -2 & 1 \end{bmatrix}$

$\begin{bmatrix} 1 & 0 & 0 \\ 0 & 1 & 0 \\ 0 & 0 & -1 \end{bmatrix} \begin{bmatrix} 1 & 0 & 2 \\ 0 & 1 & 0 \\ 0 & 0 & 1 \end{bmatrix} \begin{bmatrix} 1 & 0 & 0 \\ 0 & 1 & -1 \\ 0 & 0 & 1 \end{bmatrix}$

6. $\begin{bmatrix} 1 & 0 & 0 \\ 0 & 1 & 0 \\ 1 & 0 & 1 \end{bmatrix} \begin{bmatrix} 1 & 0 & 0 \\ 1 & 1 & 0 \\ 0 & 0 & 1 \end{bmatrix} = \begin{bmatrix} 1 & 0 & 0 \\ 1 & 1 & 0 \\ 1 & 0 & 1 \end{bmatrix}$

Sec. 3.3

1. $D(X_1, Y_1) + D(X_1, Y_2) + D(X_2, Y_1) + D(X_2, Y_2)$.
2. $D(X_1, Y_1, Z_1) + D(X_1, Y_1, Z_2) + D(X_1, Y_2, Z_1) + D(X_1, Y_2, Z_2) + D(X_2, Y_1, Z_1) + D(X_2, Y_1, Z_2) + D(X_2, Y_2, Z_1) + D(X_2, Y_2, Z_2)$.
3. Property 1: We let $Z = [z_1 \quad z_2]$. Then $D(aX + bY, Z) = D[(ax_1 + by_1 \quad ax_2 + by_2), (z_1 \quad z_2)] = (ax_1 + by_1)z_2 - (ax_2 + by_2)z_1 = a(x_1 z_2 - x_2 z_1) + b(y_1 z_2 - y_2 z_1) = aD(X,Z) + bD(Y,Z)$. Similarly for linearity in the second argument.
   Property 2: $D(X,X) = x_1 x_2 - x_2 x_1 = 0$.
   Property 3: $D(E_1, E_2) = 1(1) - 0(0) = 1$.
4. $D(X,Y,Z) = x_1 y_2 z_3 + x_2 y_3 z_1 + x_3 y_1 z_2 - x_3 y_2 z_1 - x_1 y_3 z_2 - x_2 y_1 z_3$
5. As in Prob. 3.

Sec. 3.4
1. (a) 2, -6, -12; (b) 12, 14, 168.
2. (a) Independent; (b) independent.
3. $\{(1,2,-1,2), (3,2,-1,2), (0,0,1,0), (0,0,0,1)\}$.
4. No, since the determinant of the coefficient matrix is -19.
5. No, since the determinant of the coefficient matrix is -3.
6. No. $(3,4,1)$ is in the space and $(1,0,0)$ is not.
7. $\begin{bmatrix} x \\ y \\ z \end{bmatrix} = \begin{bmatrix} -\frac{3}{5} & 1 & \frac{1}{5} \\ -\frac{2}{5} & 1 & \frac{4}{5} \\ \frac{4}{5} & -1 & -\frac{3}{5} \end{bmatrix} \begin{bmatrix} 7 \\ 6 \\ -4 \end{bmatrix} = \begin{bmatrix} 1 \\ 0 \\ 2 \end{bmatrix}$
8. (a) $\begin{bmatrix} -12 & 3 & 7 \\ 6 & -1 & -3 \\ -2 & 1 & 1 \end{bmatrix} = \operatorname{adj} A = 2A^{-1}$

   (b) $\begin{bmatrix} -1 & 0 & 1 \\ -1 & -1 & 1 \\ 7 & 4 & -6 \end{bmatrix} = \operatorname{adj} A = A^{-1}$

   (c) $\begin{bmatrix} -7 & 3 & -5 \\ 9 & -4 & 6 \\ 11 & -5 & 8 \end{bmatrix} = \operatorname{adj} A = -A^{-1}$

Sec. 3.5
1. (a) No; (b) Yes; (c) No; (d) No; (e) Yes.
2. $af_1(x)f_2(y) + bf_2(x)f_1(y)$, with f_1, f_2 linear or, more simply, rxy with $r \in \mathbf{R}$. The only alternating form is the zero form and occurs when $r = 0$.
3. For any two real numbers r_1, r_2, there exist real numbers c_1, c_2 (not both 0) such that $c_1r_1 + c_2r_2 = 0$.

Sec. 3.6
1. D would be a function of the desired type.
2. Both T_1 and T_2 must be singular, and so $\det T_1 = \det T_2 = 0$.
3. $T'f(\alpha_1, \alpha_2, \ldots, \alpha_n) = f(T\alpha_1, T\alpha_2, \ldots, T\alpha_n)$. If $T = 0$, then $T'f = 0$, and so $\det T = 0$. A similar argument for $T = I$.
4. (a) 1; (b) 1.
5. 0.
6. (a) -2; (b) 0.
7. Let $f_1(x,y) = x$, $\alpha_1^0 = (1,0)$, $\alpha_2^0 = (0,1)$. Then $f_1(\alpha_1^0) = 1 \neq 0$. If we take $f_1^0(x,y) = y$, then $f_1^0(\alpha_1^0) = 0, f_1^0(\alpha_2^0) \neq 0$. Now define $f[(x_1,x_2),(y_1,y_2)] = f_1(y_1,y_2)f_1^0(x_1,x_2) - f_1(x_1,x_2)f_1^0(y_1,y_2) = y_1x_2 - x_1y_2$, which is nonzero and alternating.

Chapter 4

Sec. 4.1
1. (a) $\sqrt{6}$; (b) $\sqrt{3}$; (c) $\sqrt{29}$.
2. (a) $\sqrt{30}$; (b) $\sqrt{29}$; (c) $\sqrt{69}$.
3. (a) -1; (b) 4.
4. (a) -2; (b) 1.
5. -10.
6. 3.

ANSWERS 351

Sec. 4.2
1. (a) 5; (b) 4.
2. (a) $\frac{1}{4}$; (b) $\frac{3}{4}$, -1.
3. (a) $\sqrt{5}$; (b) $\sqrt{3}$; (c) $\sqrt{15}$.
4. (a) $1/(5\sqrt{2})$; (b) $\sqrt{\frac{6}{7}}$; (c) $-\frac{1}{3}\sqrt{\frac{2}{3}}$.
5. (a) $\sqrt{17}$; (b) $\sqrt{5}$; (c) $\sqrt{19}$.
6. $\mathcal{L}\{(1,2)\}$.
7. $\mathcal{L}\{(2,0,-1),(0,2,1)\}$.
8. (a) $1/\sqrt{5}$; (b) $\sqrt{\frac{13}{3}}$.
9. $23/4$, $\sqrt{\frac{31}{5}}$, $\sqrt{1{,}423/210}$.

Sec. 4.3
2. Yes.
3. The form of our inductive hypothesis is "if . . . then," and so it is not a *declarative* statement for all n.
4. $(-1,2)$, $(6,3)$.
5. $\{(1,2,1),(-2,1,0),(1,2,-5)\}$, having chosen $(1,0,0)$ as the original third basis vector.
6. $\{(2,3,-1),(1,-2,-4),(2,-1,1)\}$, having chosen $(1,0,0)$ as the original third basis vector.
7. $\{(2,1,1),(4,-13,5),(-3,1,5)\}$.
8. $\{(1,1,1,1),(-9,3,7,-1),(1,-12,7,4)\}$.

Sec. 4.4
1. (a) The plane through the origin perpendicular to the geometric vector α, where $\alpha \leftrightarrow X$.
   (b) The line through the origin perpendicular to the plane determined by α and β, where $\alpha \leftrightarrow X$ and $\beta \leftrightarrow Y$.
2. The line through the origin that is perpendicular to the line through the origin determined by α, where $\alpha \leftrightarrow X$.
3. All orthogonal decompositions are direct-sum decompositions, but not conversely. In addition, an orthogonal decomposition is meaningful only in an inner-product space.
4. No. [See Example 3 in Sec. 4.2; also Theorem 7.1 (Chap. 4).]
5. (a) 1; (b) $\frac{40}{9}$; (c) $\frac{1}{13}$.
6. 1.
7. (a) 4; (b) -5.
8. -3 (standard); -27 (other basis).
9. (a) $\mathcal{L}\{(1,0,1),(0,1,1)\}$; (b) $\mathcal{L}\{(0,-2,1)\}$.

Sec. 4.5
3. (a) $\begin{bmatrix} 1 & 0 & 0 \\ 0 & -\sqrt{3}/2 & -\frac{1}{2} \\ 0 & -\frac{1}{2} & \sqrt{3}/2 \end{bmatrix}$ (b) $\begin{bmatrix} \cos\theta & \sin\theta & 0 \\ -\sin\theta & \cos\theta & 0 \\ 0 & 0 & -1 \end{bmatrix}$
4. Since it is nonsingular and preserves inner products.
6. If $X = (3,0,4)$, $TX = (\frac{2}{3}, -\frac{5}{3}, \frac{14}{3})$ and $(TX, TX) = 25 = (X, X)$.
7. If $X = (3,0,4)$, $TX = (\frac{5}{3}, \frac{1}{3}, \frac{8}{3})$, $(TX, TX) = 10$.
8. Let $X = (3,0,4)$ and $Y = (2,2,1)$. Then $TX = (\frac{2}{3}, -\frac{5}{3}, \frac{14}{3})$, $TY = (-\frac{1}{3}, \frac{4}{3}, \frac{8}{3})$, and

$$\frac{(\mathsf{T}X,\mathsf{T}Y)}{|\mathsf{T}X|\,|\mathsf{T}Y|} = \frac{10}{(5)(3)} = \frac{(X,Y)}{|X|\,|Y|}$$

9. $\begin{bmatrix} \frac{1}{2} & -\frac{1}{2} & 1/\sqrt{2} \\ \frac{1}{2} & \frac{5}{6} & 1/(3\sqrt{2}) \\ -1/\sqrt{2} & 1/(3\sqrt{2}) & \frac{2}{3} \end{bmatrix}$

Sec. 4.6
1. With $X = (x_1, x_2, x_3)$, $f_Y(X) = -x_1 + 2x_2 + x_3$; $f_Y(1,-1,1) = -2 = f_X(-1,2,1)$.
2. $Y = (2, -2, 2, -2)$.
3. $Y = (-\sqrt{2}, -2\sqrt{2})$.
4. (a) $\mathsf{T}^*X = 2X$; (b) $\mathsf{T}^*X = -X$; (c) $\mathsf{T}^*X = 0$.
5. (a) $\mathsf{T}^*E_1 = (-1,1)$, $\mathsf{T}^*E_2 = (2,-3)$; (b) $\mathsf{T}^*E_1 = (1,-2)$, $\mathsf{T}^*E_2 = (3,4)$.
6. (a) $\begin{bmatrix} 2 & 3 \\ -1 & 2 \end{bmatrix}$ (b) $\begin{bmatrix} 2 & 1 & -1 \\ 1 & 1 & 2 \\ -1 & 2 & 1 \end{bmatrix}$ (c) $\begin{bmatrix} 2 & 1 & 1 & 3 \\ 0 & 2 & 2 & 0 \\ 1 & 0 & 3 & 0 \\ 1 & -1 & -1 & 2 \end{bmatrix}$
7. $\mathsf{T}^*(x,y) = (x\cos\theta + y\sin\theta, -x\sin\theta + y\cos\theta) = [x\cos(-\theta) - y\sin(-\theta), x\sin(-\theta) + y\cos(-\theta)]$.
8. Let $\mathsf{T}_1 = (\mathsf{T} + \mathsf{T}^*)/2$ and $\mathsf{T}_2 = (\mathsf{T} - \mathsf{T}^*)/2$. Then $\mathsf{T} = \mathsf{T}_1 + \mathsf{T}_2$, where $\mathsf{T}_1 = \mathsf{T}_1^*$ and $\mathsf{T}_2 = -\mathsf{T}_2^*$.

Sec. 4.7
1. The standard inner product is the special case of the A-inner product with $A = I$.
2. (a) 0; (b) -2; (c) 3.
3. (a) Yes; (b) No.
4. (a) $\begin{bmatrix} 1 & 0 & 0 \\ 0 & 1 & 1 \\ 0 & 0 & 1 \end{bmatrix} \begin{bmatrix} 1 & 0 & 0 \\ 0 & 1 & 0 \\ 0 & 1 & 1 \end{bmatrix} = \begin{bmatrix} 1 & 0 & 0 \\ 0 & 2 & 1 \\ 0 & 1 & 1 \end{bmatrix}$
(b) $x_1^2 + 2x_2^2 + x_3^2 + 2x_2x_3$.
(c) 9.
5. (a) $2x_1y_1 + x_1y_2 + 2x_1y_3 + x_2y_1 - x_2y_2 + x_3y_1 + x_3y_2 + x_3y_3$.
(b) $2x_1y_1 - x_1y_2 + x_2y_2 + 2x_2y_3 + x_3y_1 + x_3y_2 + 3x_3y_3$.
6. (a) $2x_1^2 + 3x_2^2 - x_3^2 + 8x_1x_3 + 4x_2x_3$.
(b) $x_1^2 - 2x_2^2 + x_3^2 + 2x_1x_2 + 4x_2x_3$.
7. (a) $\begin{bmatrix} 3 & 3 \\ 0 & 4 \end{bmatrix}$ (b) $\begin{bmatrix} 3 & \frac{3}{2} \\ \frac{3}{2} & 4 \end{bmatrix}$
8. (a) $\begin{bmatrix} 1 & -2 & 1 \\ 0 & 1 & 4 \\ 0 & 0 & -6 \end{bmatrix}$ (b) $\begin{bmatrix} 1 & -1 & \frac{1}{2} \\ -1 & 1 & 2 \\ \frac{1}{2} & 2 & -6 \end{bmatrix}$
9. (a) $\begin{bmatrix} 3 & 4 & 0 \\ 0 & 0 & -5 \\ 0 & 0 & 1 \end{bmatrix}$ (b) $\begin{bmatrix} 3 & 2 & 0 \\ 2 & 0 & -\frac{5}{2} \\ 0 & -\frac{5}{2} & 1 \end{bmatrix}$

10. It is the standard inner product, since $A = I$.

Sec. 4.8
1. $\begin{bmatrix} 3 & 1 & 2 \\ 1 & 5 & -4 \\ 2 & -4 & 6 \end{bmatrix}$

2. $\begin{bmatrix} 8\pi^3/3 & -4\pi^4 & 4\pi^2 \\ -4\pi^4 & 32\pi^5/5 & -16\pi^3/3 \\ 4\pi^2 & -16\pi^3/3 & 8\pi \end{bmatrix}$

3. $11y - 5x = 0$.
4. $x + y = 0$.
5. $\sqrt{\frac{2}{7}}$.
6. $\sqrt{\frac{5}{2}}$.

Chapter 5

Sec. 5.1
1. (a) Bilinear; (b) not bilinear; (c) bilinear; (d) not bilinear.
2. For $X, Y \in V_n$, $n > 2$, $\det(X,Y)$ is not defined.
3. (a) $\begin{bmatrix} 2 & 0 & -3 \\ 0 & 2 & -5 \\ 4 & 0 & 1 \end{bmatrix}$ (b) $\begin{bmatrix} 4 & 0 & 0 \\ 0 & 2 & 0 \\ 0 & 0 & 1 \end{bmatrix}$

    (c) $\begin{bmatrix} 3 & 0 & 0 & 1 \\ 0 & 0 & -2 & 0 \\ 0 & 0 & 0 & 0 \\ 0 & 1 & 0 & -2 \end{bmatrix}$ (d) $\begin{bmatrix} 1 & 0 & 0 & 0 \\ 0 & 1 & 0 & 0 \\ 0 & 0 & 1 & 0 \\ 0 & 0 & 0 & 1 \end{bmatrix}$

4. (a) $2x_1y_1 - x_1y_2 + x_1y_3 + 3x_2y_2 - x_3y_1 + x_3y_3$.
   (b) $2xx' + xy' - xz' + 2yy' + zx' - 3zy' + zz'$.
5. $x_1y_1 + x_1y_2 + x_1y_4 + 2x_2y_1 + x_2y_3 - x_2y_4 + x_3y_1 + 2x_3y_3 - x_3y_4 - x_4y_1 + x_4y_2 + x_4y_3$.
6. (a) -12; (b) -3.
7. If $\{\alpha_i\}$, $\{\beta_i\}$ and $\{\alpha_i'\}$, $\{\beta_i'\}$ are the old and new bases of V and W, respectively, then:

   $\alpha_1' = \alpha_1 - \alpha_2 + 2\alpha_3$ $\beta_1' = 2\beta_1 + \beta_2 + \beta_3$
   $\alpha_2' = \alpha_1 + \alpha_2 + \alpha_3$ $\beta_2' = \beta_1 + \beta_3$
   $\alpha_3' = 2\alpha_1 + \alpha_3$ $\beta_3' = \beta_2 + \beta_3$

8. $13x_1y_1 + 2x_1y_2 + 5x_1y_3 + 5x_2y_1 + x_2y_3 + 7x_3y_1 + 5x_3y_3$.
9. $5x_1'y_1' + 2x_1'y_2' - 5x_1'y_3' + x_2'y_1' + 5x_2'y_2' - 5x_2'y_3' - x_3'y_1' - 2x_3'y_2' - 7x_3'y_3'$.

Sec. 5.2
1. $(X,Y) = x_1y_1 + x_2y_2 + \cdots + x_ny_n$, and $Q(X) = x_1^2 + x_2^2 + \cdots + x_n^2 = |X|^2$.
2. An inner product is a bilinear function whose matrix is positive; hence the inner-product form is polar to a positive quadratic form.

3. (a) $2x^2 - 2xy + 2xz + 2y^2 + z^2$;
 (b) $x_1^2 + 4x_1x_2 + 8x_1x_3 + 2x_2^2 - 4x_2x_3$.
4. (a) $[x \ y \ z] \begin{bmatrix} 2 & -\frac{3}{2} & 0 \\ -\frac{3}{2} & 1 & -2 \\ 0 & -2 & -4 \end{bmatrix} \begin{bmatrix} x \\ y \\ z \end{bmatrix}$

 (b) $[x_1 \ x_2 \ x_3] \begin{bmatrix} 1 & 2 & 0 \\ 2 & -3 & 0 \\ 0 & 0 & 1 \end{bmatrix} \begin{bmatrix} x_1 \\ x_2 \\ x_3 \end{bmatrix}$

5. (a) $[x \ y \ z] \begin{bmatrix} 2 & -3 & 0 \\ 0 & 1 & -4 \\ 0 & 0 & -4 \end{bmatrix} \begin{bmatrix} x \\ y \\ z \end{bmatrix}$

 (b) $[x_1 \ x_2 \ x_3] \begin{bmatrix} 1 & 4 & 0 \\ 0 & -3 & 0 \\ 0 & 0 & 1 \end{bmatrix} \begin{bmatrix} x_1 \\ x_2 \\ x_3 \end{bmatrix}$

6. (a) $[x \ y \ z] \begin{bmatrix} 1 & 2 & 4 \\ 2 & 2 & -2 \\ 4 & -2 & 0 \end{bmatrix} \begin{bmatrix} x' \\ y' \\ z' \end{bmatrix}$

 $[x_1 \ x_2 \ x_3] \begin{bmatrix} 1 & 2 & 1 \\ -3 & 2 & -2 \\ 4 & 3 & 0 \end{bmatrix} \begin{bmatrix} x_1' \\ x_2' \\ x_3' \end{bmatrix}$

 (b) $2xx' - \frac{3}{2}x'y - \frac{3}{2}xy' + yy' - 2z'y - 2y'z - 4zz'$; $x_1x_1' + 2x_1x_2' + 2x_2x_1' - 3x_2x_2' + x_3x_3'$.
7. If A is symmetric, then $(P'AP)' = P'A'P = P'AP$; hence $P'AP$ is symmetric.
8. (a) $\begin{bmatrix} 8 & 4 & -5 \\ 5 & 4 & 7 \\ 18 & -2 & -1 \end{bmatrix}$ (b) $\begin{bmatrix} 1 & 2 & -10 \\ 3 & 9 & 0 \\ 13 & -9 & 27 \end{bmatrix}$
9. $x_1y_1 - x_1y_2 - x_2y_1 + 3x_2y_2 + 4x_3y_3$.
10. $2x_1y_1 + 2x_1y_2 + 2x_2y_1 + \frac{1}{2}x_2y_3 + \frac{1}{2}x_3y_2 - 3x_3y_3$.
11. (a) $-3x'^2 - 6x'y' - 10x'z' + 2y'^2 - 24y'z' + 5z'^2$;
 (b) $3x'^2 + 14x'y' - 8x'z' + 14y'^2 - 6y'z' - 5z'^2$.

Sec. 5.3 1. (a) $x_1^2 + x_2^2 - x_3^2$, where $x_1 = \sqrt{2}\,(x - y)$, $x_2 = y + 3z$, $x_3 = \sqrt{13}\,z$;
 (b) $x_1^2 + x_2^2 - x_3^2$, where $x_1 = x + 3z$, $x_2 = \sqrt{2}\,(y + z)$, $x_3 = \sqrt{11}\,z$.
2. (a) $x^2 + y^2 + z^2 - w^2$, where $x = x_1 - x_2/2$, $y = \sqrt{7}\,x_2/2$, $z = \sqrt{3}\,x_3 + x_4/(2\sqrt{3})$, $w = x_4/(2\sqrt{3})$; (b) $x^2 - y^2 - z^2$, where $x = x_1 + 2x_2$, $y = x_1 - x_3$, $z = 2x_2 + x_3$.

3. $3y_1^2 + 6y_1y_2 + 9y_2^2$. If the old and new bases are $\{\alpha_1,\alpha_2\}$ and $\{\beta_1,\beta_2\}$, respectively, then $\beta_1 = 2\alpha_1 + \alpha_2$, $\beta_2 = \alpha_1 - \alpha_2$.
4. Different bases are used for the space.
5. If $\{\beta_1,\beta_2,\beta_3\}$ is the new basis, then $\beta_1 = \alpha_1$, $\beta_2 = \frac{1}{2}\alpha_1 + \frac{1}{2}\alpha_2 - \frac{1}{4}\alpha_3$, $\beta_3 = \frac{1}{2}\alpha_3$.

Sec. 5.4

1. (a) $\begin{bmatrix} 0 & 1 & 0 \\ 1 & 0 & 0 \\ 0 & 0 & 1 \end{bmatrix}$ (b) $\begin{bmatrix} 0 & 1 & 0 \\ 0 & 0 & 1 \\ 1 & 0 & 0 \end{bmatrix}$

2. (a) $\begin{bmatrix} 1 & 0 & 0 & 0 \\ 0 & 0 & 1 & 0 \\ 0 & 1 & 0 & 0 \\ 0 & 0 & 0 & 1 \end{bmatrix}$ (b) $\begin{bmatrix} 0 & 0 & 1 & 0 \\ 1 & 0 & 0 & 0 \\ 0 & 0 & 0 & 1 \\ 0 & 1 & 0 & 0 \end{bmatrix}$

4. (a) diag $\{1,1,-1,-1\}$; (b) diag $\{1,1,-1,-1,0,0\}$.
5. If the old basis is $\{\alpha_1,\alpha_2,\alpha_3\}$, then the desired basis is $\{\beta_1,\beta_2,\beta_3\}$ where $\beta_1 = (1/\sqrt{3})\alpha_1$, $\beta_2 = \frac{1}{2}\alpha_2$, $\beta_3 = \frac{1}{3}\alpha_3$.
6. (a) diag $\{1,1,0\}$; (b) diag $\{1,-1,-1\}$.
7. (a) $x_1'^2 + x_2'^2 - x_3'^2$; (b) $x'^2 + y'^2 - z'^2$.

Sec. 5.5

1. Every vector is an eigenvector belonging to the eigenvalue 2; by Theorem 5.1, no vector can belong to distinct eigenvalues.
2. Every real number would be an eigenvalue, with 0 as one of its associated eigenvectors.
3. 1, $\mathcal{L}\{(1,0,0,0),(0,0,0,1)\}$; 3, $\mathcal{L}\{(0,1,0,0),(0,0,1,0)\}$.
4. The expansion of $|A - xI|$ is a finite sum of terms of the form $a_k x^k$, for real a_k and positive integer k, and so is a polynomial.
5. (a) $-x^3 + 4x^2 - 6x + 1$; (b) $-x^3 + x^2 + x - 1$.
6. (a) $x^2 - 4x - 5$; $-1,5$.
   (b) $x^2 - 5x + 10$; no real eigenvalues.
   (c) $-x^3 + 5x^2 - 2x - 8$; $-1,2,4$.
7. (a) 1; $S_1 = V$; (b) 1,3; $S_1 = \mathcal{L}\{(-1,1)\}$, $S_3 = \mathcal{L}\{(1,0)\}$.
8. (a) Only real eigenvalue is -1; $S_{-1} = \mathcal{L}\{(0,0,1)\}$.
   (b) Only real eigenvalue is 1; $S_1 = \mathcal{L}\{(0,1,0)\}$.
9. (a) Eigenvalues are $-1,8$; $S_{-1} = \mathcal{L}\{(1,-2,0),(1,0,-1)\}$, $S_8 = \mathcal{L}\{(2,1,2)\}$.
   (b) Eigenvalues are $1, 1+\sqrt{5}, 1-\sqrt{5}$; $S_1 = \mathcal{L}\{(1, 0, 2)\}$, $S_{1+\sqrt{5}} = \mathcal{L}\{(-2,-\sqrt{5},1)\}$, $S_{1-\sqrt{5}} = \mathcal{L}\{(-2,\sqrt{5},1)\}$.
10. $(-3,-2,9)$.

Sec. 5.6

1. (a) For example: $P = \begin{bmatrix} 1/\sqrt{2} & -1/\sqrt{2} \\ 1/\sqrt{2} & 1/\sqrt{2} \end{bmatrix}$

   (b) For example: $P = \begin{bmatrix} \frac{1}{2} & -\sqrt{3}/2 \\ \sqrt{3}/2 & \frac{1}{2} \end{bmatrix}$

(c) For example: $P = \begin{bmatrix} \sqrt{5}/3 & -\frac{2}{3} \\ \frac{2}{3} & \sqrt{5}/3 \end{bmatrix}$

2. For example: $P = \begin{bmatrix} 1/\sqrt{2} & 1/\sqrt{3} & 1/\sqrt{6} \\ 0 & -1/\sqrt{3} & 2/\sqrt{6} \\ 1/\sqrt{2} & -1/\sqrt{3} & -1/\sqrt{6} \end{bmatrix}$

3. For example: $P = \begin{bmatrix} 0 & -\frac{1}{2} & \sqrt{3}/2 \\ 0 & \sqrt{3}/2 & \frac{1}{2} \\ 1 & 0 & 0 \end{bmatrix}$

4. $4x^2 - 4y^2 + 4z^2$, using $P = \begin{bmatrix} 0 & \frac{1}{2} & \sqrt{3}/2 \\ 0 & -\sqrt{3}/2 & \frac{1}{2} \\ 1 & 0 & 0 \end{bmatrix}$

5. The definition of an eigenvalue.
6. No. Only the canonical matrix, with diagonal entries in a prescribed order, is unique.

Chapter 6

Sec. 6.1 1. (a) $(x-1)(x+2)$; $(x-1)^2(x+2)$.
   (b) $(x+2)(x^2+x+1)$; $(x+2)^2(x^2+x+1)$; $(x+2)(x^2+x+1)^2$; $(x+2)^2(x^2+x+1)^2$.
   (c) $(x-1)(x^2+1)$.
   (d) $(x-2)(x^2-x-1)$; $(x-2)^2(x^2-x-1)$; $(x-2)^3(x^2-x-1)$.

2. (a) $(x+2)(x+1)$; $(x+2)(x+1)^2$.
   (b) $(x-1)(x+1)(x+2)$; $(x-1)(x+1)(x+2)^2$.

4. $x^3 + 2x^2 + 2x + 1$.
5. (a) $x^2 - 5x + 4$; (b) $x^2 - 3x - 18$.
6. (a) $x^3 - 2x^2 - x + 2$; (b) $-x^3 - 2x^2 + 4x + 8$.
8. The matrix satisfies the equation $x - aI = 0$.

Sec. 6.2 1. If $A = \begin{bmatrix} 1 & 0 \\ 0 & 1 \end{bmatrix}$, $P^{-1}AP = A$ for any nonsingular 2×2 matrix P. Hence the only matrix similar to A is A itself.

2. The diagonal elements of D are its eigenvalues, and D and A have the same eigenvalues.

3. For matrix $\begin{bmatrix} 2 & 0 \\ 0 & 2 \end{bmatrix}$, two linearly independent eigenvectors in V_2, associated with the eigenvalue 2, are $(1,0)$ and $(0,1)$.

4. (a) Yes; (b) No; (c) Yes; (d) No.
5. (a) For example: $P = \begin{bmatrix} 1 & 1 \\ -2 & 1 \end{bmatrix}$

ANSWERS 357

(b) For example: $P = \begin{bmatrix} 0 & 0 & 1 \\ 1 & 2 & 1 \\ 1 & 1 & 0 \end{bmatrix}$

6. (a) $m(x) = x^2 - 2x + 3$, which is irreducible.
   (b) $m(x) = (x - 2)(x - 1)^2$, which has a repeated factor.
   (c) $m(x) = (x - 1)^2$, which has a repeated factor.

Sec. 6.3

1. (a) $-1 - 3i$; (b) 1; (c) -4.
2. (a) $(35i + 51)/60$; (b) $(62i + 73)/60$.
3. (a) $A' = \begin{bmatrix} 2 & i \\ 1-i & 3i \end{bmatrix}$ $\bar{A} = \begin{bmatrix} 2 & 1+i \\ -i & -3i \end{bmatrix}$

    $A^* = \begin{bmatrix} 2 & -i \\ 1+i & -3i \end{bmatrix}$

    (b) $A' = \begin{bmatrix} i & 2i & 2 \\ 1 & 2 & -i \\ -i & 1 & 1+i \end{bmatrix}$ $\bar{A} = \begin{bmatrix} -i & 1 & i \\ -2i & 2 & 1 \\ 2 & i & 1-i \end{bmatrix}$

    $A^* = \begin{bmatrix} -i & -2i & 2 \\ 1 & 2 & i \\ i & 1 & 1-i \end{bmatrix}$

4. (a) $\begin{bmatrix} \frac{2}{3} & -i/3 \\ i/3 & \frac{1}{3} \end{bmatrix}$ (b) $\begin{bmatrix} 1 & -i & -i \\ -i/2 & -1 & -\frac{1}{2} \\ 0 & 1 & 1 \end{bmatrix}$

5. Any complex matrix $C = A + iB$, where $A = (C + C^*)/2$ and $B = (C - C^*)/2$ are hermitian.
6. (a) Eigenvalues are 0, 3; (b) eigenvalues are 1, $(1 \pm \sqrt{29})/2$.
7. (a) If A is the given matrix, $A^* = \begin{bmatrix} (1-i)/\sqrt{3} & -i/\sqrt{3} \\ -i/\sqrt{3} & (1+i)/\sqrt{3} \end{bmatrix}$,
    and $AA^* = A^*A = I$.
   (b) If A is the given matrix, $A^* = \text{diag } \{\bar{c}_1, \bar{c}_2, \ldots, \bar{c}_n\}$, and $AA^* = A^*A = \text{diag } \{c_1\bar{c}_1, c_2\bar{c}_2, \ldots, c_n\bar{c}_n\} = I$ if and only if $|c_i| = 1$, $i = 1, 2, \ldots, n$.
8. $\sigma_x^2 = I$ $\sigma_x^* = \sigma_x$ $\sigma_x\sigma_x^* = \sigma_x^2 = I$
   $\sigma_y^2 = I$ $\sigma_y^* = \sigma_y$ $\sigma_y\sigma_y^* = \sigma_y^2 = I$
   $\sigma_z^2 = I$ $\sigma_z^* = \sigma_z$ $\sigma_z\sigma_z^* = \sigma_z^2 = I$
9. $\sigma_x\sigma_y - \sigma_y\sigma_x = \begin{bmatrix} 2i & 0 \\ 0 & -2i \end{bmatrix} = 2i\sigma_z$

   $\sigma_y\sigma_z - \sigma_z\sigma_y = \begin{bmatrix} 0 & 2i \\ 2i & 0 \end{bmatrix} = 2i\sigma_x$

   $\sigma_z\sigma_x - \sigma_x\sigma_z = \begin{bmatrix} 0 & 2 \\ -2 & 0 \end{bmatrix} = 2i\sigma_y$

Sec. 6.4 1. (a) $T = E_1 + 2E_2 - E_3$; diag $\{1,2,-1,1\}$, diag $\{1,2,-1,2\}$, diag $\{1,2,-1,-1\}$.
(b) $T = iE_1 - iE_2 + 2E_3$; diag $\{i,-i,2,i\}$, diag $\{i,-i,2,-i\}$, diag $\{i,-i,2,2\}$.
2. Let $P^{-1}AP = B$, and show that $BB^* = B^*B$, using $P^* = P^{-1}$.
3. $\begin{bmatrix} (3^{25}-1)/2 & (3^{25}+1)/2 \\ (3^{25}+1)/2 & (3^{25}-1)/2 \end{bmatrix}$. (Find P such that $P^{-1}AP =$ diag $\{-1,3\}$.)
4. (a) Yes; (b) Yes; (c) No; (d) Yes.
5. $V = V_1 \oplus V_2 \oplus V_3$, where V_1 is the null space of $T - I$, V_2 is the null space of $T - iI$, V_3 is the null space of $T + iI$.
6. (a) $A = \begin{bmatrix} 1 & 1 \\ 1 & 1 \end{bmatrix}$ or any real symmetric matrix;
(b) $A = \begin{bmatrix} 1 & i \\ -i & 1 \end{bmatrix}$
7. (a) Eigenvalues of both A and A^* are 0, 2; (b) eigenvalues of both A and A^* are 0, 2.
8. $m(x) = (x-1)(x-i)(x-1-i)$, $T = E_1 + iE_2 + (1+i)E_3$; $m(x) = (x-1)(x+i)(x-1+i)$, $T^* = E_1 - iE_2 + (1-i)E_3$.
9. $\begin{bmatrix} 0 & i \\ 1 & 0 \end{bmatrix}$
10. Check that $AA^* = A^*A$ and $BB^* = B^*B$; compare $(A+B)(A+B)^*$ with $(A+B)^*(A+B)$ and $(AB)(AB)^*$ with $(AB)^*(AB)$.
11. Verify that $(aT + bT^*)(aT + bT^*)^* = (aT + bT^*)^*(aT + bT^*)$.
12. For example:
$$iA + A^* = \begin{bmatrix} 1+i & -1 & 2-i \\ -i & 2+2i & 0 \\ -1+2i & 0 & 3+3i \end{bmatrix}$$
where $A = \begin{bmatrix} 1 & i & -2i \\ 0 & 2 & 0 \\ i & 0 & 3 \end{bmatrix}$

Sec. 6.5 1. (a) $\begin{bmatrix} 2 & 0 & 0 & 0 \\ 0 & 2 & 0 & 0 \\ 0 & 0 & 1 & 0 \\ 0 & 0 & -1 & 2 \end{bmatrix}$

(b) $\begin{bmatrix} 1 & 1 & 0 & 0 & 0 & 0 \\ 2 & -1 & 0 & 0 & 0 & 0 \\ 0 & 0 & 1 & 1 & 0 & 0 \\ 0 & 0 & 2 & -1 & 0 & 0 \\ 0 & 0 & 0 & 0 & 1 & 0 \\ 0 & 0 & 0 & 0 & -1 & 2 \end{bmatrix}$

(c) $\begin{bmatrix} 2 & 0 & 0 & 0 & 0 \\ 0 & 1 & 1 & 0 & 0 \\ 0 & 2 & -1 & 0 & 0 \\ 0 & 0 & 0 & 1 & 0 \\ 0 & 0 & 0 & -1 & 2 \end{bmatrix}$

2. (a) $W_1 = \mathcal{L}\{(1,0),(0,1)\} = V$;
   (b) $W_1 = \mathcal{L}\{(1,1)\}$, $W_2 = \mathcal{L}\{(0,1)\}$;
   (c) $W_1 = \mathcal{L}\{(3,-2)\}$, $W_2 = \mathcal{L}\{(1,1)\}$.

3. (a) $W_1 = \mathcal{L}\{(1,0,0),(0,1,0)\}$, $W_2 = \mathcal{L}\{(1,-1,1)\}$;
   (b) $W_1 = \mathcal{L}\{(1,2,1)\}$, $W_2 = \mathcal{L}\{(1,0,1)\}$, $W_3 = \mathcal{L}\{(1,0,-1)\}$.

4. D^n maps every element of P_n onto 0.

6. $\begin{bmatrix} cd & d^2 \\ -c^2 & -cd \end{bmatrix}$, with $c,d \in \mathbf{C}$.

7. (a) For example: $\begin{bmatrix} 0 & 1 & -1 \\ 0 & 0 & 1 \\ 0 & 0 & 0 \end{bmatrix}$

   (b) For example: $\begin{bmatrix} 0 & 0 & 1 \\ 0 & 0 & 0 \\ 0 & 0 & 0 \end{bmatrix}$

8. (a) Check that $T(1,1,1) = (1,2,1)$ and $T(1,2,1) = (2,3,2) = (1,1,1) + (1,2,1)$.
   (b) Check that $T(1,1,1) = (1,1,1)$ and $T(1,2,1) = (2,4,2) = 2(1,2,1)$.

9. (a) $(1,4,-1) + (1,2,0) = (2,6,-1)$;
   (b) $(3,-2,0) + (-1,0,0) = (2,-2,0)$;
   (c) $(1,2,2) + (3,-4,0) = (4,-2,2)$.

10. (a) $e^t(b_0 + b_1 t)$; (b) $c_1 e^{at} + c_2 e^{bt}$, where $a = (3+\sqrt{17})/2$ and $b = (3-\sqrt{17})/2$; (c) $c_1 e^t + c_2 e^{it} + c_3 e^{-it}$.

Sec. 6.6 1. (a) $\begin{bmatrix} 2 & 0 & 0 \\ 0 & 1 & 0 \\ 0 & 0 & 1 \end{bmatrix} + \begin{bmatrix} 0 & 1 & 0 \\ -2 & 0 & 1 \\ 0 & 2 & 0 \end{bmatrix}$

(b) $\begin{bmatrix} 1 & 0 & 0 & 0 \\ 0 & 1 & 0 & 0 \\ 0 & 0 & 1 & 0 \\ 0 & 0 & 0 & 1 \end{bmatrix} + \begin{bmatrix} 0 & 1 & 0 & 1 \\ 0 & 0 & -1 & 1 \\ 2 & -1 & 0 & 0 \\ 3 & 1 & 0 & 0 \end{bmatrix}$

(c) diag $\{A'_1, A'_2\}$ + diag $\{A''_1, A''_2\}$, where $A'_1 = \begin{bmatrix} 2 & 0 \\ 0 & 3 \end{bmatrix}$,

$A''_1 = \begin{bmatrix} 0 & 1 \\ 1 & 0 \end{bmatrix}$ and $A'_2 = \begin{bmatrix} 1 & 0 & 0 \\ 0 & -1 & 0 \\ 0 & 0 & 2 \end{bmatrix}$

$A''_2 = \begin{bmatrix} 0 & 0 & 1 \\ 2 & 0 & 1 \\ 3 & 1 & 0 \end{bmatrix}$

2. (a) Yes; (b) No; (c) No.
3. $\mathcal{L}\{(1,0,0,1)\}$.
4. For example: $T(x,y,z) = (x + z, -x + y, y)$, and generator $(1,1,1)$.
5. $\begin{bmatrix} 0 & 0 & 0 & 0 & -1 \\ 1 & 0 & 0 & 0 & 3 \\ 0 & 1 & 0 & 0 & 1 \\ 0 & 0 & 1 & 0 & -5 \\ 0 & 0 & 0 & 1 & 2 \end{bmatrix}$

6. $\begin{bmatrix} 0 & 0 & 0 & a_1 & b_1 \\ 1 & 0 & 0 & a_2 & b_2 \\ 0 & 1 & 0 & a_3 & b_3 \\ 0 & 0 & 0 & a_4 & b_4 \\ 0 & 0 & 0 & a_5 & b_5 \end{bmatrix}$; $\{\alpha, T\alpha, T^2\alpha\}$; $\begin{bmatrix} 0 & 0 & 0 \\ 1 & 0 & 0 \\ 0 & 1 & 0 \end{bmatrix}$

7. $\mathcal{L}\{(1,-1,1),(1,2,-1),(3,0,2)\} = V_3$.
8. $\alpha = x^3$; $\begin{bmatrix} 0 & 0 & 0 & 0 \\ 1 & 0 & 0 & 0 \\ 0 & 1 & 0 & 0 \\ 0 & 0 & 1 & 0 \end{bmatrix}$

9. $\begin{bmatrix} 0 & -1 \\ 1 & 0 \end{bmatrix}$

Sec. 6.7 1. (a) $(x-2)^4(x-3)^2$, $(x-2)^2(x-3)$; (b) $(x-1)^6(x-2)^2$, $(x-1)^3(x-2)^2$.

2. (a) diag $\{J_{11}, J_{12}, J_{21}, J_{22}, J_{23}\}$, where $J_{11} = \begin{bmatrix} c_1 & 0 & 0 \\ 1 & c_1 & 0 \\ 0 & 1 & c_1 \end{bmatrix}$

$J_{12} = \begin{bmatrix} c_1 & 0 \\ 1 & c_1 \end{bmatrix}$, $J_{21} = J_{22} = \begin{bmatrix} c_2 & 0 \\ 1 & c_2 \end{bmatrix}$, $J_{23} = [c_2]$; $(x-c_1)^5(x-c_2)^5$, $(x-c_1)^3(x-c_2)^2$.

(b) diag $\{J_{11}, J_{12}, J_{13}, J_{21}, J_{22}, J_{31}\}$, where $J_{11} = \begin{bmatrix} c_1 & 0 \\ 1 & c_1 \end{bmatrix}$,

$J_{12} = J_{13} = [c_1]$, $J_{21} = J_{22} = [c_2]$, $J_{31} = \begin{bmatrix} c_3 & 0 & 0 \\ 1 & c_3 & 0 \\ 0 & 1 & c_3 \end{bmatrix}$;

$-(x-c_1)^4(x-c_2)^2(x-c_3)^3$, $(x-c_1)^2(x-c_2)(x-c_3)^3$.

3. (a) $(x-2)^2$, $(x-2)^2$, $(x-3)$, $(x-3)$; (b) $(x-1)^3$, $(x-1)^2$, $(x-1)$, $(x-2)^2$.

4. (a) $\begin{bmatrix} -2 & 0 & 0 & 0 & 0 \\ 1 & -2 & 0 & 0 & 0 \\ 0 & 0 & 1 & 0 & 0 \\ 0 & 0 & 1 & 1 & 0 \\ 0 & 0 & 0 & 0 & 2 \end{bmatrix}$

(b) $\begin{bmatrix} -i & 0 & 0 & 0 & 0 \\ 1 & -i & 0 & 0 & 0 \\ 0 & 0 & i & 0 & 0 \\ 0 & 0 & 1 & i & 0 \\ 0 & 0 & 0 & 0 & 0 \end{bmatrix}$

(c) $\begin{bmatrix} -1-i & 0 & 0 & 0 & 0 \\ 1 & -1-i & 0 & 0 & 0 \\ 0 & 0 & -1+i & 0 & 0 \\ 0 & 0 & 1 & -1+i & 0 \\ 0 & 0 & 0 & 0 & 0 \end{bmatrix}$

5. (a) $\begin{bmatrix} -2 & 0 & 0 & 0 & 0 \\ 1 & -2 & 0 & 0 & 0 \\ 0 & 1 & -2 & 0 & 0 \\ 0 & 0 & 0 & -2 & 0 \\ 0 & 0 & 0 & 1 & -2 \end{bmatrix}$

$\begin{bmatrix} -2 & 0 & 0 & 0 & 0 \\ 1 & -2 & 0 & 0 & 0 \\ 0 & 1 & -2 & 0 & 0 \\ 0 & 0 & 0 & -2 & 0 \\ 0 & 0 & 0 & 0 & -2 \end{bmatrix}$

(b) $\begin{bmatrix} 2 & 0 & 0 & 0 & 0 \\ 1 & 2 & 0 & 0 & 0 \\ 0 & 1 & 2 & 0 & 0 \\ 0 & 0 & 1 & 2 & 0 \\ 0 & 0 & 0 & 0 & 2 \end{bmatrix}$ (c) $\begin{bmatrix} 1 & 0 & 0 & 0 & 0 \\ 1 & 1 & 0 & 0 & 0 \\ 0 & 1 & 1 & 0 & 0 \\ 0 & 0 & 1 & 1 & 0 \\ 0 & 0 & 0 & 1 & 1 \end{bmatrix}$

6. (a) $\begin{bmatrix} \sqrt{3} & 0 \\ 0 & -\sqrt{3} \end{bmatrix}$ (b) $\begin{bmatrix} (3+\sqrt{3}i)/2 & 0 \\ 0 & (3-\sqrt{3}i)/2 \end{bmatrix}$

(c) $\begin{bmatrix} i & 0 \\ 0 & -i \end{bmatrix}$

7. (a) $\begin{bmatrix} 1 & 0 & 0 \\ 0 & 2 & 0 \\ 0 & 0 & -3 \end{bmatrix}$ (b) $\begin{bmatrix} 1 & 0 & 0 \\ 0 & 2 & 0 \\ 0 & 0 & -3 \end{bmatrix}$

8. Every Jordan block has order 1.
9. 3.
10. (a) $a = 0$, $\begin{bmatrix} 1 & 0 & 0 \\ 0 & 2 & 0 \\ 0 & 0 & 2 \end{bmatrix}$ (b) $a \neq 0$, $\begin{bmatrix} 1 & 0 & 0 \\ 0 & 2 & 0 \\ 0 & 1 & 2 \end{bmatrix}$
11. $\begin{bmatrix} c_1 & 0 \\ 0 & c_2 \end{bmatrix}$, $c_1 \neq c_2$ $\begin{bmatrix} c_1 & 0 \\ 0 & c_1 \end{bmatrix}$ $\begin{bmatrix} c_1 & 0 \\ 1 & c_1 \end{bmatrix}$

INDEX

Numbers in parentheses indicate *Problems* and are followed by their page references.

A-inner product, 225
Absolute value, 298
Addition, of linear transformations, 74
  of matrices, 100
  of multilinear functions, 169
Additive group, 3
  endomorphism of, 3
Adjoint, 213–219
  classical, 163, 283
  definitions, 214–215, 299
Algebra, linear, 2
Alternating multilinear form, 168–171
  definition, 169
Angle, 180, 182, 188–189, 191, 265, 299
Annihilator, 85–88, 327
  definition, 85
Arrangement, 36
Arrows, 4, 8
Augmented coefficient matrix, 28
Automorphism, 64
Axes, coordinate, 17

Basis, 13–14, 51–55
  change of, 119–124
  definition, 52–53
  dual, 80
  ordered, 53
  orthogonal, 194
  orthonormal, 197
Bilinear forms, 236–281
  canonical, 241
  definition, 237
  equivalent, 239, 249
  matrix of, 237
  nondegenerate, (25) 246
  nonsingular, (25) 246
  orthogonally equivalent, 275
  polar, 247
  rank of, 241
  symmetric, 247
Bilinear functions, 236, 246
  matrix of, 237
  rank of, 241

364 INDEX

Bilinear functions, symmetric, 247
Bilinear skew-symmetric function, (20) 260
Black box, 1
Bound vector, 21

Canonical form, 241
  Jordan, 316, 327–332
  variant of, 335–337
Canonical matrix, under congruence, 250, 262
  under equivalence, 243
  under similarity, 313, 331
Cartesian coordinate system, 17, (13) 185
Cayley-Hamilton theorem, 282–284
Change of basis, 119–124
Change of basis theorem, for bilinear functions and forms, 239
  for quadratic forms, 248
Characteristic of field, 297
Characteristic polynomial, 269, 284
Characteristic value, 266
Characteristic vector, 266
Classical adjoint, 163, 283
Cofactor, 39
Column-echelon form, 242, (16) 245
Column rank, 127–128
Column space, 127
Columns of matrix, 29, 97
Commutator, (9) 302
Companion matrix, 323
Complement, 26, 60
  orthogonal, 199–203
Complex numbers, 21
  absolute value of, 298
  conjugate of, 298
  imaginary part of, 297
  real part of, 297
Complex vector spaces, 297–304
  inner product in, 298
  standard inner product in, 299
Components, 5
Congruence, 249, 250, 254, 272–273, 300
Congruence transformation, 256, 263
Congruent matrices, 242, 249, 254, 273, 300
Conjugate-bilinear form, 301
Conjugate-bilinear function, 300, (13) 303
Conjugate complex number, 298
Conjugate matrix, 298, 338
Conjugate space, 77

Conjugate transpose, 299
Conjugate vector, 337
Conjunctive matrices, 301
Conjunctivity, (13) 303
Consistent equations, 143
Contraction, 66
Contravariant vector, 134–137
Coordinate function, 82
Coordinate space, complex, 297
  real, 5, 16–21, 58
Coordinate vector, 58, 98, 119
Coordinates, 58, 98, 200
Covariant vector, 134–137
Cramer's rule, 36, 41
Cross product, (23) 186

Degree of trigonometric polynomial, 231
Descartes' rule of signs, (18) 280
Determinant rank of matrix, (24) 168
Determinants, 35–42, 173–176, 269
  definition, 38, 174
  evaluation of, 38, 160–165
  Gram, 230
  as multilinear functionals, 154–158
Diagonal quadratic form under congruence, 252–258
Diagonalizable operator, 290
Differentials, 138
Dilation, 66
Dimension, 5, 51–55, 58, 119
  definition, 52–53
Direct sum, of matrices, 314
  of vector spaces, 60, (23) 63
Directed line segments, 4, 8
  as vectors, 6–14
Direction, 178–180
Direction angles, 179
Direction cosines, 179
Distance, 190, (23) 193, 206, (25) 229, 231
Distance function, applications of, 229–233
Divisor of zero, matrix, (15) 167
Domain of mapping, 63
Dot product, 181
Dual basis, 80, 128
Dual space, 77, 79, 82, 128, 173, 213, 327
Duality, 82

$\{E_i\}$ basis, 53
Eigenfunction, (25) 272

Eigenvalue, 265–266, 270
  definition, 265
  of real symmetric matrix, 273
Eigenvector, 265–270
  definition, 265
Elementary divisor, 330–331
Elementary matrices, 146–151
  of first kind, 147
  of second kind, 147
  of third kind, 147
Elementary operations, column, 150, 161
  row, 29, 146, 161
Elementary science, application to, 8, 11
Endomorphism, 3
Equality, of elements, 100
  of line segments, 8
Equations, equivalent systems of, 29
  homogeneous, 28
  of line segment, (18, 20) 68
  linear, 28–33, 35–41
  nonhomogeneous, 28
  normal, 233
  rank of system of, 143
Equivalence, 240–241, 243, 250
  matrix, (18) 245
Equivalence relation, (13, 14) 15, (8) 125, 243, (18) 245, 250, 272, 300
Equivalent bilinear forms, 239, 249
Equivalent matrices, 240–241, 300
Equivalent quadratic forms, 249
Equivalent systems of equations, 29
Euclidean algorithm, 315
Euclidean geometry, applications to, 11–14
Euclidean space, 186
Euclidean-space isomorphism, (26) 213
Euler coefficients, 231
Even polynomial, (22) 27
Extension field, 297

Field, 3
  extension, 297
Finitely generated space, 52
Forms, alternating multilinear, 168–171
  bilinear, 236–243
  column-echelon, 242
  conjugate-bilinear, 301, (13) 303
  diagonal, under congruence, 252–258
  Hermite normal, 30, 150
  hermitian, 301, (13) 303

Forms, Jordan canonical, 316, 327–332
  nondegenerate or nonsingular bilinear, (see Note) 246
  polar bilinear, 247
  positive quadratic, 224, 246
  quadratic, 224, 246–250
  row-echelon, 30
  zero, 254
Fourier coefficients, 231
Fourier series, 231, (17) 234
Free vector, 21
Full linear group, of $n \times n$ matrices, 112
  of operators, 94
Function, 63
  alternating, 169
  bilinear, 236, 246
  bilinear skew-symmetric, (20) 260
  conjugate-bilinear, 300, (13) 303
  coordinate, 79, 82
  determinant, 155, 158
  hermitian, 300, (13) 303
  linear, 236
  multilinear, 169
  scalar, 182
  vector, (23) 186
Functional, linear, 77, 79–83, 128, 155, 213–219, 236
  multilinear, 155

Gauss reduction, 28
Generator, 25, 45
Geometric vectors, 4, 8–21
  addition of, 9
  basis of, 13–14
  definition, 8
  equality of, 8, (13) 15
  linear combination of, 12
  multiplication by scalars of, 10
  parallel, 9
Geometry, applications to, 11–14
Gradient, 138
Gram determinant, 230
Gram-Schmidt process of orthogonalization, 193–197, 299
Gramian matrix, (22) 199, 230
Groups, additive, 3
  additive abelian, 3, 75
  full linear, of matrices, 208, (11) 211
  of operators, 94, 207

Groups, isomorphism of, 101
   multiplicative, 90
   multiplicative abelian, 91
   orthogonal, 207

Hermite normal form, 30, 150
Hermitian form, 301, (13) 303
Hermitian function, 300, (13) 303
Hermitian matrix, 299
Hilbert space, 305
Hom, 77
Homomorphism, 77
Hyperplane, (23) 90

Idempotent operator, (21) 79
Identical elements, 100
Identity, of group, 90
   matrix, 111
   operator, 70, 75
   of ring, 76
Inconsistent equations, 28, 143
Index of nilpotency, 323
Inertia, law of, 263
Inner product, definition of complex, 298
   definition of real, 186
   standard, on complex spaces, 299
   on real spaces, 181
Inner product spaces, 178–235
Input, 1
Intersection of spaces, 26
Invariant subspace, 313–318
   definition, 314
Invariants, 300
   of normal matrices under similarity, 313
   numerical, 239, 241, 250
   under similarity, 331
   of symmetric matrices under congruence, 260–263, 300
Inverse, group element, 90
   matrix, 112, 122, 146–151, 148
Inverse mapping, 91
Inversion, 37
Invertible linear transformation, 92
Invertible mapping, 92
Invertible operator, 93
Involutary matrix, (8) 302
Isometry, (21) 199, 206

Isomorphic mapping, 16
   of groups, 102
   of rings, 106
   of vector spaces, 16
Isomorphism, 16–17, 57
   euclidean-space, (26) 213
   group, 101

Jordan block, 330
Jordan canonical form, 316, 327–332
   description of, 329–330
   variant of, 335

Kernel, 65, (15) 68
   of operator, 293
Kronecker delta, 80

Lagrange polynomials, 309, (21) 311
Latent value, 266
Latent vector, 266
Law of inertia, 263
Leading entry, 30
Least squares, 232
Legendre polynomials, (14) 198
Length, 178–179, 182, 188–189, 191, 231, 265
Line of best fit, 233
Line segments, directed, 4, 8, 178
   equality of, 8
   equation of, (18, 20) 68
   as vectors, 7–14, 178
Linear algebra, 2
Linear combination, 12–13, 25
Linear coordinate system, 17
Linear dependence of vectors, 45, 165
Linear equations, solution of, 28–33, 35–42, 140–151
   systems of, 28–33, 35–42, 140–151
Linear functionals, 77, 79–83, 128, 155, 173, 213–219
Linear operators, 93, 110–111
   characteristic polynomial of, 270
   definition, 93
   idempotent, 308
   orthogonal, 275
Linear systems of equations, 28–33, 35–42, 140–151

INDEX 367

Linear transformations, 2, 63–139, 298
  definition, 65
  matrix of, 96–103
  nonsingular, 90–95
  rank of, 127–128
Lorentz transformation, (28) 304

Mapping, 1, 63
  distance-preserving, 206, 208
  inverse, 91
  invertible, 92
  one-to-one, 91
  onto, 63, 93
  surjective, 63
Matrices, 96–134
  addition of, 100
  column, 135
  congruent, 242, 249, 273, 300
  conjunctive, 301
  direct sum of, 314
  elementary, 146–151
  equal, 100
  equivalent, 240–241, 300
  full linear group of, 112
  of linear transformations, 96–103
  multiplication by scalars of, 100, 103
  as multiplicative systems, 106–116
  operations on, 97–103
  orthogonally similar, 273
  partition of, 114
  Pauli spin, (8) 302
  product of, 108
  ring of, 111
  row, 135
  row-equivalent, 148
  similar, 123, 273, 289–294, 301
Matrix, augmented coefficient, 28
  characteristic polynomial of, 269
  of coefficients, 28
  companion, 323
  conjugate, 298, 338
  diagonal, 112, 254, 273, 289–294
  elementary, of first kind, 147
    of second kind, 147
    of third kind, 148
  gramian, (22) 199, 230
  hermitian, 299
  identity, 111
  inversion of, 112, 122

Matrix, involutary, (8) 302
  Jordan canonical, 324, 330–331
  of linear transformation, 98
  minimum polynomial of, 283
  negative, 263
  nilpotent, (15) 167, 317
  nonsingular, 112
  normal, 306
  order of, 99
  orthogonal, 208, 224
  positive, 224
  rank of, 129–130
  ring of, 111
  row-equivalent, 29
  scalar, 112, 282
  square, 38, 99
  symmetric, (25) 134, 135, 254, 273
  trace of, (20) 133
  transformation, 98
  of transition, 119–120
  transpose of, 119, 121
  triangular, (26) 153, 162–163
  unitary, 300
  zero, 100
Matrix adjoint, 299
Metric, 178, 182, 191
Minimum polynomial, 283, 285, 293, 330
Module, 305
Multilinear functionals, 155
Multiplication by scalars, 3
  of linear transformations, 76
  of matrices, 103

n-linear form, 169
  skew-symmetric, 169–170
  symmetric, 169
Negative matrix, 263
Nilpotency, index of, 323
Nilpotent matrix, 317
Nilpotent operators, (11) 78, 317, 321–324
Nonsingular linear transformation, 90–95
  definition, 92
Nonsingular operator, 93
Nonsingular matrix, 112
Norm, 188, (30, 31) 193, 299
Normal equations, 233
Normal matrix, 306
  invariants under similarity, 313

Normal operators, 306–309
  definition, 306
Null space, 65, 70, 142
Nullity, 72

Odd polynomial, (22) 27
One-to-one correspondence, 16
One-to-one mapping, 91, 93
Onto mapping, 63, 93
Operations, 3
  binary, 3
  elementary column, 150, 151, 242
  elementary row, 29, 146
  on linear transformations, 73–77
  on matrices, 97–103
Operator, linear, 93, 98, 110–111
  adjoint, 213–215, 299
  characteristic polynomial of, 284
  definition, 65, 98
  diagonalizable, 290
  hermitian, 299
  idempotent, (21) 79
  identity, 70, 75
  invariant of, (1) 67
  inverse, 91
  invertible, 93
  kernel of, 293
  minimum polynomial of, 285
  nilpotent, (11) 78, 317, 321–324
  nonsingular, 93
  normal, 282, 306–309
  orthogonal, 206, 275, 299
  positive, 221–227
    definition of, 222, 224
  self-adjoint, 218, (19) 220, 222, 299
  symmetric, 290
  transpose, 130, 217, 327
  unitary, 299
Order of matrix, 99, 262
Origin, 17
Orthogonal basis, 194
Orthogonal complement, 199–203
  definition, 202
Orthogonal group, 207
Orthogonal matrix, 208, 224
Orthogonal operator, 206, 299
Orthogonal projection, 200–201, 230, 308
Orthogonal reduction, quadratic forms, 272–278

Orthogonal subspace, 189, 202
Orthogonal transformation, 206–210
  definition, 206
Orthogonality, 182, 194, 199
Orthogonalization, Gram-Schmidt process of, 193–197
Orthogonally similar matrices, 273
Orthonormal basis, 197
Output, 1

Parallel geometric vectors, 9
Parallelogram law, 9–10, (22) 186, (26) 193, 299
Parameters, 33
Parity, 37, (14) 159
Partition of matrices 114
Pauli spin matrices, (8) 302
Permanent, (20) 159
Permutations, 36–37
  even, 37
  identity, 37
  inverse, 37
  odd, 37
  product of, 37
  symmetric group of, 37
Points, 19–21
Polar bilinear form, 247
Polarization identity, (12) 250, 299
Polynomials, 4
  characteristic, 269
  even, 27
  Lagrange, 309, (21) 311
  Legendre, (14) 198
  minimum, 283, 293, 330
  odd, 27
Positive matrix, 224
Positive operator, 221–227, (20) 311
  definition, 222, 224
Positive quadratic form, 224, 246
Primary decomposition, 313–318
Primary decomposition theorem, 315
Principal-axis theorem, 273–274
Principal value, 266
Principal vector, 266
Product, 2
  A-inner, 225
  cross, (23) 186
  dot, 2
  of group elements, 90

Product, inner, 2, 178, 298
  scalar, 2
  T-inner, 225
Projections, 60, (14) 68, (11) 73, (15) 204, (20) 296, 308
  orthogonal, (20) 192, 200–201, 230, 308
Pythagorean theorem, 189–190

Quadratic forms, 224, 246–281
  definition, 247
  diagonal, under congruence, 252–258
  equivalent, 249
  nondegenerate, (24) 252
  orthogonal reduction of, 272–278
  orthogonally equivalent, 275
  positive, 224, 246, (19) 260
  signature of, 261

Range, of linear transformation, 65, (15) 68, 70
  of mapping, 63
Rank, 72, 127–131, 241, 243, 250, 262
  column, of matrix, 127–128
  determinant, of matrix, (24) 168
  of linear transformation, 128
  row, of matrix, 127, 129
  of similar matrices, 130
  of systems of equations, 143
Rank space, 65
Real line, 19–21
Real vector space, 2, 297
Reflection, 207
Relation, (13) 15
  congruence, 242–243
  equivalence, 243
  similarity, 243
Rigid motion, 206, 208
Rings, 76
  isomorphic, 106
Rotations of plane, 64, 206–208
Row of matrix, 29, 97
Row-by-column rule, 108
Row-echelon form for matrix, 30
Row-equivalent matrices, 29, 148
Row rank of matrix, 127, 129
Row space of matrix, 127

Scalars, 2, 3, 5
Schwarz inequality, 190
Segre characteristic, 330–331
Self-adjoint operator, (19) 210, 218, **222**, 299, 305
Self-dual vector space, (25) 85
Sets, 1
Shear, 66
Signature, 261–262
Similar matrices, 123, 273, 301
  definition, 123
Similarity, (8) 125, 243, 272–273, 282–**338**
Skew-symmetric n-linear form, 169
Solution space, 142
Solutions of linear equations, 28
  complete set of, 142
  one-parameter family of, 33
  three-parameter family of, 33
  trivial, 143
Spaces, column, of matrix, 127
  coordinate, 16–21, 297
  definition of complex vector, 297
  definition of real vector, 2
  dual, 77, 79, 82, 128, 173, 213
  general euclidean, 186–191
  Hilbert, 305
  inner product, 178–235
  null, 65
  rank, 65
  row, of matrix, 127
  solution, 142
  zero, 5
Spanning vectors, 25, 52
Spectral resolution, 309
Spectral theorem, 294, 305–309
  statement of, 307, 309
Spectrum, 309, 330–331
Square matrix, 99
Standard inner product, **181, 187**
Subfield, 297
Subset, proper, 46
Subspace, 24–26
  invariant, 313–318
  orthogonal, 189, 202
  T-cyclic, 321–324
  T-invariant, 313–318
Sum of spaces, 26
  direct, 60, (23) 62
Surjective mapping, 64
Symmetric bilinear form, 247

Symmetric billnear form, signature of, 261
Symmetric bilinear function, 247
Symmetric group, 37
Symmetric matrix, (25) 134, 135, 254, 273
   invariants under congruence of, 260–263
Symmetric n-linear form, 169
Symmetric operator, 290
Systems, 1
   algebraic, 73, 100
   equivalent, 29
   isomorphic, 17, 57
   linear, 1
   linear coordinate, 17
   of linear equations, 140–177
      rank of, 143

T-cyclic subspace, 321–324
   definition, 321
T-inner product, 225
T-invariant subspace, 314
Tensor, 136–137
Trace, (20) 133
Tranformations, linear, 2, 63–139, 298
     congruence, 256, 263
     definition, 65
     distance-preserving, 206
     invertible, 92
     Lorentz, (28) 304
     nonsingular, 90–95
     operations on, 73–77
     orthogonal, 206–210
     product of, 74
     sum of, 74
     transpose, 128
     zero, 70, 75
Transition matrix, 119–120
Translation, 206, 208
Transpose linear transformation, 128
Transpose matrix, 119, 121
Transpose operator, 128–130, 174, 217, 327
Triangle inequality, 191, 299
Triangle law, 9–11
Trivial solution, 28

Unit point, 17, 53
Unit vector, 197
Unitary matrix, 300
Unitary operator, 299
Unitary vector space, 299

Value, characteristic, 266
   latent, 266
   principal, 266
Vector quantities, 8
Vector spaces, complex, 297–304
   conjugate, 77
   definition of real, 2, 134
   dimension of, 5, 51–55, 58, 119
   dual, 77, 79, 82
   examples of real, 3–5
   finite-dimensional, 1–62
   of geometric vectors, 4
   isomorphic, 16–17, 57
   left, 24
   real coordinate, 5
   right, 24
   self-dual, (25) 85
   unitary, 299
   zero, 5
Vectors, 2
   A-orthogonal, (22) 228
   addition of, 2
   basis of plane, 13
   bound, 21
   characteristic, 266
   conjugate, 337
   contravariant, 134–137
   coordinate, 58, 98, 119
   coordinates of, 200
   covariant, 134–137
   displacement, 138
   distance between, 190, (27) 193
   free, 21
   geometric, 4, 8–21
   latent, 266
   linear dependence of, 45
   linear independence of, 46, 49, 52, 164
   multiplication by scalars of, 2–3, 76–77
   orthogonal, 182–190
   perpendicular, 182
   spanning, 25
   unit, 53, 197
   zero, 2–3

Zero form, 254
Zero matrix, 100
Zero transformation, 70, 75
Zero vector space, 5